ADVANCED COMPOSITES

for AEROSPACE, MARINE, and LAND APPLICATIONS II

Cover Photograph: Scanning electron micrograph of the tensile fracture surface of the 2014/Al_2O_3/10p composite deformed in tension showing noticeable failure by cracking of the reinforcing particulates. Further details can be found in the technical manuscript "Influence of Reinforcement Content on Tensile Response and Fracture Behavior of an Aluminum Alloy Metal Matrix Composite" by K. Manigandan, T. S. Srivatsan, Zhencheng Ren, and Jingyi Zhao.

TMS2015

144th Annual Meeting & Exhibition

March 15-19, 2015 • Walt Disney World • Orlando, Florida, USA

ADVANCED COMPOSITES
for AEROSPACE, MARINE, and LAND APPLICATIONS II

Proceedings of a symposium sponsored by

The Minerals, Metals & Materials Society (TMS)

held during

TMS2015

144th Annual Meeting & Exhibition

March 15-19, 2015

Walt Disney World • Orlando, Florida, USA

Edited by:

T. Sano • T. S. Srivatsan

Editors
T. Sano
T. S. Srivatsan

ISBN 978-3-319-48607-9 ISBN 978-3-319-48141-8 (eBook)
DOI 10.1007/978-3-319-48141-8

Chemistry and Materials Science: Professional

Published by Springer International Publishers, Switzerland, 2016
Reprint of the original edition published by John Wiley & Sons, Inc., 2015, 978-1-119-08226-2

Printed on acid-free paper

This Springer imprint is published by Springer Nature
The registered company is Springer International Publishing AG
The registered company address is: Gewerbestrasse 11, 6330 Cham, Switzerland

TABLE OF CONTENTS
Advanced Composites for Aerospace, Marine, and Land Applications II

Advanced Processing Techniques

Metal Matrix Composites

Composite Microstructure and Mechanical Property Characterization

Advanced Composites and Syntactic Foams

Carbon Fiber Reinforced Composites and Modeling & Simulations

Poster Session

PREFACE

The use of advanced composites has increased in recent years, often replacing heavier monolithic materials for structural applications. The high demand for composites is also seen in the transportation and defense industries. To explore the growth of composite knowledge and to promote communication of recent composite research for aerospace, marine, and land applications, the international symposium titled "Advanced Composites for Aerospace, Marine, and Land Applications II" was held during the TMS 2015 Annual Meeting & Exhibition in Orlando, Florida, USA, March 15-19, 2015. This symposium was the second in the series on the topic of advanced composites and was well-represented with abstracts from United States and international engineers and scientists from academia, research laboratories, and industries. This volume contains many of the papers presented in the five-session symposium, sponsored by the Composite Materials Committee of TMS (The Minerals, Metals & Materials Society). Over 46 abstracts were approved for presentation at this symposium and were divided into five sessions:

Session 1:	Advanced Processing Techniques
Session 2:	Metal Matrix Composites
Session 3:	Composite Microstructure and Mechanical Properties
Session 4:	Advanced Composites and Syntactic Foams
Session 5:	Carbon Fiber Reinforced Composites and Modeling & Simulations

In these five sessions, recent advances in composite materials processing, characterization, properties, and modeling specifically for aerospace, marine, or land applications were presented. Each presenter brought new knowledge and ideas to the symposium, and we, the symposium organizers, extend our warmest thanks and appreciation to the presenters and participants. We also extend our most sincere thanks and appreciation to the elected and governing representatives of TMS for their understanding and acknowledgment of our interest, and timely approval of our request to organize this intellectually stimulating symposium.

Profuse gratitude and appreciation are extended to Ms. Trudi Dunlap (Programming Manager at TMS) for her sustained attention, assistance, interest, involvement, and timely participation stemming from understanding. Ms. Dunlap ensured a timely execution of the numerous intricacies related to both orchestration and layout of this symposium from the moment following its approval through the compilation and publication of this proceedings publication. Special thanks and much deserved appreciation is extended to Ms. Patricia Warren (Programming and Proceedings Specialist at TMS) for her patience, understanding, and much valued and desired attention as well. Also, the timely publication of this bound volume would not have been possible without the cooperation of the authors and the publishing staff headed by the dedicated Content Senior Manager, Mr. Matt Baker at TMS.

We hope that this bound volume will provide the readers new perspectives and directions in their composite research. The growth and demand of composite materials will only continue and we hope that we have assisted the aerospace, marine, and land engineering and science community in reaching composite material research and application goals.

Dr. Tomoko Sano
Weapons and Materials Research Directorate
Lightweight & Specialty Metals Branch
U.S. Army Research Laboratory
RDRL-WMM-F
Aberdeen Proving Ground, MD 21005

Dr. T. S. Srivatsan
Professor
Division of Materials Science and Engineering.
Department of Mechanical Engineering
The University of Akron
Akron, Ohio 44325, USA

SYMPOSIUM ORGANIZERS

Tomoko Sano

Dr. Tomoko Sano, Materials Engineer, U.S. Army Research Laboratory, received her M.S. in Materials Science and Engineering at Carnegie Mellon University in 2001 and her Ph.D. in Materials Science and Engineering at Carnegie Mellon University in 2005. After a postdoctoral fellowship at the U.S. Army Research Laboratory, she was appointed the Technical Assistant to the Director of the Weapons and Materials Research Directorate, where she was awarded three special act awards for the technical execution of directorate level assignments and programmatic actions. She is currently a materials engineer in the Materials and Manufacturing Science Division of the Weapons and Materials Research Directorate of the U.S. Army Research Laboratory, Aberdeen Proving Ground, Maryland. Her expertise and interests are in mesoscale microstructural and micromechanical characterization, and she conducts research to correlate processing, microstructure, and properties.

Dr. Sano has been the first author of over 20 technical publications including 2 book chapters, has been the first editor of three books, and presented over 20 conference presentations. She is an active member of the TMS Composite Materials Committee, serving as the *JOM* liaison in 2012 and 2014.

T.S. Srivatsan

Dr. T.S. Srivatsan is Professor of Materials Science and Engineering in the Department of Mechanical Engineering at The University of Akron. He received his graduate degrees (Master of Science in Aerospace Engineering in 1981) and Doctor of Philosophy in Mechanical Engineering in 1984) from Georgia Institute of Technology. Dr. Srivatsan joined the faculty in the Department of Mechanical Engineering at The University of Akron in August 1987. Since joining, he has instructed undergraduate and graduate courses in the areas of (a) Advanced Materials and Manufacturing Processes, (b) Mechanical Behavior of Materials, (c) Fatigue of Engineering Materials and Structures, (d) Fracture Mechanics, (e) Introduction to Materials Science and Engineering, (f) Mechanical Measurements, (g) Design of Mechanical Systems, and (h)Mechanical Engineering Laboratory. His research areas currently span the fatigue and fracture behavior of advanced materials to include monolithic(s), intermetallic, nanomaterials and metal-matrix composites; processing techniques for advanced materials and nanostructure materials; inter-relationship between processing and mechanical behavior; electron microscopy; failure analysis; and mechanical design. He has authored/edited/co-edited 55 books in areas cross-pollinating mechanical design; processing and fabrication of advanced materials; deformation, fatigue and fracture of ordered intermetallic materials; machining of composites; failure analysis; and technology of rapid solidification processing of materials. He serves as co-editor of *International Journal on Materials and Manufacturing Processes* and is on the editorial advisory board of other journals in the domain of materials science and engineering. His research has enabled him to deliver over 200 technical presentations in national and international meetings and symposia; technical/professional societies; and research and educational institutions. He has authored and co-authored over 650 archival publications in international journals, chapters in books, proceedings of national and international conferences, reviews of books, and technical reports. In recognition of his efforts and contributions and their impact on furthering science, technology, and education he has been elected Fellow of the ASM International, Fellow of American Society of Mechanical Engineers (ASME), and Fellow of American Association for the Advancement of Science (AAAS). He has also been recognized as Outstanding Young Alumnus of Georgia Institute of Technology and Outstanding Research Faculty at the College of Engineering at The University of Akron. He offers his knowledge in research services to the U.S. Government (U.S. Air Force and U.S. Navy), National Research Laboratories, and industries related to aerospace, automotive, power-generation, leisure-related products, and applied medical sciences .

SESSION CHAIRS

Dr. Sampath Vedamanickan
Indian Institute of Technology Madras

Dr. Jonathan Spowart
Air Force Research Laboratory

Dr. Alicia Esther Ares
IMAM (CONICET-UNaM)

Dr. Christopher Muhlstein
Georgia Institute of Technology

Mr. Paul Moy
U.S. Army Research Laboratory

Dr. Timothy Walter
U.S. Army Research Laboratory

Mr. Gundolf Kopp
German Aerospace Center (DLR)

Dr. David Saylor
Food and Drug Administration

Dr. Tomoko Sano
U.S. Army Research Laboratory

Dr. T. S. Srivatsan
The University of Akron

ADVANCED PROCESSING TECHNIQUES

Advanced Composites for Aerospace, Marine, and Land Applications II
Edited by: T. Sano and T.S. Srivatsan
TMS (The Minerals, Metals & Materials Society), 2015

Effect of Processing Parameters on the Microstructure of Mechanically Alloyed Nanostructured Al-Mn Alloys

Kristopher A. Darling, Anthony J. Roberts, James E. Catalano Mark A. Tschopp, and Laszlo J. Kecskes

U.S. Army Research Laboratory
Aberdeen Proving Ground, Maryland, USA

Keywords: aluminum, manganese, Al, Mn, high-energy ball milling, mechanical alloying, precipitation, microstructure, solid solubility

Abstract

High-energy mechanical alloying was utilized to study the alloy formation between Al and Mn. A modified SPEX mill was designed to investigate the role of milling energy and milling power on enhancing the solubility of Mn in Al. Despite having a large influence on alloying kinetics (an increased rate of 3.25×) no further grain refinement or degree of alloying was observed with increasing milling energy. It is hypothesized that alloying in this particular binary system takes place and is controlled by the increasing interfacial surface area created between the Mn particles and the Al matrix. However, extensive comminution eventually likely leads to an auto-catalytic exothermic reaction between the Mn particles and Al matrix, leading to the formation of the equilibrium phase Al_6Mn. This reaction reduces the degree of and, in turn, frustrates further alloying, highlighting the practical limits of creating solid solutions in this system through mechanical alloying.

1. Introduction

Nanocrystalline materials have increased strengths over their polycrystalline counterparts because the reduced grain sizes result in 1) Hall-Petch strengthening through dislocation pileup at the grain boundaries and 2) a change in deformation mechanisms at the smallest grain sizes [1]. Despite their significant advantage in mechanical properties over coarser grain sizes, the main setback for nanocrystalline metals continues to be their inability to mitigate grain growth during material processing or subsequent applications. In fact, the thermal stability is especially important when bulk nanocrystalline materials are produced via the powder processing route, in which high homologous temperatures are required during consolidation of the nanostructured powder [2-3]. In many cases, this has motivated a number of studies where various solutes are added to pure nanocrystalline materials in an attempt to thermodynamically or kinetically hinder grain boundary motion (e.g., adding Ta to Cu [4] or Zr to Fe [5]. Furthermore, a number of different thermodynamic-based models have arisen to provide guidance for selecting solutes in a number of nanocrystalline systems [6-10].

Nanocrystalline Al is one such system of interest. Wrought and cast Al alloys are used for a wide array of applications because of their combination of moderate strength and low density. However, improving the strength of nanocrystalline alloys could have a wide-reaching effect for structural applications. Some recent studies have investigated the addition on transition metals to Al to form binary alloys. The low diffusivity of these transition metals can help form a solid solution and enhance their thermal stability, but these element additions are often associated with low equilibrium solubility concentrations. Various processing routes (e.g., rapid solidification [11], evaporation/inert condensation [12-13], electrodeposition [14-15], powder metallurgy [16-20]) have extended their solubility, which has enhanced the strength by refining the nanocrystalline grain size, dispersing nanoscale precipitates, forcing solute into solution, and pinning dislocations through other mechanisms.

This paper focuses on adding Mn to nanocrystalline aluminum to enhance strength and reduce the grain size. Ruan and Schuh first found hardness values up to 5 GPa (corresponding to a tensile strength that exceeds 1.5 GPa) with nanocrystalline Al-Mn alloys produced via controlled electrodeposition [14-15]. These remarkable properties were caused by a high concentration of Mn in solution (content 8.2 at.%) combined with an engineered, duplex structure composed of a high volume fraction of an fcc nanocrystalline icosahedral phase surrounded by a Mn-rich amorphous phase. They show that a significant advantage of the Al-Mn system is that the microstructure (and corresponding properties) can be systematically tuned with very small changes in the Mn content [14]. While electrodeposition allows for a precise control of chemistry, phase composition, and grain size, it is limited in its ability to form thicker films and bulk parts.

An alternative method for producing bulk nanocrystalline Al-Mn materials is high energy mechanical alloying. Because mechanical alloying is often utilized in conventional powder processing, scalability from laboratory results to bulk nanocrystalline specimens is less of a concern than with the inherent limitations of the electrodeposition or vapor deposition processes. In a recent study, it was shown that high energy mechanical alloying led to a substantial increase in the Mn solubility limit in α-Al [21]. The solubility limit, after mechanically alloying in a

SPEX mill for 200 hours, was found to be 3.1 at.% – a five-fold increase over the equilibrium solubility limit – with the residual Mn content existing as a micrometer to ultrafine-grained particle dispersion. While the ability to force Mn into solution was encouraging, the overall strength of the nanocrystalline Al-Mn alloy was found to be mainly a result of the reduction in grain size of the matrix. Unfortunately, while similar trends to those obtained by Ruan and Schuh were obtained via mechanical alloying, this process was unable to reproduce either the same Mn solubility (8 at.%) or strength levels reported with the electrodeposition process.

In the present work, the authors present the influence of milling energy and milling time on the microstructure of the resultant Al-Mn system. It is discovered that the extent of alloying via mechanical alloying is limited to approximately 3 at.% Mn, which results in a Vickers microhardness of approximately 4.0 GPa. The results of this study show how milling parameters affect the kinetics in alloy formation and highlight some potential reasons for the terminal solid solubility of Mn in the Al-Mn binary system. Understanding and optimizing the processing route via mechanical alloying will enable an ultrahigh-strength, bulk nanocrystalline Al-Mn material for structural applications.

2. Experimental Procedures

The Al-Mn powder was produced by mixing appropriate amounts of Al and Mn elemental powders to obtain the desired Al-8 at.% Mn composition. Alfa Aesar (Ward Hill, NY) supplied both precursor powders (-325 mesh, 99.9% purity). The ball-to-powder mass ratio in each vial was 6.67:1 with a total powder charge of 7.5 g. Additionally, prior to milling, 1.5 wt.% stearic acid was added to each vial before sealing it in an argon atmosphere containing less than 1 ppm oxygen and H_2O. Stearic acid was used as a process control agent to prevent cold welding of the powder charge to the milling media.

The powders were loaded into hardened steel vials containing 440C stainless steel balls. Mechanical alloying was then performed at various intervals up to 100 hours in a modified SPEX 8000D mill capable at running higher than twice the factory set number of revolutions per minute (i.e., up to 2400 RPM). At each time interval, approximately 300–400 mg of powder was extracted for analysis. Specifically, for this effort, the number of RPMs was increased by a factor of 1 (no change), 1.25 (+25%), and 1.5 (+50%). This increases the milling kinetic energy and milling power (i.e., the number of impacts/unit time). The mechanical milling resulted in loose, un-agglomerated powders with a particle size range of 20–200 μm.
X-ray diffraction of the as-milled powders was performed with an X'Pert PRO PANalytical MPD x-ray diffractometer using CuKα (λ = 0.1542 nm) radiation. Typical scans were run from 20–120° 2θ, 0.0065° increment, and 15 s dwell time. X-ray analysis was performed using Scherrer and Stokes-Wilson analyses for grain size and strain determination [22].

3. Results

Mechanical alloying of elemental Al and Mn powders produced a nanoscale composite with a fine-scale Mn dispersion within the Al matrix. In general, an increase in milling time resulted in an improved Mn dispersion, because of the continual breakdown and fragmentation of the larger Mn particles. Residual larger Mn particles (~1 μm in diameter) were still observed in the Mn particle dispersion, but their volume fraction decreased with milling time, in agreement with previous reports [21].

Figure 1 shows the effect of milling energy and milling time on reducing the grain size of the α-Al matrix (see Figure 1A) and, the Mn particle size dispersion (see Figure 1B). Additionally, Figures 2A and 2B show the effect of milling energy and milling time on the cumulative mean lattice strain stored within the Al matrix and Mn dispersoids, respectively. Here the milling energy is given as a factor (1 or 1.25), which indicates the increase in RPM over that of the factory set value of 980 (SPEX 8000D). Furthermore, unless otherwise stated, open data points refer to data in which precipitation of a secondary intermetallic phase occurred. This phenomenon will be discussed later in the manuscript. Figure 1A shows the reduction in grain size for the Al matrix. The precursor elemental Al powder was premilled for 20 hours to increase its hardness relative to that of the precursor Mn powder [21]. Since the Al was pre-milled, its starting grain size was already 30 nm with an induced lattice strain of 0.12. Over the course of milling, the mean grain size of the Al matrix reduced to approximately 11 nm. While the increased milling energy did not affect the equilibrium grain size, it did significantly alter the rate/kinetics of grain refinement by as much as a factor of 3. The open data points, as observed in Figures 1A and 1B, indicate that precipitation also occurs much earlier, which is attributed to the increased alloying rate. This same relationship of elevated milling energy increasing the kinetics of refinement is also observed for the Mn particle size (Figure 1B). Subsequent experiments investigated whether further grain size reduction could be achieved utilizing energy factors up to 1.5, but little improvement was observed.

Figures 2A and 2B show the lattice strain as a function of milling time and milling energy for the Al matrix and Mn precipitates, respectively. The strain in the Al matrix monotonically increases as milling time is increased to levels as high as ~0.35 strain, which exceeds that reported for pure nanocrystalline Al milled in a similar manner [23]. At an energy factor of 1.25, the strain accumulation rate is much higher than for the baseline milling energy. The observed strain evolution in the Mn particles trends similarly with increased strains for higher milling times. However, in the case of Mn particles, the increased milling energy resulted in a lower Mn strain at higher milling times.

Figure 3 shows the change in lattice parameter as a function of milling time and milling energy, as calculated using Cohen's method in combination with a least squares method [21]. The increased milling energy results in a greater Al lattice parameter. For the 25% increase in milling RPM, there is a corresponding increase in the alloying rate (slope) by a factor of 3.25. Interestingly, there exists a limit below which the lattice parameter for Al can no longer be decreased without precipitation occurring. Thus, the amount of Mn in the supersaturated Al-matrix can be approximated by knowing the change in lattice parameter as a function of solute content. The Mn content of the as-milled Al-Mn solid solution is quantified using the analysis of Jones [24], wherein the Al lattice parameter is reported as a function of the Mn content. Since Mn has a smaller atomic radius than Al, the more Mn in solution, the lower the Al lattice parameter. So, the change in the Al lattice parameter suggests that, regardless of milling energy or the rate at which the lattice parameter can be expanded, the as-milled Al-Mn powders have a terminal supersaturated solid solution (~3 at.% Mn). The remaining 5 at.% Mn is assumed to reside as dispersed Mn particles. It is also observed that upon precipitation of the Al-Mn intermetallic at longer milling times, the lattice parameter increases slightly, which suggests that the precipitation process removes Mn solute from solution.

Further analysis of the X-ray data results in the supersaturated Al-Mn grain size as a function of dissolved Mn solute content (Figure 4). The open data points denote the first three hours of milling, which results in little, if any, alloying. In contrast, increasing the milling energy resulted in faster rates of alloying. For a given grain size, a higher dissolved Mn solute content is indicated for increased milling energies. Note that there is a simple linear relationship between dissolved Mn content and grain size, too. While there is a difference in the rate of alloying occurring at the initial milling times as well as throughout the milling process, the point at which precipitation occurs for both milling energies is the same grain size and dissolved Mn content.

4. Discussion and Conclusions

High-energy mechanical alloying was utilized to study the alloy formation between Al and the transition metal Mn, the result being the formation of a high-strength nanocrystalline Al-Mn solid solution Mn composite. Previous investigations into this particular binary system by the authors revealed that high energy mechanical alloying could produce a solubility of approximately 3 at.% Mn in Al. The overall microhardness of the composite microstructure, ~4 GPa, and found to only be marginally increased by the presence of the Mn dispersion; rather, the main strengthening mechanism in the samples was determined to be related to the decreasing grain size of the matrix. The experimental results also suggest a simple linear relationship between the Mn solute concentration and grain size and, therefore, a potential alternative hardening mechanism in the alloy. It was suggested that if the solubility of Mn in Al could be further increased, still higher mechanical properties could be attained. However, it was postulated that this would not be possible while utilizing the current milling equipment, namely a SPEX 8000D shaker mill. In this work the authors created a modified SPEX mill capable of running at even high energies and studied the effect of milling energy on alloy formation. The results of this work by the authors have the implications for mechanical alloying of the Al-Mn binary system and are as follows. A higher RPM resulted in not only a higher kinetic milling energy but also a simultaneous increase in the milling power (energy/unit time). Therefore, even low increases in RPM can result in a very noticeable increase in alloying kinetics to a point. Despite an increase in alloying rate (a factor of 3.25 times the original rate), further grain refinement and a higher dissolved Mn content could not be attained. Therefore, no significant enhancement in mechanical strength was reported. However, slight differences in the dissolved Mn content were noted, specifically a higher amount of dissolved Mn content was observed at larger grain sizes. This may have led to the slight differences in the observed Hall-Petch behavior for Al-Mn alloys studied here.

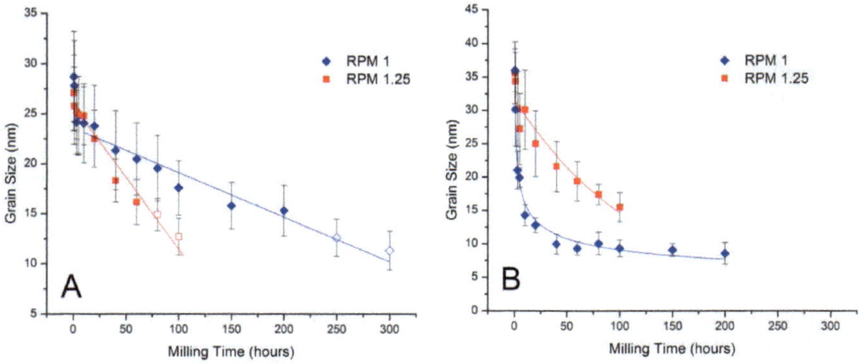

Figure 1. **Average grain size for the Al-Mn samples as a function of milling energy and time**

A: **plot of the mean grain size of the Al-Mn matrix**

B: **plot of the mean Mn particle size**

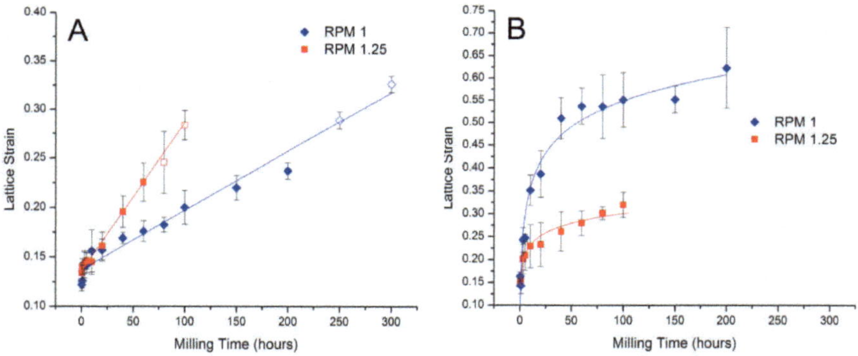

Figure 2. **Mean lattice strain for the Al-Mn samples as a function of milling energy and time.**

A: **plot of the stored lattice strain in the Al-Mn matrix**

B: **plot of the stored lattice strain in the Mn particles.**

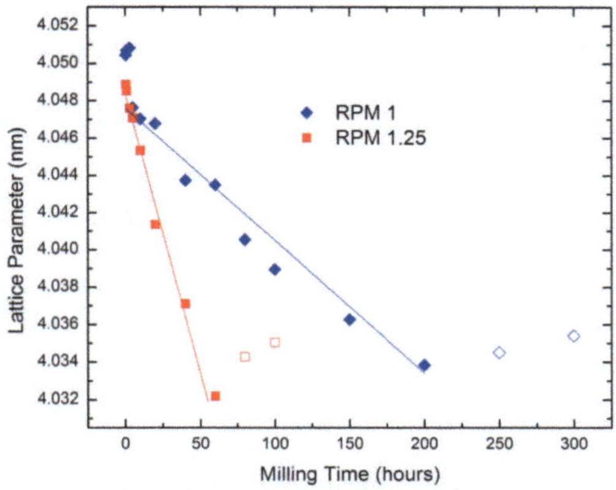

Figure 3. **X-ray diffraction results for the mechanically alloyed Al-Mn samples as a function of milling energy. The lattice parameter as a function of milling time for both milling energies shows a linear increase, until precipitation is observed in the pre-milled Al-Mn and non-pre-milled Al-Mn alloys.**

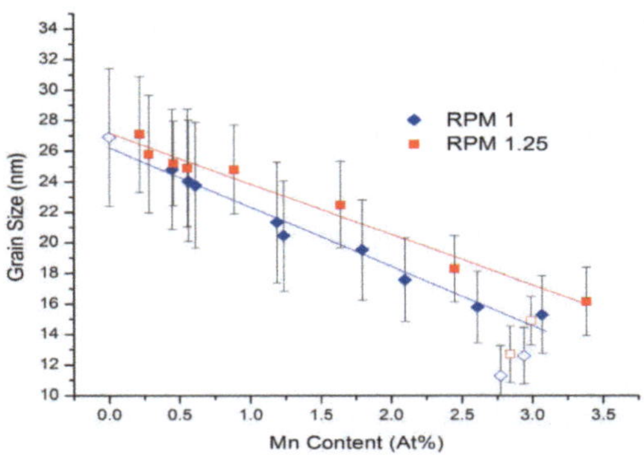

Figure 4. **X-ray diffraction results for the mechanically alloyed Al-Mn samples as a function of milling energy. The mean grain size of the Al-rich matrix as a function of dissolved Mn content is also shown to have a simple linear relationship until precipitation is observed.**

References

1. MA Meyers, A Mishra, and DJ Benson, ***Prog. Mat. Sci.,*** 51 (2006) 427.

2. K.A Darling, MA Tschopp, RK Guduru, WH Yin, Q Wei, and LJ Kecskes, ***Acta Mat***., 76 (2014) 168.

3. MA Tschopp, HA Murdoch, LJ Kecskes, and KA Darling, ***JOM***, 66 (2014) 1000.

4. KA Darling, AJ Roberts, Y Mishin, SN Mathaudhu, and LJ Kecskes, ***J. Alloy Comp***. 573, 142 (2013).

5. KA Darling, BK VanLeeuwen, JE Semones, CC Koch, RO Scattergood, LJ Kecskes, and SN Mathaudhu, ***Mat. Sci. Eng. A***, 528 (2011) 4365.

6. T Chookajorn, HA Murdoch, and CA Schuh, ***Science***, 337 (2012) 951.

7. HA Murdoch, CA Schuh, ***Acta Mat.,*** 61 (2013) 2121.

8. KA Darling, MA Tschopp, BK VanLeeuwen, MA Atwater, and ZK Liu, ***Comp. Mat. Sci***., 84 (2014) 255.

9. M Saber, H Kotan, CC Koch, and RO Scattergood, ***J. App. Phys***., 113 (2013) 063515.

10. M Saber, H Kotan, CC Koch, and RO Scattergood, ***J. App. Phys***., 114 (2013) 103510.

11. AInoue, M Watanabe, HM Kimura, F Takahashi, A Nagata,and T Masumoto, ***Mat. Trans. JIM,*** 33 (1992) 723.

12. MN Rittner, JR Weertman, JA Eastman, KB Yoder, and DS Stone, ***Mat. Sci. Eng. A***, 237 (1997) 185.

13. PG Sanders, M Rittner, E Kiedaisch, JR Weertman, H Kung,and YC Lu, ***Nanostruct. Mat.,*** 9 (1997) 433.

14. SY Ruan, CA Schuh***, Acta Mat***., 57 (2009) 3810.

15. S Ruan, KL Torres, GB Thompson, and CA Schuh, ***Ultramicroscopy,*** 111 (2011) 1062.

16. SS Nayak, M Wollgarten, J Banhart, SK Pabi, and BS Murty, ***Mat. Sci. Eng. A***, 527 (2010) 2370.

17. M Krasnowski, T Kulik, ***Intermetallics,*** 18 (2010) 47.

18. M Krasnowski, T Kulik, ***Mat. Chem. Phys***., 116 (2009) 631.

19. TT Sasaki, T Ohkubo, and K Hono, ***Acta Mat***., 57 (2009) 3529.

20. KV Rajulapati, RO Scattergood, KL Murty, Z Horita, TG Langdon, and CC Koch, ***Metall. Mat. Trans.*** A, 39A (2008) 2528.

21. KA Darling, AJ Roberts, L Armstrong, D Kapoor, MA Tschopp, LJ Kecskes, and SN Mathaudhu, ***Mat. Sci. Eng. A,*** 589 (2014) 57.

22. BD Stock, SR Cullity, Elements of X-Ray Diffraction, Upper Saddle River, NJ, 2001.

23. KV Rajulapati, Thesis "Synthesis and Mechanical Properties of Two Phase Nanostructured Al Based Composites" p 81, http://www.lib.ncsu.edu/resolver/1840.16/4561

24. H Jones, ***J. Mat. Sci. Let.,*** 1 (1982) 405.

Advanced Composites for Aerospace, Marine, and Land Applications II
Edited by: T. Sano and T.S. Srivatsan
TMS (The Minerals, Metals & Materials Society), 2015

Al-NiTi METAL MATRIX COMPOSITES FOR ZERO CTE MATERIALS: FABRICATION, DESIGN, AND MODELING

Adam Hehr[1], Xiang Chen[2], Joshua Pritchard[1], Marcelo J Dapino[1], Peter M. Anderson[2]

[1]Department of Mechanical & Aerospace Engineering, The Ohio State University, 201 W. 19th Avenue; Columbus, OH 43210, USA
[2]Department of Materials Science & Engineering, The Ohio State University, 2041 College Rd; Columbus, OH 43210, USA

Keywords: Composite, Shape Memory Alloy, Characterization, Modeling

Abstract

Al-NiTi composites fabricated via ultrasonic additive manufacturing (UAM) provide a light-weight solution for low thermal expansion applications. It is shown that the thermal expansion of Al 6061 can be reduced by over 50% by incorporating a 13% volume fraction of NiTi fibers. This reduction in thermal expansion occurs from the contraction of the NiTi fiber during heating, thereby offsetting the thermal expansion of the Al matrix. Al-NiTi composites are made possible by low temperature UAM process. Successful implementation of these composites requires a careful design approach that includes the processing characteristics as well as the thermo-mechanical response of the shape memory fibers and matrix. This is achieved using a NiTi microstructure based FEA model implemented that captures the underlying thermomechanical response of the NiTi fibers and calculates the complex stress state within the composite.

Introduction

Ultrasonic additive manufacturing (UAM) is a solid-state consolidation process where thin foils of similar or dissimilar metals are ultrasonically joined in a layer-by-layer process to form gapless, 3D metal parts [1]. Periodic machining is utilized during the UAM process to achieve complex designs, features, and to embed various objects into the structure, such as reinforcing fibers. A schematic of the UAM process is shown in Figure 1. Due to the physics of ultrasonic welding, metallic bonding takes place at temperatures far below metallic melting temperatures [2]. Thus, temperature-sensitive metals or components such as nickel titanium (NiTi) shape memory alloys can be integrated into metallic structures. The weld power of current UAM systems has increased nearly an order of magnitude over the past five years. The increase in weld power and available down force remedies poor interfacial properties observed in early UAM builds [3].

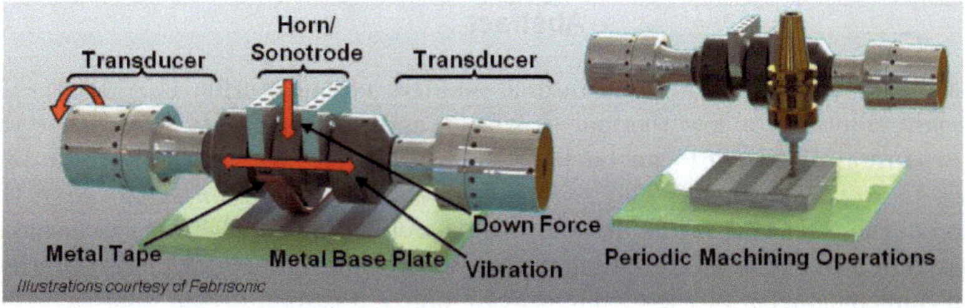

Figure 1: UAM process for developing low temperature metal-matrix composites.

The low formation temperature and additive nature of UAM enables the manufacture of NiTi-Al composites. NiTi fibers can be integrated into the Al matrix to counteract thermal expansion upon heating. Specifically, when the composite is heated, the strain recovery of the NiTi fibers counters the expansion of the aluminum matrix. This combination creates a light, stiff, and thermally-stable material for engineering applications. Previous efforts have shown a 60 % reduction in the average coefficient of thermal expansion for Al 3003 up to 100°C [4]. Yet, metallic bonding between the aluminum matrix and NiTi fibers was not always observed. Instead, the interface was believed to be predominately supported by mechanical coupling in the form of a friction fit since the preexisting oxide layer on the fiber was still present after welding [5]. The bonding between the Al matrix and NiTi fibers was improved by removing the preexisting oxide layer; the results were evaluated tested in a recent interface study using Al 6061 [6]. Fiber pullout testing and Energy Dispersion X-ray Spectroscopy (EDS) were used to test the Al-NiTi interface strength and bond character, respectively. Fiber pullout testing showed that the matrix failed for all surface finishes under investigation, including those with the preexisting oxide. On the other hand, EDS supported the possibility of metallic bonding between the Al matrix and NiTi fibers when the preexisting oxide was removed prior to welding since little to no oxide concentration was observed at the NiTi-Al interface after welding. Lastly, this study showed that circular fiber geometries exhibited strong consolidation character because the matrix was able to flow in and around the fibers during fabrication. The resulting interface exhibited strong resistance to ply delamination.

The focus of this paper is to test these recent NiTi fiber consolidation and bonding observations and to develop an accurate model that describes composite behavior. Multi-layer NiTi-Al composites were fabricated and tested with Al 6061 and mechanically polished NiTi fibers. The corresponding simulations utilized a microstructural constitutive model of NiTi implemented in ABAQUS. This microstructural model describes underlying NiTi physics, i.e., texture effects, martensite variant interaction, and crystal plasticity. Overall, it can simulate a variety of engineering materials and predict the stress redistribution and local thermo-elastic response within the composite. Previous models of NiTi-Al composites utilized a phenomenological approach implemented in MATLAB [7]. Although they require little computational power, these models captured only the axial stress state of the NiTi fiber and did not predict the interface or shear stress. The simulations reported in this work are calibrated to single fiber mechanical tests using a crystallographic description of the transformation process [8].

Methods

Material Stabilization (Training) and Characterization

Mechanically polished 0.381 mm (0.015") diameter NiTi fibers were obtained from Nitinol Devices & Components, Inc. for this study. The fibers had an austenite start temperature above room temperature, were trained to be straight. A stabilization process was applied to the fibers prior to composite construction [9] in order to minimize changes in flow stress and inelastic strain when the NiTi fibers were incorporated into the Al matrix. The stabilization process involved cycling the material 10 times at 75°C to ensure that the material was in the super-elastic state, i.e., above the Austenite finish temperature. These cycles were performed using a TestResourses 131R1000-6 tensile frame with MTS Screw Action Grips with serrated faces and a thermal chamber.

The stabilized NiTi fibers were then characterized using isothermal tensile testing and differential scanning calorimetry (DSC). Isothermal cyclic tensile tests were performed on a 158.8 mm (6.25 in) long piece of NiTi fiber at 65°C, 75°C, and 85°C to quantify the stress as a function of temperature. The nominal strain rate was 0.8%/min to minimize internal heating during the test in accordance with ASTM F2516, Standard Test Method for Tension Testing of Nickel-Titanium Super-Elastic Materials. The nominal strain was determined by dividing measured displacement by the fiber's length.

The austenitic transformation temperatures were measured using a TA Instruments 2920 Differential Scanning Calorimeter. The sample was first cooled to near −10°C to ensure that the NiTi fiber was in the martensitic state or below the martensite finish (M_f) temperature. The sample was then heated to 100°C at 10°C/min while simultaneously measuring the reference heat flow. The transformation temperatures were determined by fitting lines to both sides of the transformation peak and finding the intersections between the plateaus as specified in ASTM F2004, Standard Test Method for Transformation Temperature of Nickel-Titanium Alloys by Thermal Analysis. This DSC system was not used to measure the martensitic transformation temperatures because reliable results for cooling could not always be achieved.

Composite Design, Fabrication, and Testing

A 9 kW UAM system was utilized (Figure 2). The metal matrix used in this study was Al 6061-H18.[1] Foils were 0.152 mm (0.006 in) in thickness and 25.4 mm (1 in) in width. Aluminum 6061 was chosen for its high strength to weight ratio and close compatibility with UAM.

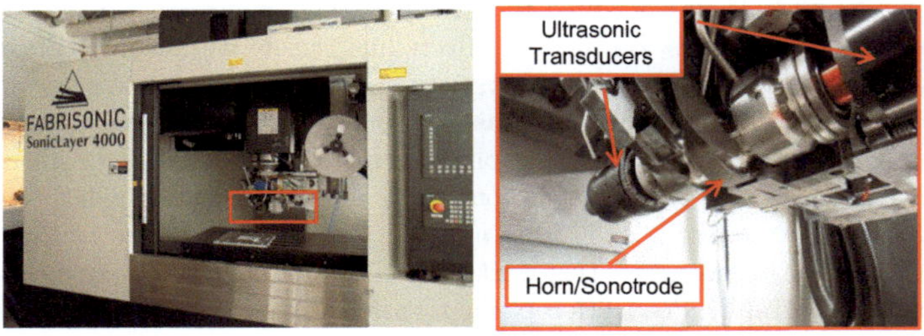

Figure 2: State-of-the-art UAM system fitted with a 9 kW weld head (inset), a 3-axis CNC mill with 25 HP, 8000 RPM spindle, and integrated 40 W laser.

The composite design for this study placed NiTi fibers away from shared interface boundaries to minimize delamination risk. The upper and lower portion of a fiber did not share the same common boundary with the upper and lower portions of a neighboring fiber, see Figure 3(a). A 0.397 mm (0.016 in) diameter ball-nose end mill was used to cut a 0.356 mm (0.014 in) deep channel in the Al matrix to assist in the encapsulation and placement of the NiTi fibers during consolidation. The channel dimension was 0.025 mm (0.001 in) less than the NiTi fiber diameter to enhance scrubbing and plastic flow of the Al around the fiber and into the pocket, Figure 3(b). Welding was conducted on 101.6 mm $\times$ 76.2 mm $\times$ 4.76 mm (4.0 in $\times$ 3.0 in $\times$ 0.188 in) Al 6061-T6 build plates that were constrained in place using a custom base plate, fabrication fixture, and vacuum chuck. Table 1 shows the parameters that were used to weld the aluminum tapes. These parameters were obtained from a recent design of experiments for optimal welding parameters of Al 6061-H18 [10]and they have been shown to be sufficient in pilot NiTi-Al welds.

[1]Al 6061-H18 was fabricated by work hardening annealed 6061 material

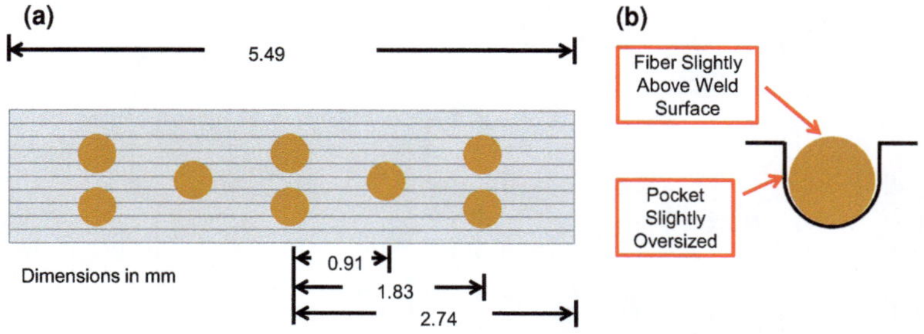

Figure 3: Composite design: (a) fibers were strategically placed to eliminate shared interface boundaries and minimize possible delamination; (b) pocket design for fiber placement.

Table 1: Ultrasonic welding parameters used in study.

Parameter	Value
Temperature	22°C (70°F)
Force	6000 N
Amplitude	34.6 μm (74%)
Speed	84.6 mm/sec (200 in/min)

Removal of the completed composite from the surface of the build plate was done in multiple steps: (i) CNC milling to reduce the composite geometry to the desired dimensions, (ii) manual milling to remove the build plate from under the composite, and (iii) a slow speed cutting with a high volume coolant flow to remove the end of the composite attached to the remaining build plate without significant heat generation. Two composites were fabricated. Each measured approximately 5.49 mm × 74.00 mm × 1.27 mm (0.22" × 2.91" × 0.05"). The NiTi volume fractions for the two composites were 13% and 13.8% respectively. Slight manufacturing variations in the removal process caused the volume fraction to be slightly different.

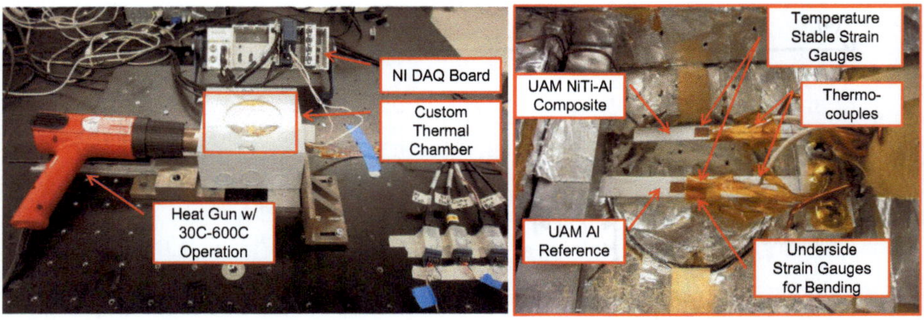

Figure 4: Thermal test utilized to measure composite CTE. Important test elements have been detailed.

Thermal testing was performed on the composites using a custom 15.24 cm × 15.24 cm (6.0" × 6.0") thermal chamber with foil-faced fiberglass insulation, Figure 4. The top of the chamber was equipped with a circular viewing window. A Milwaukee Variable Temperature Heat Gun supplied heated air to the chamber, and this air was deflected upon entering the chamber with an aluminum baffle to produce even heating of the samples. Composite CTE measurements were made by mounting strain gauges on a reference piece of UAM Al 6061 and composite so that the relative strain difference between the two materials could be determined [11]. The composite and the reference sample were mounted within the chamber in a free-stress, cantilevered condition. Micro Measurement WK 13031CF350 strain gages were mounted to the top and bottom sides of both samples, approximately 3.8 cm (1.5") from the fixed end. The temperatures of each sample were measured using type-J thermocouples with Omegatherm 201 thermal conductivity paste at the composite thermocouple interface to ensure accurate measurements. The outputs of the strain gauges and thermocouples were connected to a National Instruments data acquisition system, which interfaced with a LabVIEW VI for recording data. Before each sample was tested, the strains were zeroed using the calibration features within LabVIEW. Each test was performed by heating the sample from room temperature to approximately 100°C using the slowest speed setting on the heat gun (near 45 min. to reach 100°C). Samples were then cooled back to room temperature by setting the heat gun to 30°C and allowing the chamber to equilibrate with the room once the heat gun could no longer provide cooling (near 1 hr).

Composite Modeling

Numerical simulations of the composite were conducted to better understand the behavior of the fabricated Al-NiTi composite, especially the matrix-fiber interaction that determines its performance. The primary challenges in modeling are to accurately capture the response of isolated NiTi-fibers and to capture the physics of stress redistribution within the fiber-matrix composite. A microstructural finite element constitutive model [8] was utilized to meet both challenges. This model is implemented as a User-defined MA-

Terial (UMAT) subroutine in the commercial finite element package ABAQUS [12]. The NiTi fiber response was calibrated to the stress-strain response at three different temperatures and the critical transformation temperatures determined from Differential Scanning Calorimetry (DSC). The full list of calibrated parameters of the NiTi model is summarized in Table 2. The phase transformation between austenite and 24 correpondence variant pairs (CVPs, a.k.a., martensite plates) is controlled by the driving force per unit volume described by

$$f_t = \mathbf{b}_t \cdot \left((\mathbf{F}^e)^{\mathrm{T}} \mathbf{F}^e \mathbf{T}^* \right) \mathbf{m}_t - \frac{\lambda_{\mathrm{T}}}{\Theta_{\mathrm{T}}}(\Theta - \Theta_{\mathrm{T}}) - \sum_{u=1}^{N_T} \mathbf{h}_{tu} v_u$$

where t (= 1 to 24) denotes the martensite plate type; $\mathbf{b}_t$ and $\mathbf{m}_t$ are the shear magnitude and shear plane normal vectors associated with the transformation from austenite to martensite plate type t, $\mathbf{T}^*$ is the 2nd Piola-Kirchhoff stress, and $\mathbf{F}^e$ is elastic part of the deformation gradient; λ_{T} is the transformation latent heat, Θ_{T} is the thermodynamic equilibrium temperature, Θ is the current temperature; v_u is the volume fraction of plate type u, and the $\mathbf{h}_{tu}$ matrix specifies the interaction energy among the plates. In terms of physical significance, the first term on the right-hand side expresses the mechanical contribution to the driving force of transformation from austenite to CVP-type t; the second term expresses the thermal contribution; and the third term expresses the inter-CVP interaction. Evolution of volume fraction of each martensite CVP (v_t) is controlled by the following consistency condition

$$-f_c \leq f_t \leq f_c$$

where f_c is the critical force/unit volume for transformation. This will be determined through calibration. For further details regarding the development and implementation of the SMA model, the readers are referred to Ref. [13] and [8].

Table 2: Material parameters of the NiTi fiber (calibrated) and Al matrix (from [4]).

Category	**Parameters and their calibrated/adopted values**
Elastic and thermal properties of the fiber	Elastic constants of austenite (GPa): $C_{11(\mathrm{A})} = 130$; $C_{12(\mathrm{A})} = 98$; $C_{44(\mathrm{A})} = 34$; Elastic constants of martensite (GPa): Same as austenite Thermal expansion coefficients (10^{-6}/K): $\alpha_{\mathrm{A}} = 11$; $\alpha_{\mathrm{M}} = 6.6$
Phase transformation properties of the fiber	Equilibrium transformation temperature (K): $\Theta_{\mathrm{T}} = 303$; Latent heat of transf./unit vol. (J/cm^3): $\lambda_{\mathrm{T}} = 151$; Hardening matrix $\mathbf{h}$: $h_{tu} = \begin{cases} C_{44(\mathrm{A})}/1500 \ (t = u) \\ C_{44(\mathrm{A})}/3000 \ (t \neq u) \end{cases}$; Critical driving force force (J/cm^3): $f_c = 6.0$; Number of martensite CVPs: $N_{\mathrm{T}} = 24$; Max. principal shear strain by one CVP: $\lvert\mathbf{b}_t\rvert = 0.1308$;
Elastic and thermal properties of Al	Young's modulus (GPa): $E_{\mathrm{Al}} = 68.0$; Poisson's ratio: $\nu_{\mathrm{Al}} = 0.33$; Thermal expansion coefficients (10^{-6}/K): $\alpha_{\mathrm{Al}} = 23.6$

Since the matrix-fiber interface is expected to yield or fracture much earlier than either of the pure phases, plastic deformation of the fiber is not incorporated into the modeling in this work. This is accomplished by setting the reference plastic slip rate to zero in the NiTi model. The fiber is meshed with eight-node, coupled-temperature-displacement, reduced-integration elements (C3D8RT) in the ABAQUS element library. Each element is treated as a separate austenite grain, and assigned a crystallographic orientation within a random variation of magnitude of $\pm 5°$ from $[110]_{B2}$, approximating the RD texture reported for the same Nitinol product in Ref. [14], as suggested by the manufacturer. A total of 800 C3D8RT elements (and therefore, grains) are used for the fiber. The texture data used in the simulation has been found to significantly influence the predicted composite performance, a result in agreement with previous works [15].

In the composite simulation, only one-eighth of a fiber–containing Representative Volume Element (RVE) is modeled by applying a mirror-plane condition to the $x = 0$, $y = 0$ and the $z = 0$ planes. Comparison with the full model showed that no significant difference results from such a simplification. Figure 5 shows the RVE, which contains 13 vol.% of fiber (shaded part). Considering the mirror plane perpendicular to the fiber axis, this RVE corresponds to a length-to-diameter ratio of 20:1, which is in agreement with the experimental geometry. The aluminum matrix is modeled as purely elastic, with properties given in Table 2. To further simplify the computation, a uniform temperature over the whole model is assumed at any given instant so that heat conduction within the sample is not explicitly modeled. Also, the matrix-fiber bonding is assumed to be perfect, i.e., no-slip.

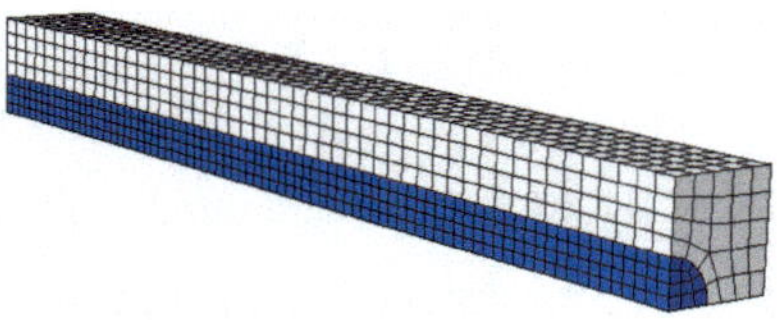

Figure 5: The finite element model of the composite, using one-eighth of a fiber-containing RVE. Three mirror-plane conditions are applied – two of them through the center axis of the fiber (shown in blue), while the remainder is perpendicular to the fiber.

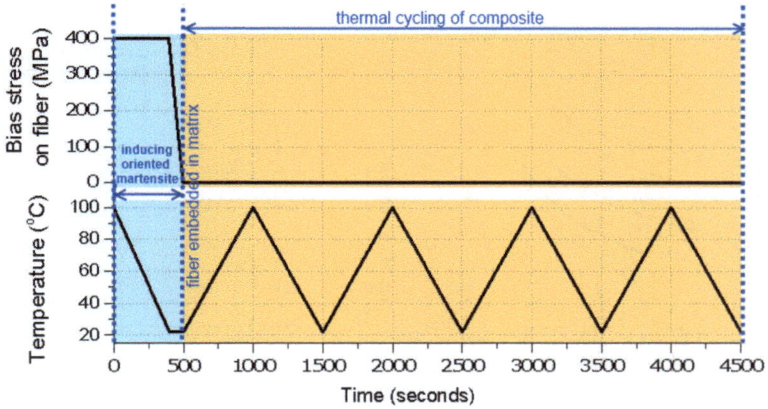

Figure 6: Simulated mechanical and thermal loadings: the NiTi fiber is first cooled under a tensile stress of $\sigma_{\text{bias}} = 400$ MPa in the absence of Al matrix to induce oriented martenite, and then unloaded at 22°C (blue regime). This is followed by embedding the fiber in matrix at t = 500s. The heating-cooling cycle of the composite from 22°C to 100°C is repeated 4 times (orange regime) to study the evolution of performance.

Two simulation efforts are documented here. The first one is the calibration of NiTi fiber parameters against isothermal tensile loading-unloading curves at three different temperatures: 65°C, 75°C and 85°C, as well as the DSC curve measured during heating of the NiTi fiber at its pre-deployment state. This is followed by simulations of the composite at working condition. The double Y-axis plot of Fig. 6 shows the loading history of the deployment (blue regime) and that of the subsequent thermal cycles (orange regime). A total of four thermal cycles between 22°C and 100°C are simulated.

Results

NiTi Fiber Performance

Characterization results from isothermal tensile testing and DSC are shown in Figure 7. It should be noted that, as described in the **Methods** section, the behavior shown in Figure 7(a) is achieved after mechanically cycling the NiTi fiber multiple times in a super-elastic state, thereby stabilizing the response of the as-received material. This is due to the crystallographic defects induced during the stabilization process, such as transformation-induced dislocations [16]. Since the focus of this work is the composite behavior, the NiTi model is calibrated to the stabilized behavior. The resulting parameters are listed in Table 2. As shown in Figure 7, the calibrated NiTi fiber model captures both the uniaxial tension and DSC behavior reasonably well.

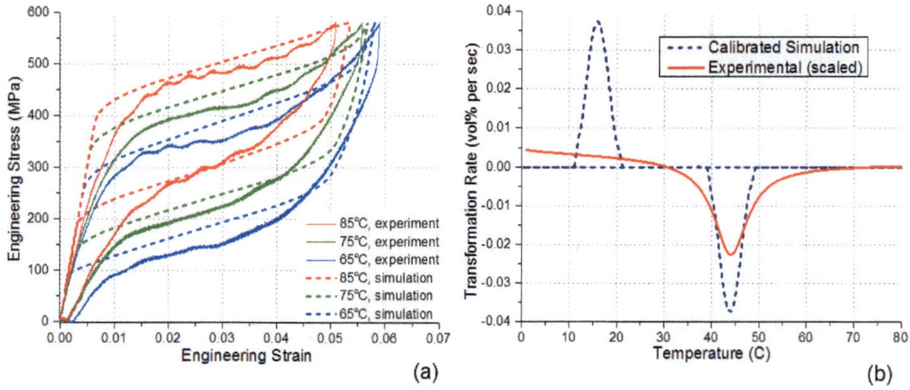

Figure 7: Model calibration: (a) against isothermal stress-strain curves; (b) against DSC

Experimental Composite Performance

Figure 8 shows that empirical strain-temperature results for both NiTi-Al composites and pure UAM Al-6061. The composites begin transforming near 40°C and exhibit compression from the NiTi. The first cycle demonstrates enhanced compression and curvature during heating, which may be linked to the collapse of voids within the UAM build. Lastly, the NiTi-Al composites exhibit CTEs less than 50% of Al-6061.

Figure 9 shows the optical images of the end of the composite before and after heating integrity of the fiber-matrix interface. The composite end was imaged to assess the shear transfer taking place between the NiTi fiber and Al matrix. Damage in the composite due to testing is likely to initiate at the composite end. The image is inverted since the image is not of the same sample but is of the same cross section, i.e., some material was removed between samples with a saw. As a result, key fibers have been boxed. Some voids are present in the composite prior to testing, yet no obvious cracking can be seen. After testing, the composite remains reasonably crack free.

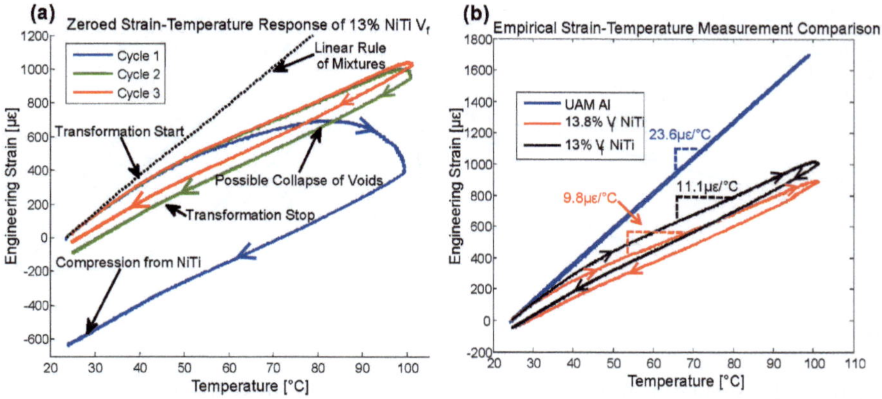

Figure 8: Empirical strain-temperature results: (a) 13% NiTi volume fraction composite for 3 cycles; (b) comparison of Al (no NiTi), composite with 13% volume fraction and composite with 13.8% volume fraction. Fit slopes from 40 to 100°C are given.

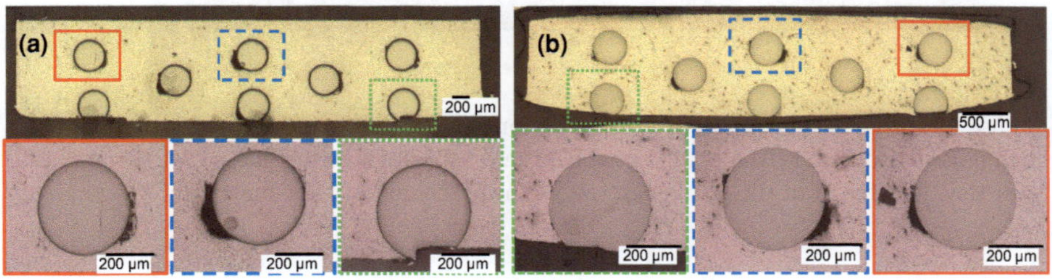

Figure 9: End cross section of 13.8% NiTi composite: (a) the composite before heating, some void presence can be seen with no obvious cracking; (b) the composite after 3 heating cycles, little to no cracking formed during thermal cycling.

Simulated Composite Performance

The composite strain-temperature response from the FE simulations is shown in Figure 10(a). The composite strain is measured and averaged over the middle cross-section plane of the matrix, thus avoiding end effects (or St. Venant effects, Figure 11).

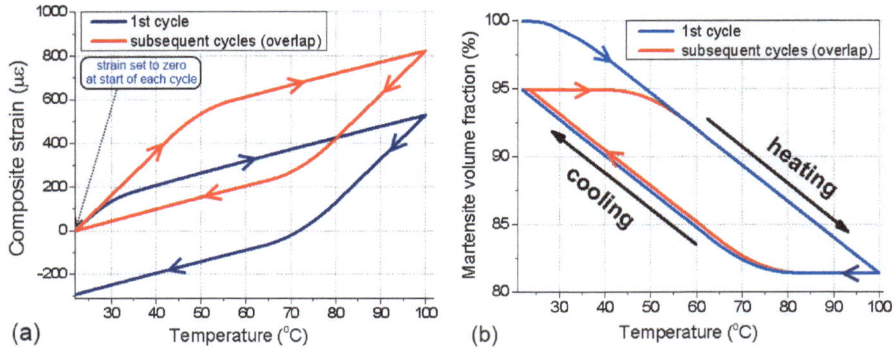

Figure 10: FE simulation results: (a) Strain vs. temperature curve for the first 4 cycles, of which the latter three effectively overlap; (b) Evolution of martensite volume fraction in the first 4 cycles.

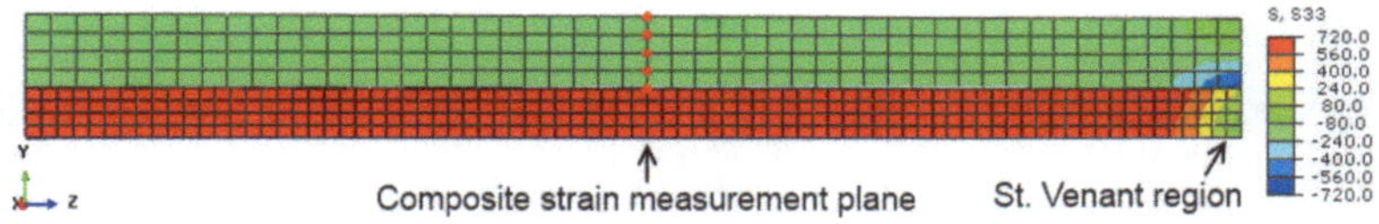

Figure 11: Axial normal stress (σ_{zz}) distribution at $T = 100°C$ during the 1st cycle. The highlighted nodes show the plane on which the composite strain is measured. It can be seen that it is far from the St. Venant region near the end of fiber-matrix interface.

Several observations can be made regarding the comparison between the simulation result in Figure 10(a) and experimental result in Figure 8(a). First, the pronounced difference between the 1st-cycle and subsequent-cycle behaviors is partially captured in the FE simulation, the reason of which will be discuss later. Second, the transition in apparent CTE of the composite in the temperature regime of 40 - 60°C is also captured for subsequent cycles. This corresponds to the onset of the martensite-to-austenite transformation, as shown in Figure 10(b). Third, the contraction above 70°C in Figure 8(a) is absent in Figure 10(a), which implies that the FE model does not capture these physics. Fourth, the simulation slightly under-predicts the maximum strain at 100°C, under-predicts the strain hysteresis for the 1st cycle, and over-predict the strain hysteresis for "subsequent" cycles. These discrepancies are likely interconnected, as will be discussed in the next section.

Discussion

Evolution of Fiber State

The thermo-mechanical mechanisms underlying the composite behavior are discussed first. The driving force for the change of CTE of the composite arises from axial contraction of the NiTi fiber. This contraction arises from the phase transformation from an oriented martensite (B19') phase to a high temperature austenite (B2) phase. Constrained by the Al matrix, this phase transformation competes with the conventional thermal expansion of the Al matrix as the composite is heated. This generates an internal stress field that reduces the amount of expansion in the Al matrix and, at the same time, limits the shape recovery of the NiTi fiber.

The evolution of the NiTi fiber state is schematically shown in Figure 12. The fiber starts at the fully reoriented martensite state (Point A) at the beginning of the 1st-cycle. During heating, it enters in the martensite-to-austenite (M-A) transformation region, but the transformation is only partial (to Point B), as revealed by the martensite volume fraction of 82% at $T = 100°C$ in Figure 10(b). As the sample is cooled from 100°C, the martensitic (A-M) transformation occurs. However, it is again incomplete by the end of the cooling, achieving only 95% martensite. Therefore, the fiber still lies in the A-M (blue) region at the end of the 1st cycle, as represented by Point C in Figure 12.

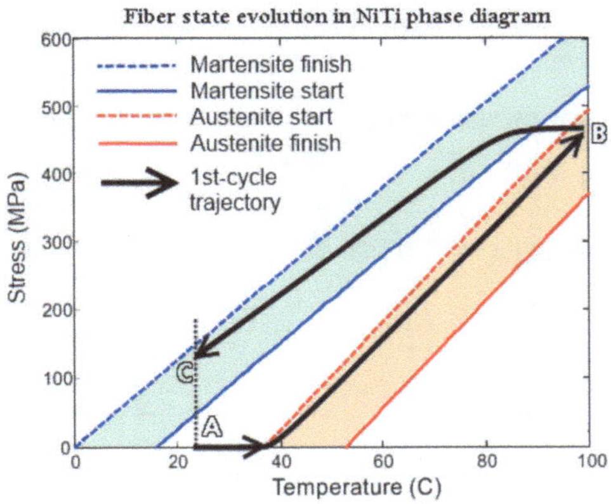

Figure 12: Stress-Temperature phase diagram of the NiTi fiber. The upper (blue) region corresponds to the austenite-to-martensite (A-M) transformation region, and the lower (red) region corresponds to the martensite-to-austenite (M-A) transformation region. No transformation happens in the region between these, and the volume fraction of phases here totally depends on previous path. The 1st-cycle trajectory shown here is based on the present finite element simulations.

The partial A-M transformation is found to be the mechanism behind the compressive strain at the end of the 1st cycle shown in Figure 10(a). It arises from the difference in

A-M mechanical driving forces during the reorientation stage (blue regime in Figure 6) and that during in situ cooling in the composite. Since this difference is eliminated with the completion of the 1st cycle, subsequent cycles shift the fiber state between Point C and Point B in Figure 12, therefore they overlap in the strain-temperature curve in Figure 10(a) and nearly overlap in Figure 8(a).

Not-captured Experimental Features

The largest difference between experiment (Figure 8(a)) and simulation (Figure 10(a)) is the contraction at the high temperature end of the 1st cycle. The mechanisms considered in the FE model does not capture this feature. Since there is experimental evidence that the voids in matrix are generated in the UAM process, it is possible that collapse of these voids may be responsible for this behavior. This hypothesis agrees with three known facts: i) The level of internal stress in the composite increases sharply with temperature at $T > 50°C$. Consequently, the contraction is observed at the most stressed state and agrees with void collapse; ii) The contraction does not occur in subsequent cycles. iii) The compressive strain at the end of the 1st cycle is significantly larger in the experiments (Figure 8a) compared simulations (Figure 10a). All of these support the hypothesis that the aluminum matrix plastically deforms through the collapse of voids.

To further evaluate the void collape hypothesis, the 13.8% NiTi composite was cross sectioned and polished at the strain gauge location to examine void presence. Cross sections before and after testing is shown in Figure 13. The significant reduction in void volume fraction from thermal cycling supports the collapse-of-void hypothesis.

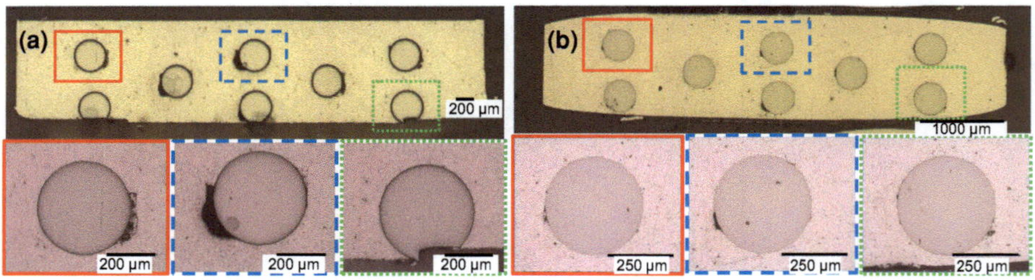

Figure 13: Comparison of end and strain gauge cross sections of the 13.8% NiTi composite: (a) the composite before heating, with significant voids; (b) the composite after 3 heating cycles, showing reduced void volume fraction.

The collapse of voids and axial plastic contraction of the matrix during the first cycle may explain the under-estimation of composite strain and over-estimate of strain hysteresis in subsequent cycles. The decrease in the natural length of the matrix during the first cycle will make the axial stress in fiber lower than in the FE simulations (~110MPa). In Figure 12, a lower stress than that at Point C will reduce the volume fraction of stress-induced martensite volume fraction at 22°C to less than the 95% predicted value shown in Figure 10b). The reduced martensite volume fraction will decrease the available martensite for transformation. Since the ability of contraction in fiber is determined by the amount of transformation that occurs during a thermal cycle, a lower martensite fraction at 22°C means lower constraint on the thermal expansion of the Al matrix. This can explain why the observed strain in subsequent cycles is higher in the experiments compared to simulations. Also, a lower martensite fraction at room temperature implies that transformation during subsequent thermal cycles is also lower, which agrees with a smaller strain hysteresis.

Conclusions

NiTi-Al composites provide a light weight alternative compared to iron-based thermally invariant materials. This was demonstrated by fabricating Al 6061 matrix composites with approximately 14 vol% aligned commercially-available NiTi wires. The composite exhibited a reduced coefficient of thermal expansion of 11.1 $\mu\varepsilon/^{\circ}$C over the range 45-100 C, compared to 23.6 $\mu\varepsilon/^{\circ}$C for the matrix. A state-of-the-art microstructural NiTi model implemented in ABAQUS was developed to capture key composite and NiTi physics. The simulations were calibrated to single fiber stress-strain and differential calorimetry experimental data. The predicted composite response qualitatively captured the experimental results, including an initial axial contraction of the composite 10^{-4} during the first thermal cycle and a reduced coefficient of thermal expansion during subsequent cycles. Differences between the experiments and simulations s arise from the model assumption that the matrix does not plastically deform. However, experimental observations show the collapse of voids around fibers during thermal cycling. The results suggest that the Al matrix plastically deforms during initial thermal cycling and then stabilizes. In principle, the simulations can incorporate plastic deformation and work hardening of the matrix. Overall, this combined experiment-simulation approach can be used to accelerate the development of thermally invariant, shape memory based composites.

Acknowledgments

This work is supported by the NSF I/UCRC on Smart Vehicle Concepts at The Ohio State University (www.SmartVehicleCenter.org). Support for A.H. comes from a National Science Foundation Graduate Fellowship under Grant No. 1102690. Any opinions, findings, and conclusions or recommendations expressed in this material are those of the author(s) and do not necessarily reflect the views of the National Science Foundation. X.C. and P.M.A. acknowledge the support of the NSF (Division of Materials Research, DMR 1207494), DOE (Office of Basic Energy Sciences, SC-0001258), and the Ohio Supercomputer Center (Grant PAS0676). The technical assistance of Phillip Evans (MIT Lincoln Laboratory), Walter Green (OSU), Steven Bright (OSU), and Richard Boger (Dassault Systèmes Simulia Corp) is gratefully acknowledged.

References

[1] K. Graff, Ultrasonic Additive Manufacturing, American Society for Metals International, 2011.

[2] M. Sriraman, M. Gonsera, H. Fujii, S. Babu, M. Bloss, Thermal transients during processing of materials by very high power ultrasonic additive manufacturing, Journal of Materials Processing Technology 211 (2011) 1650–1657.

[3] K. Graff, M. Short, M. Norfolk, Very High Power Ultrasonic Additive Manufacturing (VHP UAM), in: International Solid Freeform Fabrication Symposium, Austin, TX, 2011.

[4] R. Hahnlen, Characterization and Modeling of Active Metal-Matrix Composites with Embedded Shape Memory Alloys, Ph.D. thesis, The Ohio State University, Columbus, OH (2012).

[5] R. Hahnlen, M. Dapino, NiTi-Al Interface Strength in Ultrasonic Additive Manufacturing Composites, Composites Part B: Engineering 59 (2014) 101–108.

[6] A. Hehr, M. Dapino, Interfacial Shear Strength Estimates of NiTi - Al Matrix Composites Fabricated via Ultrasonic Additive Manufacturing, Composites Part B: Engineering In Review.

[7] R. Hahnlen, M. J. Dapino, Stress-Induced Tuning of Ultrasonic Additive Manufacturing, in: Proceedings of SPIE Vol 8342, San Diego, CA, 2012.

[8] S. Manchiraju, P. Anderson, Coupling between martensitic phase transformations and plasticity: A microstructure-based finite element model, International Journal of Plasticity 26 (10) (2010) 1508–1526.

[9] D. Lagoudas (Ed.), Shape Memory Alloys: Modeling and Engineering Applications, Springer, 2008.

[10] P. Wolcott, A. Hehr, M. Dapino, Optimized Welding Parameters of Al 6061 Ultrasonic Additive Manufactured Structures, JMR 29 (18).

[11] Vishay, Measurement of Thermal Expansion Coefficient Using Strain Gages, Tech. rep., Micro-Measurements, tN-513-1 (2010).

[12] ABAQUS 6.11 Documentation, Dassault Systèmes Simulia Corporation, 2011.

[13] P. Thamburaja, L. Anand, Polycrystalline shape-memory materials: effect of crystallographic texture, Journal of the Mechanics and Physics of Solids 49 (4) (2001) 709–737.

[14] K. Kim, S. Daly, The effect of texture on stress-induced martensite formation in nickel–titanium, Smart Materials and Structures 22 (7) (2013) 075012.

[15] H. Paranjape, P. M. Anderson, Texture and grain neighborhood effects on Ni-Ti shape memory alloy performance, Modelling and Simulation in Materials Science and Engineering 22 (7) (2014) 075002.

[16] M. Bowers, X. Chen, M. De Graef, P. Anderson, M. Mills, Characterization and modeling of defects generated in pseudoelastically deformed NiTi microcrystals, Scripta Materialia 78 (2014) 69–72.

Advanced Composites for Aerospace, Marine, and Land Applications II
Edited by: T. Sano and T.S. Srivatsan
TMS (The Minerals, Metals & Materials Society), 2015

CHARACTERIZATION OF THE INTERFACE OF CO-EXTRUDED ASYMMETRIC ALUMINUM-TITANIUM COMPOSITE PROFILES

N. Grittner[1], M. Engelhardt[1], B. Striewe[2], A. von Hehl[2], C. Klose[1], F. Nürnberger[1]

[1] Institut für Werkstoffkunde (Materials Science), Leibniz Universität Hannover, An der Universität 2, 30823 Garbsen, Germany

[2] Stiftung Institut für Werkstofftechnik, Badgasteiner Str. 3, 28217 Bremen, Germany

Keywords: aluminum, titanium, co-extrusion, diffusion bonding, intermetallic layer, ECAP, LACE

Abstract

The combination of different light alloys enables lightweight solutions with tailor-made properties at the macroscopic global as well as at the microscopic scale. In this context co-extrusion by Equal Channel Angular Pressing (ECAP) offers a great potential for advanced profiled structures. While titanium alloys show particular high mechanical strength and good corrosion resistance, aluminum alloys provide a considerable high specific bending stiffness along with low materials costs.

The mechanical properties of metallic compounds strongly depend on the bonding mechanisms, which are initiated during processing. In order to enable a rigid adhesive bond between Al and Ti in this study the extrusions were processed by means of ECAP that is known for reducing the activation energy for diffusion bonding. Hence, the study is focused on the general manufacturing of Al-Ti-compounds by ECAP. The second aim of the study was to investigate the development of the bonding zone during processing. Experiments were made with the material combination Al 99.5 and titanium grade 2. The bonding zones of the co-extruded samples were analyzed by scanning electron microscopy and energy dispersive X-ray analysis. The bond strength was determined by quasi-static tensile tests. Compared to the as-extruded condition an intermetallic layer was formed during heat treatment. The layer was characterized by scanning electron microscopy and electron probe micro analysis.

Introduction

The growing demands on sophisticated lightweight designs in transportation promote the market of customized materials, such as composites as well as compound materials. For the latter one attractive material combination constitutes aluminum and titanium. While aluminum alloys provide a high specific bending stiffness, titanium alloys show high tensile strength combined with an outstanding corrosion resistance. Due to the differing material properties the compound production process has to meet demanding requirements. Frequently applied metallurgical joining technologies are diffusion welding, laser beam welding, compound rolling as well as co-extrusion.

In contrast to conventional welding techniques, where one or both joint partners partially melt, diffusion welding, co-extrusion and compound rolling produce an adhesive bonding under compression at elevated temperatures utilizing diffusion [1, 2]. Due to mechanical deformation during processing compound rolling leads to an extension of contact surfaces promoting bonding mechanism like mechanical clamping and adhesion. Recently, compound rolling has become commercial relevant due to its applicability to mass production [3, 4].

The bonding of different alloys during co-extrusion is primary realized by adhesion. Due to the short extrusion duration diffusion occurs mainly during subsequent heat treatment. Apart from extrusion ratio the extrusion temperature is the crucial factor for the formation of diffusion zones and, thus, the growth of intermetallic phases. A high extrusion ratio and a high temperature promote the bonding [5]. Typical material combinations for compound extrusion are aluminum-steel and aluminum-titanium [6].

Studies on the co-extrusion using the ECAP process were already performed on the material combinations Al 99.95 - Cu 99.95 and the magnesium alloy AZ31 - Al 99,95 [7-9]. It was shown that with both material combinations simple compound profiles with the ECAP process could be produced. Both combinations of materials have directly after extrusion an extremely small diffusion zone, the thickness of the diffusion zone grew only by subsequent heat treatment. It could also be detected a positive effect on the mechanical properties of the composite.

Objectives

The study is focused on the general manufacturing of Al-Ti-compounds by ECAP and the development of the bonding zone during processing.

The co-extrusion experiments were performed with Al 99.5 combined with Ti grade 2. After co-extrusion the mechanical properties of the compounds were determined by means of tensile tests. Besides, the bonding zone was metallographical analyzed both after co-extrusion and after subsequent heat treatment.

Experiments

The aim is to develop a process for the production of asymmetric aluminum-titanium composite profiles with uninterrupted titanium reinforcing elements. For this purpose, an ECAP die with a lateral feed for titanium profiles was designed and put into operation, as seen in Figure 1. The tool design enables to feed the titanium profile laterally by an angle of 90°. In this case the titanium profile is not formed, but the aluminum is pressed laterally to the titanium. Due to two processes a connection occurs: At first a mechanical clamping takes place by pressing the aluminum into bumps of the titanium surface. This creates an intimate material contact, secondly, due to the prevailing temperature and pressure conditions in the contacting zone a diffusion process starts, which is a prerequisite for an adhesive bond.

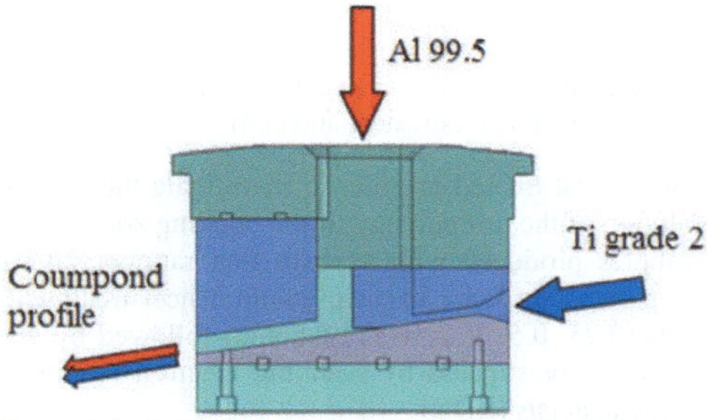

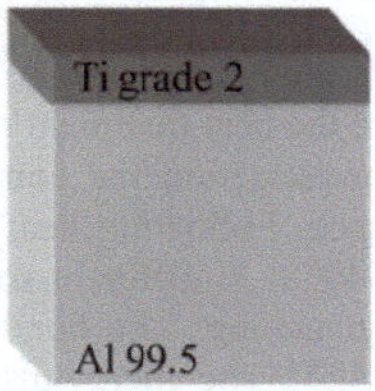

Figure 1a): ECAP die with lateral supply of Ti reinforcement b) Profile geometry

For the experiments the combination of pure titanium (ASTM grade 2) with Al 99.5 (EN AW-1050) was regarded. To influence the extrusion ratio and therefore the conditions in the contact zone, two different profiles were produced: An aluminum profile with a thickness of 70 mm with a 6 mm thick titanium reinforcement, extrusion ratio 4 and an aluminum profile with a thickness of 28 mm with a 2 mm thick titanium reinforcement, extrusion ratio 20. The process was carried out as a direct extrusion process utilizing an extrusion press (SMS Meer) with a maximum press force of 10 MN. The billet preheating temperature was set to 500 °C for 2 h; the die preheating temperature as well as the container temperature was set to 450 °C. All billets had a final diameter of 120 mm and a length of 300 mm. An extrusion speed of 2 mm/s was used for all tests. For the microstructural, mechanical and SEM investigations cross-sections were prepared by cutting transverse to the extrusion direction.

The interface was investigated longitudinal and transverse to the extrusion direction by light and scanning electron microscopy. The samples were taken from the beginning of the profile.

The adhesive strength in the contact zone of the composite was characterized by using the quasi-static tensile test. The experiments were performed on a tensile testing machine of type ElectroPuls E1000 from Instron. The tensile samples were separated by wire cutting from the profile. The geometry of the test-piece is shown in Figure 2 and has a thickness of 1 mm. Due to uneven strain across the length of the sample only the crosshead displacement was regarded. Based on the known entire cross-section of the samples the stress values were calculated to determine the bond strength. The tensile tests were performed on the samples with aluminium profile thickness of 70 mm and a titanium reinforcement thickness of 6 mm (produced with an extrusion ratio of 4). Unfortunately the samples from the profile of the aluminium profile thickness of 28 mm with the 2 mm thick titanium reinforcement (produced with an extrusion ratio of 20) could not be tested in this way due to the small dimension of the titanium reinforcement.

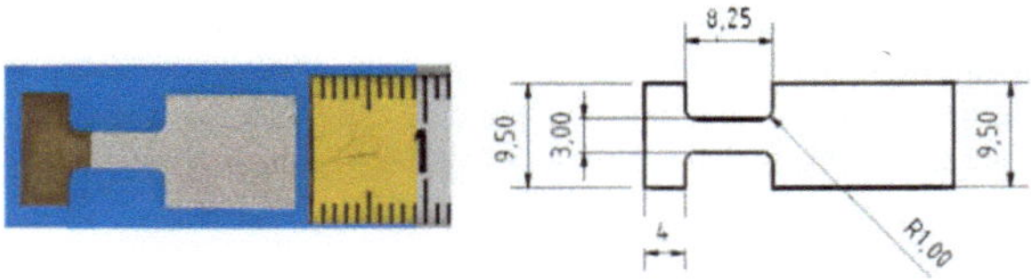

Figure 2: Tensile specimen with an aluminium profile thickness of 70 mm and a titanium reinforcement thickness of 6 mm (produced with an extrusion ratio of 4)

Additionally some samples were heat treated in order to investigate the influence of the heat treatment on the evolution of the morphology of the bonding zone. For this investigation the compound samples produced with an extrusion ratio of 20 were used. Following heat treatment parameters were chosen; solution heat treatment at 540° C at different soaking times (0.25, 0.5, 1, 2, 4 and 8 hours) followed by water quenching. The bonding zones were investigated by scanning electron microscopy (SEM) and energy dispersive X-ray analysis (EDX).

Results and Discussion

The results of the co-extrusion experiments with both extrusion ratios demonstrate that a co-extrusion of aluminium-titanium compound profiles by means of ECAP is feasible. Due to the relative movement inside the die the fed titanium profile was pulled into the die, enabling an intimate contact to the pressed aluminium. Due to different flow rates of the materials the aluminum left the die earlier, but without any gap between the joining partners as to be seen in Figure 3.

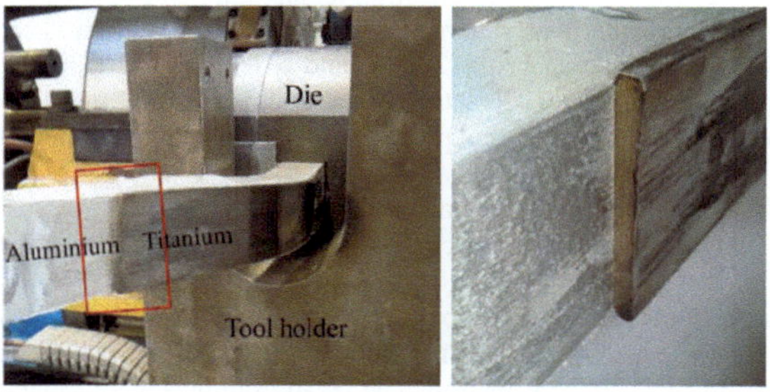

Figure 3: Compound profile consisting of pure titanium (ASTM grade 2) and Al 99.5 (EN AW-1050)

Metallography

The interface layer of the composite samples was investigated with a scanning electron microscope. The specimens were taken longitudinally to the extrusion direction. In Figure 4 and Figure 5 the interfaces of the sample produced with an extrusion ratio of 20 are displayed. There was no gap detected between the joining partners. Due to the high extrusion ratio and the resulting pressure on the interface the titanium surface shows a flaky structure, which suggests an enhancement of the mechanical interlocking. However, due to the huge local deformation micro-cracks can be sporadically detected in the titanium surface, see Figure 5.

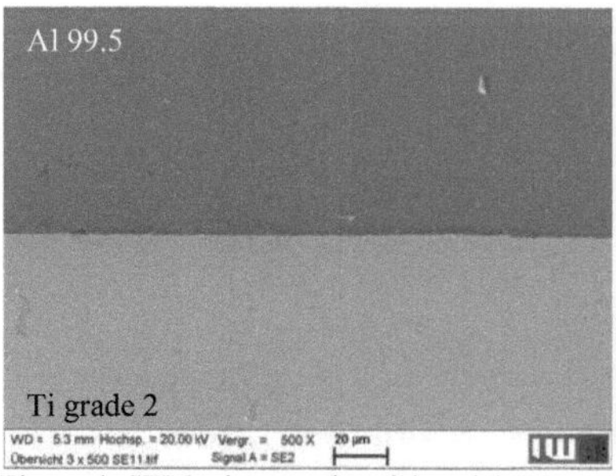

Figure 4: SEM micrographs of the of the interface of the sample produced with an extrusion ratio of 20

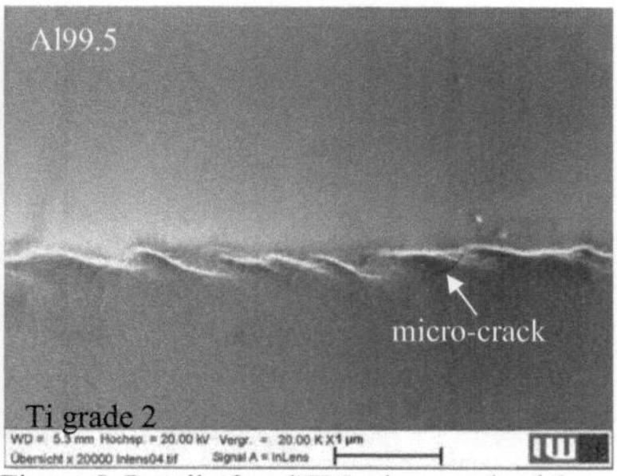

Figure 5: Detail of an SEM micrograph of the interface of the sample produced with an extrusion ratio of 20

In Figure 6 and Figure 7, the interface of the sample produced with an extrusion ratio of 4 is displayed. Even with the lower extrusion ratio of 4 no gap between the joining partners was detected. Due to the lower compression ratio the surface of the titanium is less deformed, thus, the surface of the titanium shows a more even structure. A strong mechanical interlocking is not visible. Due to the rigid joint it can be assumed that the coupling of the two materials is essentially based on intermolecular bonding.

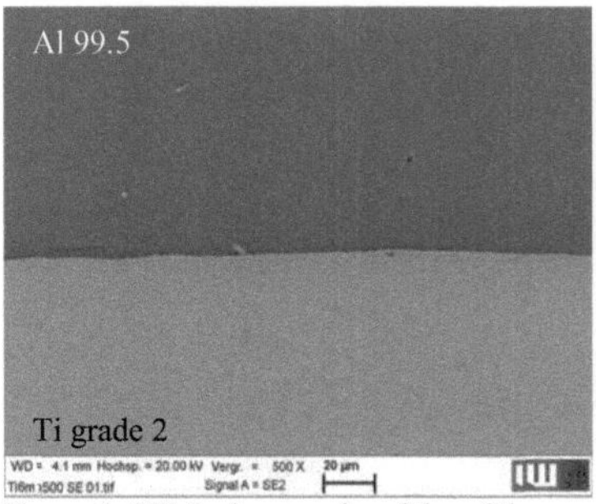

Figure 6: SEM micrographs of the of the interface of the sample produced with an extrusion ratio of 4

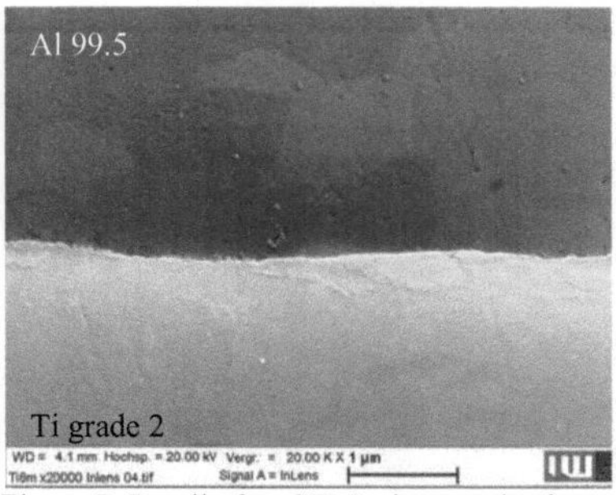

Figure 7: Detail of an SEM micrograph of the interface of the sample produced with an extrusion ratio of 4

To analyze the distribution of the elements in the interface of the compound the sample produced with an extrusion ratio of 4 was selected for an EDX line scan, as shown in Figure 8, the measurement was carried out with a relatively small excitation voltage of 10 kV, to minimize the extent of the excitation bulb in the ground of the substrate and to avoid distortion of the measurement signal largely. The result of the measurement is also shown in Figure 8. A smooth transition between the two elements was measured that may indicate local diffusion processes between the joining partners. A plateau, which usually characterizes the formation of an intermetallic phase in the interface, could not be detected.

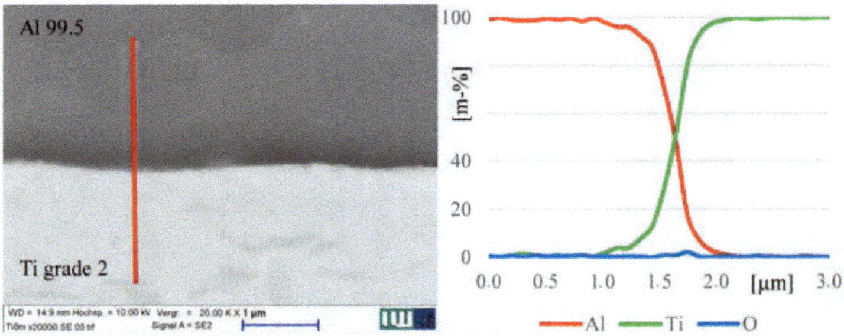

Figure 2: Position of the measurement to the analysis of the interface by EDX line scan

Tensile test

The results of the quasi-static tensile tests are shown in Figure 9. Each of the co-extruded samples mostly failed in the bonding zone. The tensile strength after extrusion is in average about 51 MPa with a standard deviation of ± 6 MPa. The tensile strength of Al99.5 ranges usually between 60 and 95 MPa [10] which means that the tensile strength of the compound material is about 10 MPa lower than that of Al99.5.

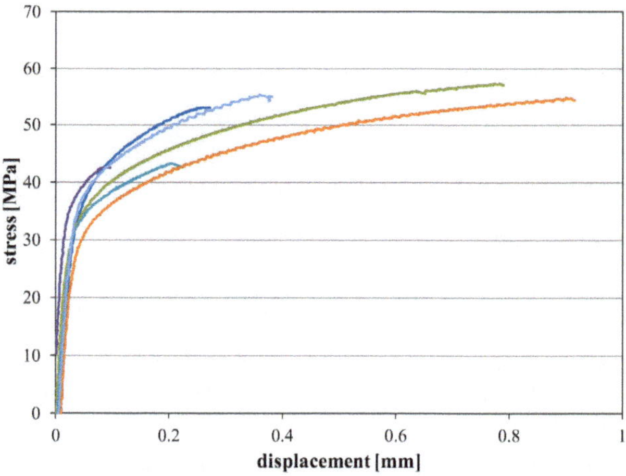

Figure 9: Tensile stress of compound samples produced with an extrusion ratio of 4

Subsequent heat treatment

After co-extrusion the samples were heat treated. The heat treated samples were metallographically prepared and finally examined by SEM. It can be concluded that the heat treatment led to a formation of a dense intermetallic layer between the aluminum and titanium. As the backscattered electron image in Figure 10 shows, a heat treatment of 8 hours at 540° C enabled a layer growth to a thickness between 0.6 and 1.3 μm.

In Figure 10 shows also the EDX spectrum of the location that is marked with an arrow in the backscattered electron image. As expected, the diffusion zone mainly consists of the two elements titanium and aluminum. From previous investigations it is known that the phase $TiAl_3$ forms preferably [11].

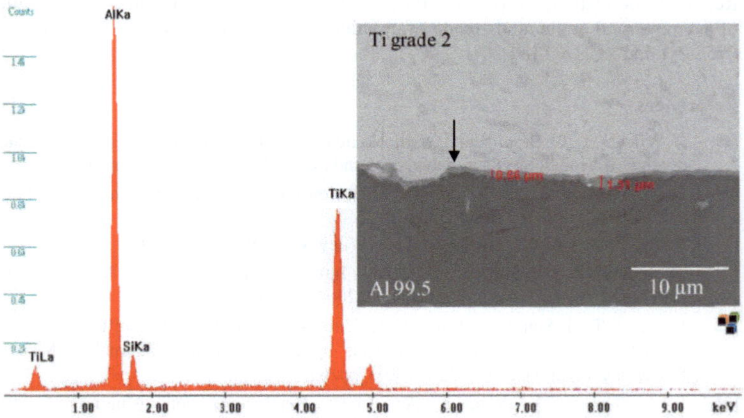

Figure 10: Backscattered electron image of a sample produced with an extrusion ratio of 20 after subsequent heat treatment (8 h; 540 °C). The arrow marks the corresponding EDX spectrum of the diffusion zone.

Summary and outlook

It was successfully demonstrated that using the newly developed **lateral angular co-extrusion process (LACE)**, a modified ECAP die with lateral feeding of reinforcements, enables the co-extrusion of asymmetric aluminum-titanium compound profiles. The co-extruded non-heat treated compounds show a firmly bonded joint at the interface. Nevertheless, the tensile specimens failed in the bonding zone throughout tensile testing. But, subsequent heat treatment produced a dense intermetallic phase by diffusion that, first, proves the existence of a close contact at the interface after co-extrusion. Second, these results demonstrate the (immense) potential of a subsequent heat treatment for the optimization of the bonding zone with its expected positive influence on the bond strength and the failure mode of the compound. The parameters for the formation of an optimum interface must be developed in further experiments.

Due to the positive results of the first trials with pure metals, further experiments using application-oriented alloys of titanium, for example Ti-6Al-4V (ASTM grade 5), and aluminum, heat-treatable and non-heat-treatable - for example AlMgSi1 (EN AW-6082) -, have to be done. In this context, the impact of a subsequent heat treatment will be investigated thoroughly. Local bonding properties will be studied according to previous works [12] as well as the compound's performance with focus on internal stresses and distortion [13]. These results of the new LAE process will be comparable with other bonding techniques for compound profiles, for example laser beam welding. All of these aspects will be part of the subsequent studies.

Acknowledgment

The authors gratefully acknowledge the financial support by DFG (German Research Foundation) for project "Process chains for co-extrusion and heat treatment of asymmetrical compounds made of aluminum and titanium" with support code HE 6758/2-1 and SCHA 1484/20-1.

References

[1] J. Wilden, J. P. Bergmann: Möglichkeiten zum Herstellen von Mischverbindungen durch Diffusionsschweißen. Schweißen und Schneiden, 2004, 56 (5), pp. 199

[2] P. G. Partridge, C. M. Ward-Close: Diffusion bonding of advanced materials. Metals and Materials, 1989, 5 (6), pp. 334

[3] J. Hirsch: Walzen von Flachprodukten. Wiley-VCH, Weinheim, 2001.

[4] M. Schmidtchen, P. Simecek, D. Hadjuk: International Conference on Metal Forming - MEFORM 2004, Freiberg, Deutschland, 2004, pp. 81-110.

[5] J. S. Lee, T. H. Son, C. S. Kang, C. H. Yun, S. C. Lim, H. C. Kwon: Fabrication and characterisation of Ti-Cu clad materials by direct extrusion; Journal of Materials Processing Technology 187-188 (2007), pp. 653-656.

[6] J. J. Theler, A. Wagner, A. Ames: Herstellung von Aluminium/ Stahl-Verbundstromschienen mit metallurgischer Bindung zwischen Aluminium und Stahl durch Verbundstrangpressen. Metallwissenschaft und Technik 30 (1976) 3, pp. 223-227.

[7] Eivahi A.R.; Taheri K.A.: A new method for producing bimetallic rods. Materials Letter 61 (2007) pp. 4110-4113.

[8] Llorca-Isern N.; Escobar A. M.; Roca A.; Cabrera J.M.: Equal Channel Angular Pressing of Cu-Al Bimetallic Rod. Materials Science Forum Vols. 706-709 (2012) pp. 1811-1816.

[9] Liu, X.B; Chen R.S.; Han E.H.: Preliminary investigations on the Mg-Al-Zn/Al laminated composite fabricated by equal channel angular extrusion. Journal of Materials Processing Technology 209 (2009) pp. 4675-4681.

[10] Hesse, W.: Aluminium-Werkstoffdatenblätter. 5. Auflage, Aluminium Verlag, Düsseldorf 2007

[11] D. Dietrich, N. Grittner, T. Mehner, M. Schaper, H.J. Maier, T. Lampke: Microstructural evolution in the bonding zones of co-extruded aluminium/titanium. Journal of Materials Science, Volume 49, Issue 6 (2014), pp. 2442-2455,

[12] Striewe, B.; Grittner, N.; von Hehl, A.; Hunkel, M.; Stelling, O. Characterisation of the bonding zone of co-extruded aluminium-titanium compounds Materials Science and Engineering Technology 43 (2012) 4, S. 315-320

[13] Barr, A.; Hunkel, M.; von Hehl, A. Determination of local material properties of laser beam welded aluminium-steel and aluminium- titanium compounds Materials Science and Engineering Technology 43 (2012) 4, S. 321-327

Advanced Composites for Aerospace, Marine, and Land Applications II
Edited by: T. Sano and T.S. Srivatsan
TMS (The Minerals, Metals & Materials Society), 2015

FABRICATION AND CHARACTERIZATION OF NIOBIUM-ALUMINUM COMPOSITES: EFFECT OF SINTERING TEMPERATURE

Lucio Vazquez, Juan Manuel Miranda, Elizabeth Garfias and Dulce Dyolotzin

Universidad Autónoma Metropolitana-Azcapotzalco,
Av. San Pablo 180
Col. Reynosa Tamaulipas
México D. F., C.P. 02200, **Mexico**

Keywords: Composites, Nb-Al, Particles, Distribution, Mechanical, Properties, Microscopy

Abstract

Objectives of this research were to measure the distribution of Nb particles in the Al matrix, (density ratio 8.6:2.7) to define the effect that sintering temperature, 390 and 550°C, had on the consolidation of the composites, and the influence on the density of the mechanical properties: hardness, elastic strength, elastic modulus, ductility, and impact toughness. Specimens, 3, 5 and 7 % wt of nanometric particles of Nb, the rest 100 meshes particles of Al, were prepared by a P/M technique. Nb – Al powders were mixed in a mill for 12 h, using a 6 mm balls load of ZrO_2, the ratio in wt ZrO_2 balls/composite powder was 10:1. The powders were transferred to a die to make cylindrical specimens, applying a pressure of 3.75 MPa, the specimens were sintered either at 390 or 550°C for 3 h, within a N_2 atmosphere. Mechanical properties do not show an reinforcement effect due to Nb presence, while Vickers hardness is about 10 points higher for composites, elastic strength decreases 10 points gradually for composites, the highest value correspond to 3 % Nb composite, Al has a hardness slightly higher than the composite with 7 % Nb. Scanning microscopy showed an homogeneous Nb particles distribution for samples sintered at 550°C, homogeneity decreased for samples sintered at 390°C.

1. Introduction

A previous work [1] on a composite Al-3Mg-Nb prepared by adding Nb powder into a ladle containing the melted aluminum- magnesium and stirring before pouring showed a nonhomogeneous distribution of Nb particles. In the present work, it was used a powder metallurgy technique, mixing the components Al-Nb in a ceramic mill followed by sintering, the authors expected to improve the distribution.
This work is part of a project to determine the conditions which render the best properties testing materials and fabrication procedure.

2. Experimental Procedure

2.1 Manufacture of samples

Experimental Materials. Aluminum used consists of powder, 100% passes through a 100 meshes sieve and are completely retained by a 200 meshes sieve. The chemical composition of the powder is 99.96 % aluminum with the impurities shown on Table 1.

Table 1. Chemical Composition of Aluminum

Element	Sb	Ba	Cd	Pb	Cr
Concentration ppm	18	40	8	9	5

Nb was a nanosize powder. Purity 99.8 % supplied by Sigma Aldrich.

Weighting of Materials. Components of the composite were weighted in an analytical balance.

Grinding. The ratio of load of balls to B4C-Al mixture was 10:1. The load composition consisted of 900 g of zirconia one cm diameter spheres. The grinding parameters used were mill speed 340 rpm, grinding time 12 h.

Pressing. The mixture Nb-Al was poured into D2 steel dies to make rounded specimens 2 cm diameter and 2 and 6 cm height or quadrangular specimens, 1 cm x 1 cm and a height of 6 cm. It was applied a 375 MPa pressure at room temperature, holding the load constant for ten minutes. A thin film of SAE 20W-50 Quaker State oil was applied to the wall dies to reduce friction, to avoid sticking of the material to the walls as well as to circumvent crumbling of the specimen during the extraction operation.

Sintering. The specimens were sintered in a furnace using a nitrogen atmosphere with a flow pressure of 1MPa. The thermal cycle consisted of 3 h at 550°C. After this time furnace was off, nitrogen flow was interrupted; samples were left within the furnace for cooling until room temperature.

2.2 Sampling Testing and Results

Composite composition. A previous work (2) showed that contents of 3, 5, and 7 % of reinforcing material allow releasing the specimens from the die without crumbling. Therefore, these compositions were adopted.

Density Density of the specimens was determined by the Archimedes method using a hydrostatic balance. Three rounded specimens 2 cm height for each composition were weighed out in a balance, then samples were weighed submerged in water and the displaced volume of water was measured. Figure 1 shows average density values as a function of composition for three samples at 390|C and 550°C, respectively.

Microhardness. Three rounded samples of 2.0 cm height for each composition were grounded and polished to mirror finishing. Vickers hardness was used with a 50 g load. Figure 2 shows graphs of average hardness vs. composition.

Compression. The specimens tested in microhardness were used to carry out the compression tests. Figures 3a, 3b and 3c, show graphs yield stress vs. composition, fracture stress vs. composition and elasticity modulus vs. composition, respectively, for the two temperatures.

Impact Test. Three specimens with a square transverse section 1cmx1cm and 5 cm long without groove, because the fragility of the composites made impossible to machine a groove according to ASTM standard. The results were useful to compare only the absorbed energy by aluminum and the three composites compositions at the two temperatures. Figure 4 shows the results for the two experimental temperatures.

Optical Microscopy. The material of all the specimens forms cells, a characteristic which was already observed in a previous work [2], Figure more evident at 100X, it is attributed to the procedure used to make the specimens: by compression, Figure 9. The distribution of B4C is homogeneous, although it seems that the reinforcement concentrates on the boundaries of the cells.

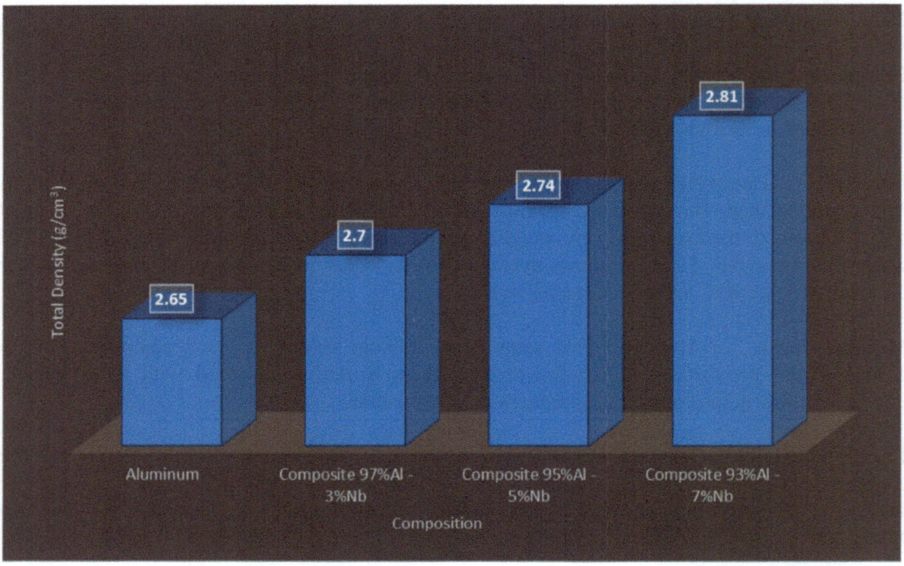

(a)

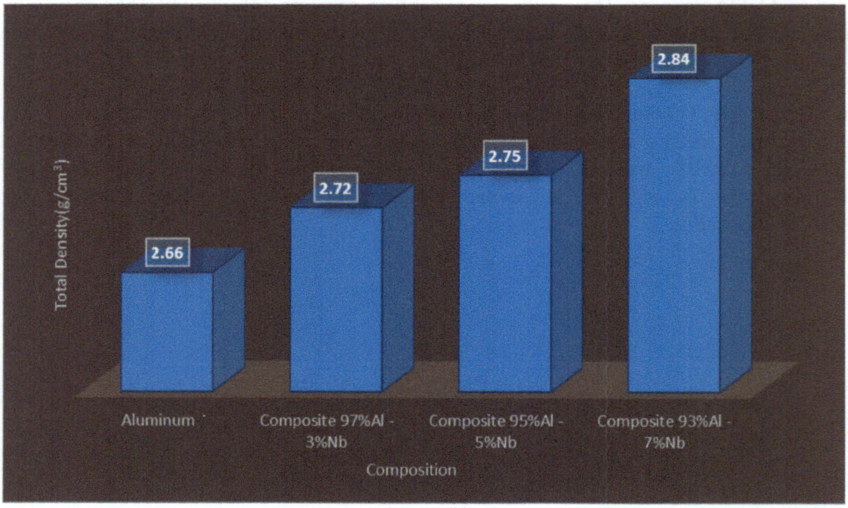

(b)

Figure 1. Density vs. composition, (a) at 390°C, (b) at 550°C.

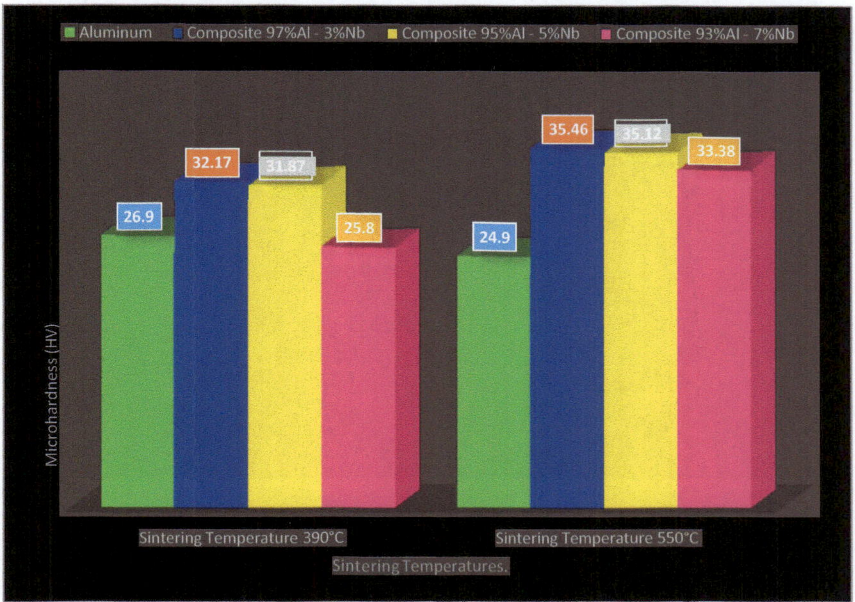

Figure 2. Microhardness vs. Composition at 390°C and 550°C

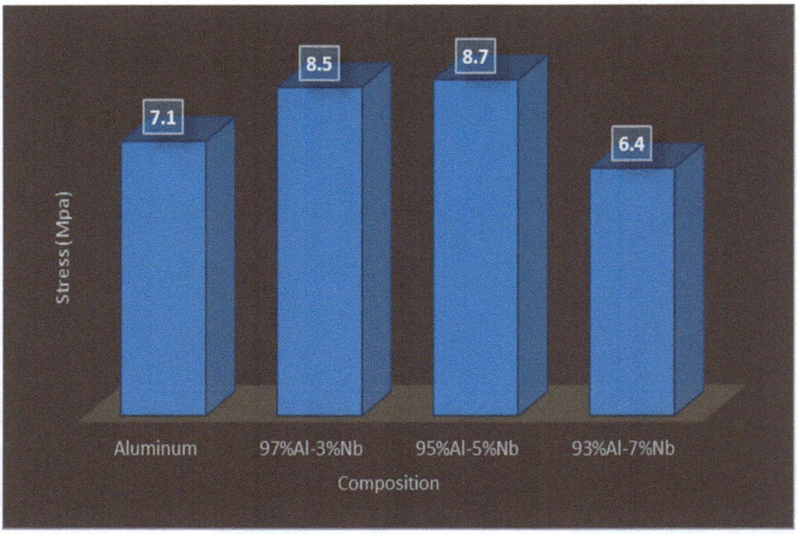

(a)

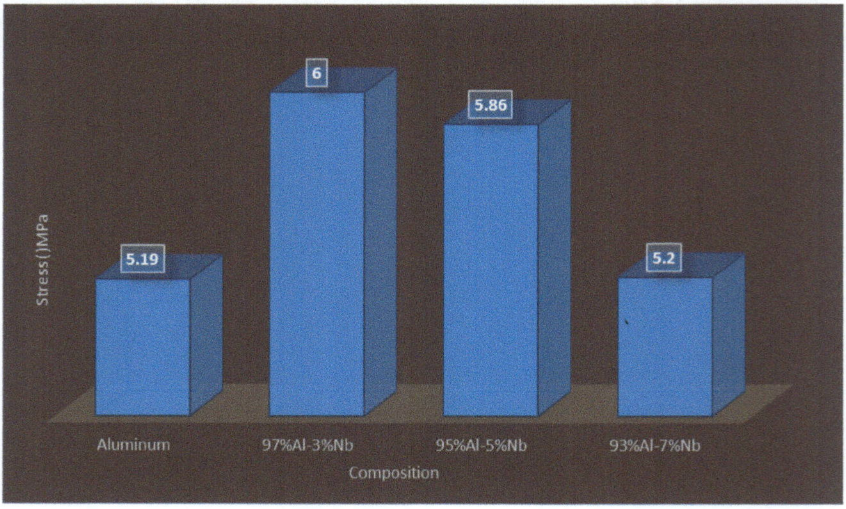

(b)

Figure 3a. Yield stress vs. Composition for sintering at a) 390°C and b) 550°C.

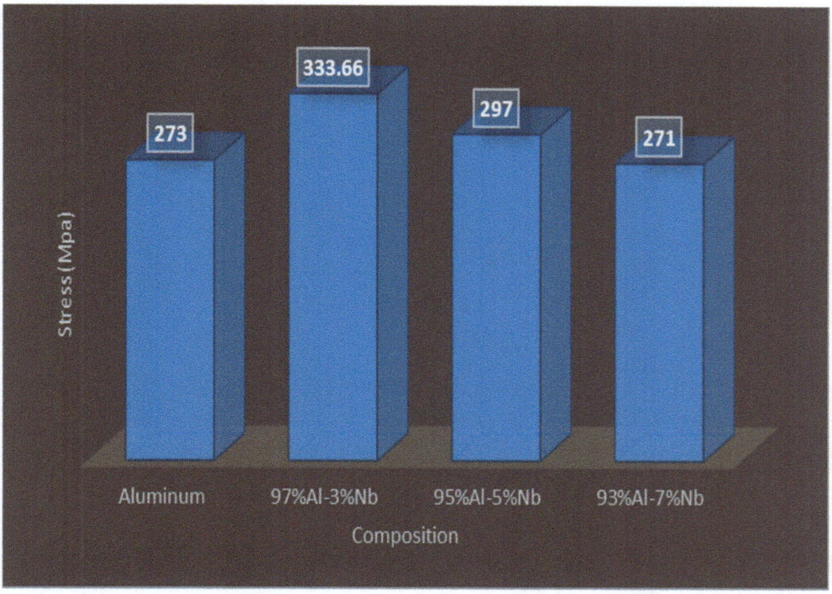

(a)

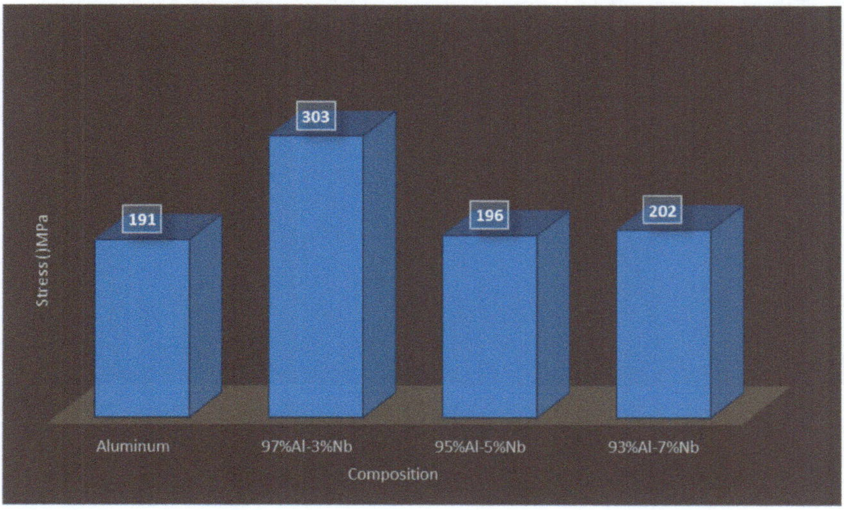

(b)

Figure 3b. **Fracture stress vs. Composition for sintered specimens at a) 390°C and b) 550°C.**

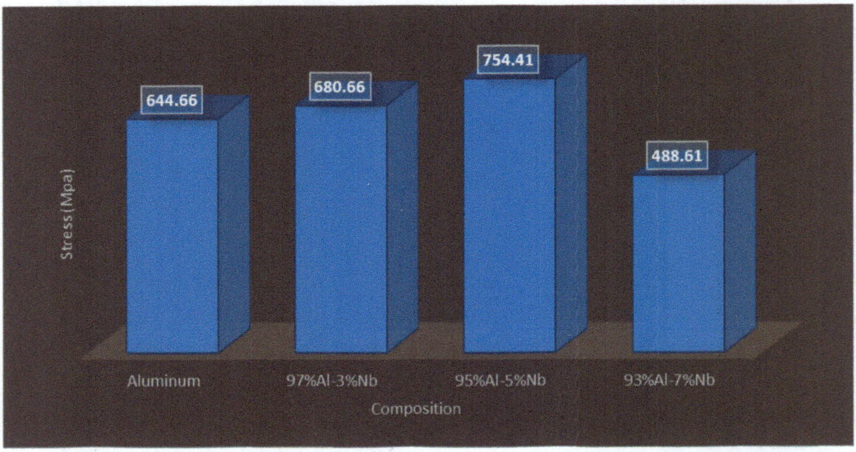

(a)

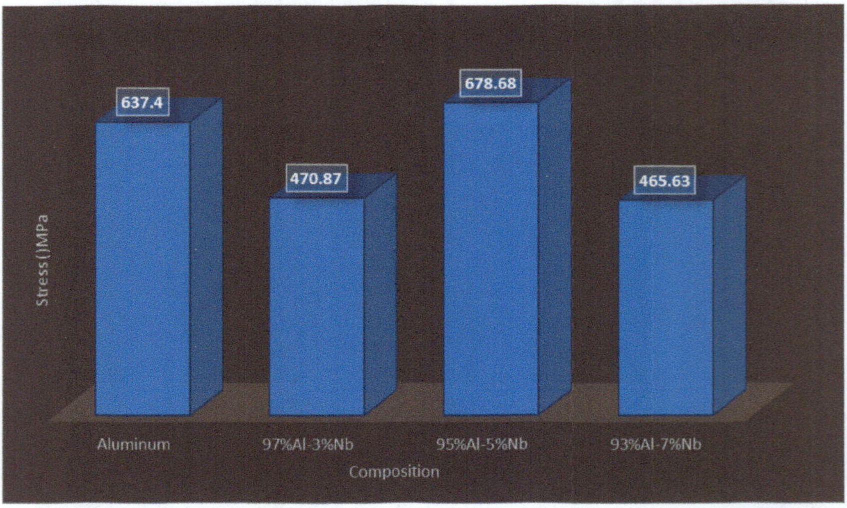

(b)

Figure 3c. Elastic Modulus vs. Composition for sintered specimens at a) 390°C and b) 550°C.

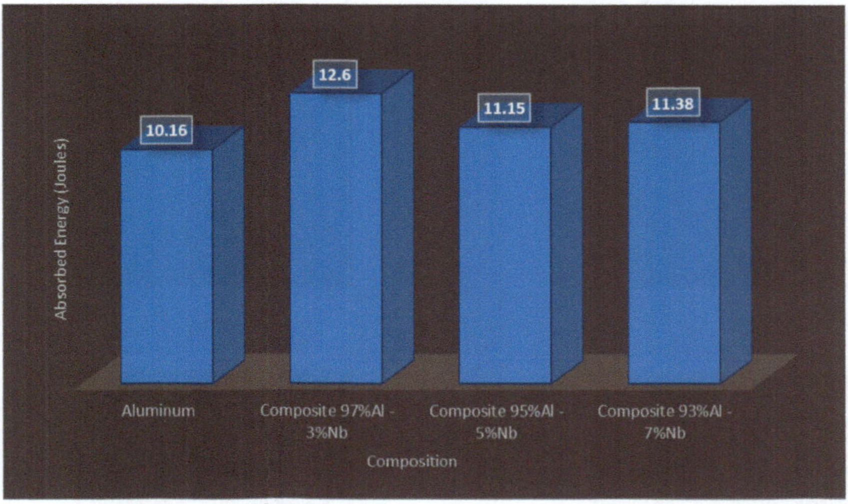

(a)

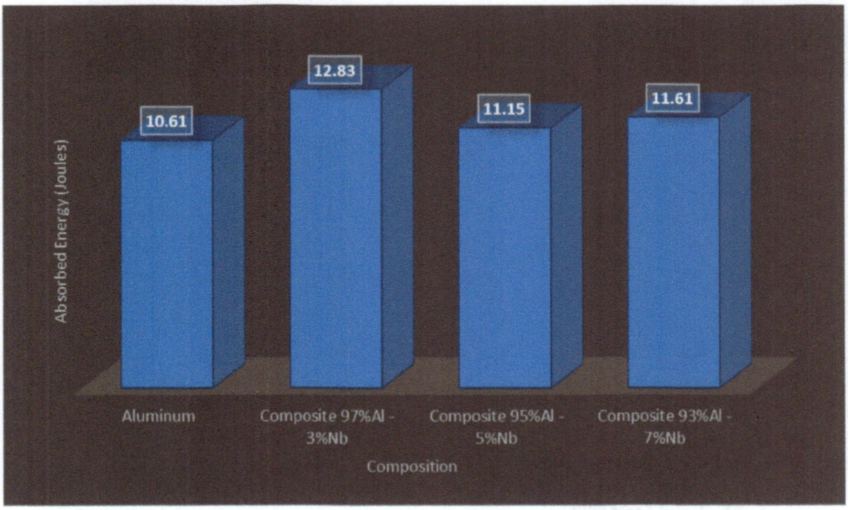

(b)

Figure 4. Absorbed energy by Impact test. Sintered composites at (a) 390°C and (b) at 550°C.

Figure 5. **Optical micrographs:**
a, c and e correspond to a sintering temperature of 390°C, while
b, d and f to 550°C.
a and b belong to 3 %, c and d to 5 %, e and f, to 7 % in weight of Nb, respectively.
100X

Scanning Electronic Microscopy (SEM).

SEM shows that the particles of Nb are distributed rather homogeneously for all the compositions at 390°C as well at 550°C, Figure 6.

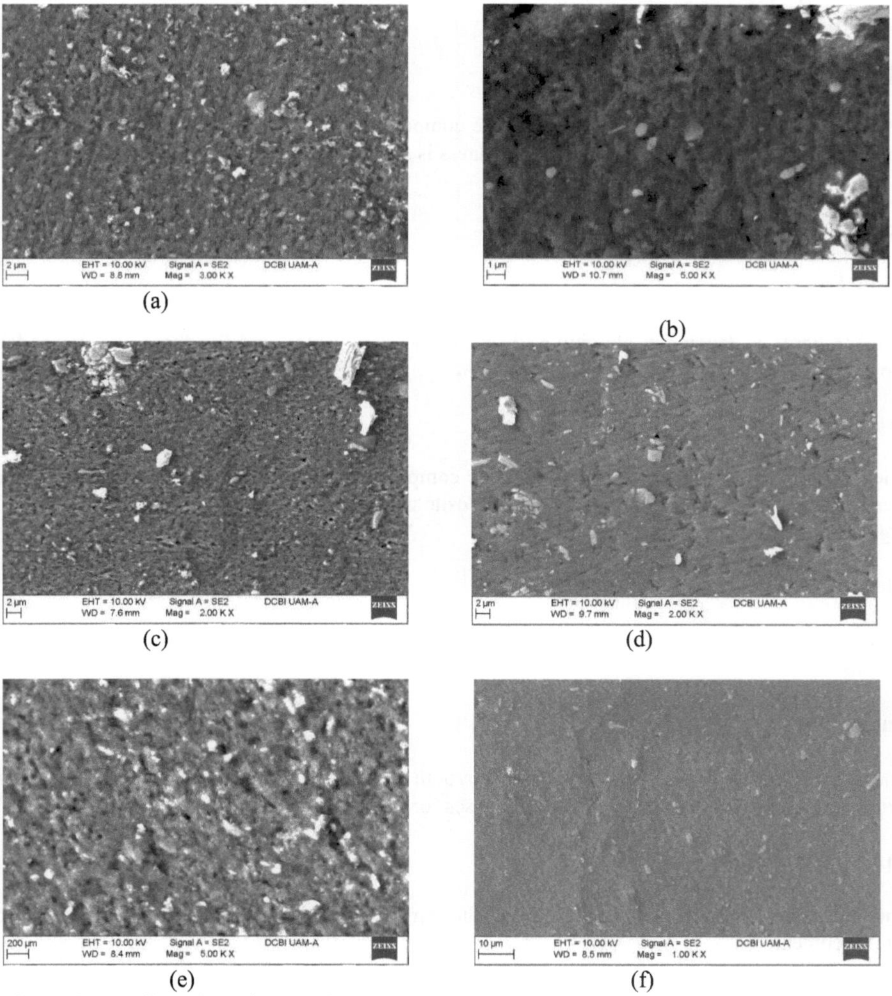

Figure 6. **Scanning micrographs:**
a, c and e correspond to a sintering temperature of 390°C, while
b, d and f to 550°C.
a and b belong to 3 %, c and d to 5 %, e and f, to 7 % in weight of Nb, respectively.

3. Analysis of Results

Density.

Aluminum presents a lower density than the composites. Density increase with Nb content, Figure 1.

Microhardness.

Aluminum displays a lower hardness than the composites. Hardness is higher for the composites sintered at 550°C, Figure 2. Hardness is slightly higher for 3% and 5% Nb content than for 7% in weight.

Compression Properties.

Yield Stress.

This property is lower for aluminum than for the composites and for these was lower for the composites sintered at 550°C, Figure 3a.

Fracture Stress.

The values of fracture stress are higher for the composites than for aluminum, and for the latter were higher at 390°C than at 550°C The composite with 7% in weight presented the lower value. Figure 3b.

Yield Stress.

It increases in the order: aluminum, 3%, and 5% of weight and is lower for the composite with 7% Nb

Impact Test.

Energy absorbed follows a similar pattern for both experimental temperatures: it is lower for aluminum, higher for 3% Nb content, it decreases for 7%, but it is similar for the latter content

Optical microscopy.

Specimens display a cellular structure for all the samples, no matter the sintering temperature or the Nb content.

4. Conclusions

Mechanical properties are better for composites than for aluminum. P/M technique allows preparing composites with 10% or less of B4C. Composites with higher content crumble during the release from the die.

Micro hardness. Hardness increases with B4C content.

Compression tests. Yield, maximum and fracture stress increase with B4C content. Strain decreases as B4C content increases.

Impact test. Toughness increases as B4C increases.

Microscopy. Boron carbide particles deposit on cell boundaries, but distribution within the cells is homogeneous.

Acknowledgements

The authors are grateful for the support that Materials Department, Universidad Autonoma Metropolitana, gave to make possible this work, using its laboratories and allowing spending time to the researchers to carry out this duty.

References

1. Lopez N., Rodriguez X., Domínguez S., Vazquez L. Cortés V., Dominguez M., Lara G., Vite M., "Manufacture and Characterization of Al-3Mg-Nb Alloys", Supplemental Proceedings, v. 3., 289 – 294, 137 Annual Meeting & Exhibition, TMS 2008.

2. Vazquez L., Hernandez L., Altamirano A., Cortés V., Garfía E., and Vite M., "Effect of Composition of B4C Aluminum Composites on Mechanical Properties and Resistance Corrosion", Proceedings Advance Composites, for Aerospace, Marine, and Land Applications, 23 – 33, Annual meeting & Exhibition, TMS 2014.

3. Goni, L., Mitxelena, I., and Coleto, I., *Development of low cost metal matrix composites for commercial applications*, ***Material Science and Technology***, v. 16, No. 7 – 8. July – August 2000, pp. 743 – 746.

4. Bhagat, R. B., *Advanced aluminum powder metallurgy, alloys and composites*, ASM Handbook, v.7, 1998, pp. 840.

5. Ruiz-Navas E. M., Da Costa, C. E., *Lopez* F. V., and Castello, J. M. T., *Mechanical alloying: a method to metallic powder composite materials*, Revista Metalúrgica, Madrid, v. 13, No. 4, 2000, pp. 279-286.

6. Torralba J. M., Da Costa, C. E. and Velasco F., *P/M aluminum matrix composites: an overview*, ***J. of materials Processing Technology***, v. 133, Issues 1-2, 2003, pp. 203-206.

7. Kaczmar, I., Pietrzak, K., and Wlosinski, W., *The production and application of metal matrix composite materials*, ***J. of Materials processing technology***, v. 106, 2000, pp. 58-67.

8. Annual Book of Standards, v. 6-2, 2010.

Advanced Composites for Aerospace, Marine, and Land Applications II
Edited by: T. Sano and T.S. Srivatsan
TMS (The Minerals, Metals & Materials Society), 2015

SPARK PLASMA SINTERING (SPS) OF CARBON NANOTUBE (CNT) / GRAPHENE NANOPLATELET (GNP)-NICKEL NANOCOMPOSITES: STRUCTURE PROPERTY ANALYSIS

Tushar Borkar[1], Hamidreza Mohseni[1], Junyeon Hwang[2], Thomas Scharf[1], Jaimie Tiley[3], Soon H. Hong[4], and Rajarshi Banerjee[1,*]

[1] *Department of Materials Science and Engineering*
University of North Texas
Denton, TX 76203, USA

[2] *Institute of Advanced Composite Materials*
Korea Institute of Science and Technology
Jeonbuk, 565-905, Korea

[3] *Materials and Manufacturing Directorate*
Air Force Research Laboratory
Dayton, OH 45433, USA

[4] *Department of Materials Science and Engineering*
Korea Advance Institute of Science and Technology
Daejeon, 305-701, Korea

Keywords: Spark plasma sintering (SPS), Molecular level mixing (MLM), Carbon nanotube (CNT), Graphene nanoplatelet (GNP), Tensile strength, Interface.

Abstract

Carbon nanotubes (CNTs) and graphene nanoplatelets (GNPs) are attractive reinforcements for lightweight and high strength metal matrix composites due to their excellent mechanical and physical properties. The CNT/Ni (DM) nanocomposites exhibiting a tensile yield strength of 350 MPa (about two times that of nickel ~ 160 MPa) and an elongation to failure ~ 30%. In contrast, CNT/Ni (MLM) exhibited substantially higher tensile yield strength (~ 690 MPa) but limited ductility with an elongation to failure ~ 8%. GNP/Nickel nanocomposites were also processed via DM followed by SPS consolidation. The Ni-1vol%GNP nanocomposite exhibited the best balance of properties in terms of strength and ductility. The enhancement in the tensile strength (i.e. 370 MPa) and substantial ductility (~40%) of Ni-1vol%GNP nanocomposites was achieved due to the combined effects of grain refinement, homogeneous dispersion of GNPs in the nickel matrix, and well-bonded Ni-GNP interface effectively transfers stress across metal-GNP interface during tensile deformation.

1. Introduction

Carbon nanotubes, originally discovered by Iijima [1-5] almost two decades ago have attracted significant interest as reinforcements in various metallic, ceramic, and polymeric matrices due to their unique atomic structure and excellent properties, such as high tensile strength, elastic modulus, low density, high aspect ratio, excellent thermal as well as electrical conductivity [6- 14]. CNT/metal nanocomposites have potentially attractive properties, such as high tensile strength, stiffness, excellent wear and corrosion resistance which make them potential candidates for structural applications including in aerospace and transportation in general [15-21]. Most of the investigations focus on CNT/polymer composites because of compatibility and flexibility of CNTs in the polymer matrix aiding the fabrication process [7, 22-23]. CNT/polymer nanocomposites typically exhibit high interfacial strength between the CNT and the polymer matrix due to the molecular level interaction [24]. CNT/metal nanocomposites have attracted less attention due to their inferior mechanical properties because of the inhomogeneous dispersion of CNTs in the metal matrix, the poor quality of metal-CNT interfaces and the low relative density of CNT/metal nanocomposites. Few researchers have reported improvement in mechanical and tribological properties of CNT/metal nanocomposites fabricated by conventional powder metallurgy techniques. Cha et. al. [24] have reported a unique method of fabricating CNT/metal nanocomposites by molecular level mixing process followed by spark plasma sintering (SPS) resulting in a uniform dispersion of CNTs in the metal matrix [24]. CNT/metal nanocomposites sintered from nanocomposite powders prepared by molecular level mixing exhibit significant improvement in mechanical properties mainly due to the homogeneous distribution of CNTs within the metal matrix and the strong interfacial bonding between CNTs and the metal matrix [25]. The process for fabricating CNT/Ni composite powders using molecular level mixing consists of four steps [24-28]. CNTs produced by thermal chemical vapor deposition (CVD) are dispersed in ethanol to make a stable suspension by functionalizing the CNT surfaces. The electrostatic repulsive force between the CNTs overcomes the van der Waals' force to obtain a stable suspension within the solvent due to functionalization of CNTs. Second, a salt contains Ni ions is dissolved in the CNT suspension. Additional ultra sonication helps the dispersion of Ni ions among the suspended CNTs and promotes reaction between the Ni ions and functional groups on CNT surfaces. The third step is to dry the mixture by heating to remove the solvent and ligands; leads to Ni ions on the CNTs are get oxidized to form powders. The fourth and final step is the calcination and reduction process to obtain chemically stable CNT/Ni composite powder.

Graphene is a two dimensional (2D) building block of carbon material with a single atomic layer of sp^2 hybridized carbon atoms densely packed in a honeycomb lattice [29-31]. In recent years, graphene has attracted considerable attention due to its excellent mechanical and physical properties, such as high young's modulus (1 TPA), high tensile strength (130 GPa), excellent thermal conductivity (5000 W/mK) $wm^{-1}k^{-1}$), high electron mobility (200,000 $cm^2v^{-1}s^{-1}$), large surface area (=2600 m^2s^{-1}), and very low density (2.2 gcm^{-2}), which make it a suitable candidate for various applications in the fields of solar energy conversion, supercapacitors, field effect transistors, lithium secondary batteries, metallization lines in devices, high speed devices associated with high mobility of carriers, spintronics, and sensors [31-42]. One of the best possible ways to employ such extraordinary properties would be the incorporation of graphene in a composite material. While the 2D material graphene is very promising due to its excellent intrinsic properties, the introduction of this 2D form into composites can be rather challenging due to the difficulties associated with handling such a 2D material. In contrast, GNPs are easier to handle and disperse in all kinds of matrices, as compared to CNTs, have potential to replace CNT reinforcements in most polymer/ metal matrix composites [32]. GNPs are widely used as reinforcements in many polymer-matrix composites due to their cost and ease of operation compared to CNTs. GNP/Polymer nanocomposites exhibited significant improvement in mechanical and electrical properties [30, 43-47]. However, very little work on GNP/metal nanocomposites has been reported, with even less discussing the controversial effect of graphene reinforcement on mechanical properties of composites. The main challenges associated with the processing and development of GNP/metal composites for practical applications are (a) to attain homogeneous dispersion of graphene into metal matrix due to large density differences between metal and GNPs, (b) to form a strong metal-graphene interfacial bonding which effectively transfers the load across the interface during mechanical deformation, and (c) to retain chemical as well as structural integrity of graphene (due to high pressure and high temperature processing conditions of composites) [48]. Bartolucci et al. fabricated aluminum-graphene nanocomposites via milling, hot isostatic pressing, and hot extrusion, which exhibited reduction in hardness as well as tensile strength of the composites due to formation of Al_4C_3 (aluminum carbide) which lowers the interfacial bonding and load transfer across the interface [37]. On the other hand, Latief et al fabricated Al-GNP composites via powder metallurgy route exhibited improved mechanical properties and corrosion resistance [34]. Kuang et al. electrodeposited nickel- graphene composite coatings that have shown about 15% improvement in thermal conductivity and about 4 times enhancement in hardness over that of pure nickel [31].

Spark plasma sintering (SPS) is a novel tool for processing of metal matrix composites at lower temperatures and shorter processing times. In the SPS process, a local high temperature state is generated when sparks discharge in a gap or at the contact point between the particles of powder causing evaporation and melting on the surface of powder particles, thus helping in the formation of fully dense nanocomposites [49-50]. This initiation of the spark discharge in the gap between particles is assisted by fine impurities and gases on and between the surfaces of the particles. SPS is a high-speed powder consolidation (sintering) technology capable of processing conductive and non-conductive materials, and it utilizes uniaxial force and on-off DC pulse energizing. This high frequency transfers and disperses the joule heat phenomenon throughout the specimen, resulting in a rapid and thorough heat distribution, ensuring high homogeneity and consistent precipitate densities. Consolidation of nanopowders is possible in SPS without excessive grain growth due to the high heating rates and lower sintering times and temperatures involved [49-63]. SPS is effective for any powder material applications, but interest is especially high for nano-crystalline structures, because of high surface area per volume. "Net" or "Near- net" shapes are also possible with SPS directly from powder to finished product, having symmetrical and simple geometry in one step. However, for complex shapes, further machining may be required. This paper focuses on a comparison of CNT/Ni nanocomposites prepared via two routes. The first of these involves dry milling of nickel and CNT powders, followed by SPS consolidation of the dry mixed powders. The second approach involves SPS consolidation of powders processed via the novel, recently proposed, molecular-level process. Also, this report mainly emphases on the effect of graphene nanoplatelet volume fraction on microstructure and mechanical properties of GNP/Ni nanocomposites. The microstructure and mechanical properties of both types of nanocomposites have been compared and contrasted.

2. Experimental details

Multiwalled CNTs fabricated by chemical vapor deposition (CVD) with an average dia. of about 10 nm and average length of few μm were obtained from Hanhwa Nanotech Co. Ltd., South Korea. For CNT/Ni nanocomposite (ball milled), CNT/Ni nanocomposite powder synthesized by mixing nickel powder and CNTs through high energy ball milling process for 24 hrs. at 400 rpm. The volume fraction of CNTs in both the composite powder was about 5%. To fabricate CNT/Ni composite powders by molecular level mixing process, the CNTs were purified and functionalized by using hydrofluoric (HF), acid solution mixture of H_2SO_4/HNO_3. The functionalized CNTs (diameter of 5–10 nm, length of 500 nm–2 μm) were cleaned with distilled water and dried. The CNT–Ni precursor was prepared through a chemical reduction process, using hydrazine hydrate as reducing agent, sodium hydroxide as the oxidation agent and ethylene glycol as a solvent. The functionalized CNT were dispersed in ethylene glycol solution by ultra- sonication and $Ni(C_2H_3O_2)_2 \cdot 4H_2O$ (Aldrich, 98% purity) was added in this solution. The CNT solution was mixed with a Ni salt-solution and 2 M-NaOH solution and hydrazine was injected into the mixed solution. The mixed CNT–Ni solution was heated at 70°C for 30 minutes to complete the CNT/Ni composite powder fabrication. Fabricated CNT/Ni composite powders were separated from the remaining solution and subsequently washed with ethanol and dried. Subsequently, these powders were reduced in a H_2 + CO gas mixture for 2 hrs, at 400°C to process the final CNT/Ni composite powders [25]. The graphite nanoplatelets (GNPs) fabricated by exfoliation process from graphite with an average thickness of approximately 6-8 nm and average length of few μm were commercially procured from XG Science, South Korea. GNP/Ni nanocomposite powders were synthesized by mixing nickel powder and GNPs through high- energy ball milling process for 24 hrs. with 400 rpm. The volume fraction of GNPs in GNP/Ni composite powder was ~ 1%, 2.5%, and 5%. The CNT-GNP/Ni composite powders were precompacted in a graphite mold under a pressure of 5 MPa. The precompacted powders were sintered by SPS system at a temperature of 800° C for 5 min under controlled argon atmosphere under a pressure of 80 MPa. The heating rate was maintained at 100° C/ min.

CNT-GNP/Ni nanocomposites were characterized in FEI-Quanta Nova-SEM. The microhardness of CNT-GNP/Ni composite was measured using a standard Vickers microhardness tester under a load of 1.96 N for 15 sec. Average of 10 readings were taken into account. X-ray diffraction analysis of nanocomposites was performed using (1.54 Cu Kα) line of Rigaku Ultima III X-Ray diffractometer. Conventional and high-resolution transmission electron microscopy (TEM) analysis was carried out using a FEI TECHNAI F20 Field emission gun (FEG) TEM operating at 200 keV. TEM foils were prepared via jet polishing techniques and then ion milled using Gatan Duo Mill using 4 keV for the further milling. To assess tensile properties of the composites mini-tensile testing was carried out on a custom-built mini-tensile tester. The tensile samples were prepared using Electric discharge machining (EDM). The gauge length, width, and thickness of the gage section of the samples were 5.0 mm, 1.0 mm, and 1.35 mm, respectively. Each sample was ground and polished (in the order of 600, 800, 1200 grit size silicon carbide (SiC) papers and final polishing on 0.04 μm colloidal silica before tensile testing.

The test was carried out at room temperature at a crosshead speed of 0.02 mm/min. Total three tensile samples of each condition were tested for quantitative analysis.

3. Results and Discussion

3.1 Carbon Nanotube (CNT) /Nickel Nanocomposites:

Microstructure of CNT/Ni composites:

Back-scattered SEM images of the SPS processed pure nickel and CNT/Ni nanocomposites (DM and MLM) are shown in Fig. 1. Pure nickel (Fig. 1(a) and (b)) exhibits a uniform grain structure without any porosity with an average grain size about 45 μm. The full and rapid densification of powder compacts without any substantial grain growth is one of the major advantages of SPS processing that is a result of sintering by joule heating and the spark plasma generated by the pulsed high electric current passing through the compact. CNT/Ni (DM) nanocomposite shows a substantially refined grain structure (Fig. 1(c)), as compared with the pure Ni, even though both samples were prepared with the same initial size of Ni powder. Within the Ni matrix, there appears to be a uniformly distributed second phase or possibly porosity, as shown in Fig. 1(c). However, a higher magnification SEM image, shown in Fig. 1(d), confirms that the regions exhibiting a darker contrast are actually CNT bundles of 1-2 μm in size, rather than porosity. Therefore, based on the comparison of microstructures of the SPS processed pure Ni and CNT/Ni composite (DM), it is apparent that addition of CNTs into the nickel matrix helps in grain refinement by possibly inhibiting grain growth and providing nucleation site for new grains during recrystallization. Fig. 1(e) and (f) show SEM images from the CNT/Ni (MLM) nanocomposites, that also exhibits a uniform grain structure. Previous investigations of such CNT/Ni nanocomposites processed via molecular level mixing and SPS have clearly shown that there is a uniform distribution of individual CNTs within the nickel matrix [25].

X-ray diffraction analysis:

The XRD patterns for the SPS processed pure nickel and CNT/Ni nanocomposites are shown in Fig. 2. The XRD pattern for pure nickel shows peaks corresponding to the (111), (200), and (220) crystallographic planes of nickel. Both the dry-milled as well as molecular level mixed CNT/Ni nanocomposites show an additional (0002) carbon peak. Presence of carbon peak in both the CNT/Ni nanocomposites confirms the presence of carbon nanotubes. In all three cases, the maximum intensity peak is the Ni(111) peak. The Ni(111)/Ni(200) intensity ratio for SPS processed pure Ni is 2.17, while that for dry-milled and SPS processed is 2.45, and finally for molecular level mixed and SPS processed is 2.25. Comparing these three values with the standard ratio expected for randomly oriented Ni grains in the International Center for Diffraction Data (ICDD) files obtained from Joint Committee of Powder Diffraction Standards (JCPDS) of 2.38, it is apparent that there is no significant change in texture resulting from the introduction of CNT's in the nickel matrix. This is in contrast to previous publications on electro-deposited CNT/Ni nanocomposites where a clear increase in the Ni(200) peak intensity has been reported, indicating a change in texture due to the addition of CNTs [64-65].

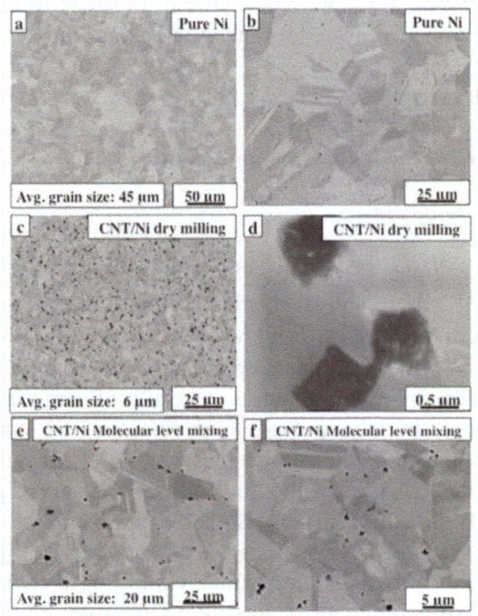

Figure 1: **Backscatter SEM images of (a) and (b) Pure Nickel, (c) and (d) CNT/Ni Dry Milling, (d) and (e) CNT/Ni Molecular level mixing**

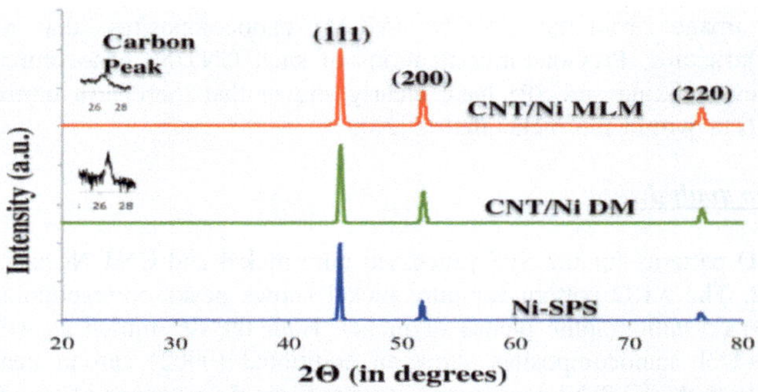

Figure 2: **XRD pattern obtained from Pure Nickel, CNT/Ni Dry milled, and CNT/Ni Molecular Level Mixed nanocomposites.**

Microhardness and Tensile behavior of CNT/Ni composites:

The Vickers microhardness of the CNT/Ni nanocomposites are shown in Table 1. There is a considerable enhancement in the microhardness with the addition of CNTs to the nickel matrix. The hardness of CNT/Ni (MLM) nanocomposites is 98.6 HV, and that of the CNT/Ni (DM) nanocomposite is 94.5 HV, both of these values are about 1.5 times higher than that of pure SPS processed nickel. The yield strength and elongation (tensile) properties of SPS processed pure nickel and both types of CNT/Ni nanocomposites are also listed in Table

1. The corresponding engineering stress-strain curves for these materials are shown in Fig. 3. The tensile yield strength of CNT/Ni (DM) nanocomposite is 350 MPa, which is about 2 times higher than that of pure nickel (160 MPa). Even more striking is that the tensile yield strength of the CNT/Ni (MLM) nanocomposite is 690 MPa, about 4 times that of pure Ni and 2 times that of CNT/Ni (DM) nanocomposite. However, comparing the ductility of all three types of nanocomposites, while pure Ni exhibits about 50% elongation to failure, the elongation values for the CNT/Ni (DM) and CNT/Ni (MLM) are 30% and 8% respectively. Comparing the stress-strain behavior (Fig. 3), it appears that while the CNT/Ni (DM) nanocomposite exhibits a modulus similar to that of SPS processed pure Ni, the CNT/Ni (MLM) nanocomposite exhibits a substantial higher modulus.

Table 1: Microhardness, yield strength and elongation properties of pure nickel and CNT/Ni nanocomposites.

Sample	Hardness (HV)	Yield Strength (MPa)	Ductility/Elongation (%)
Pure Nickel	77.52±3	160±5	50
CNT/Ni (DM)	94.5±6	350±4	30
CNT/Ni (MLM)	98.6±4	690±10	8

Since the details of the deformation behavior of the CNT/Ni MLM nanocomposite have been discussed in a recent paper [24], these are not being repeated in the present paper. However, it is important to note the contrasting aspects of deformation between the two types of CNT/Ni nanocomposites being compared in the present study. While the CNT/Ni (DM) nanocomposite exhibits a classical strain-hardening behavior all the way from yield to failure, similar to pure Ni, the CNT/Ni (MLM) nanocomposite exhibits the sharper yield point accompanied by a drop in stress and subsequently strain-hardening to a limited degree before failure. Figs. 4(a), (b), and (d) are secondary electron SEM micrographs showing the fracture surfaces of the pure Ni, CNT/Ni (dry-milled) and CNT/Ni (molecular level mixed) respectively. From these images it is evident that while pure Ni and CNT/Ni (DM) exhibit similar fracture surfaces, the CNT/Ni (MLM) nanocomposite exhibits a somewhat different fracture surface. The CNT/Ni (DM) nanocomposite fracture surface (Fig. 4(b)) exhibits typical cup and cone type failure characteristics, showing dimples of the order of few microns in size. This is indicative of a very ductile deformation behavior in agreement with the greater than 30% elongation observed for this nanocomposite. An EDS carbon map, corresponding to the fracture surface shown in Fig. 4(b), is shown in Fig. 4(c). Based on this map the location of the CNT bundles on the fracture surface can be determined and a few examples of these bundles have been marked with dotted circles in both Figs. 4(b) and (c). Contrastingly, the CNT/Ni (MLM) nanocomposite fracture surface (Fig. 4(d)), exhibits much smaller dimples of the order of sub-micron to at most a micron. Additionally, regions containing CNTs have been marked with circles in the higher magnification image of the fracture surface shown in Fig. 4(e). This is indicative of some strain localization and eventual failure at a lower level of strain. This is agreement with the observed lower ductility of the CNT/Ni (MLM) nanocomposite as compared to the CNT/Ni (DMA) nanocomposites.

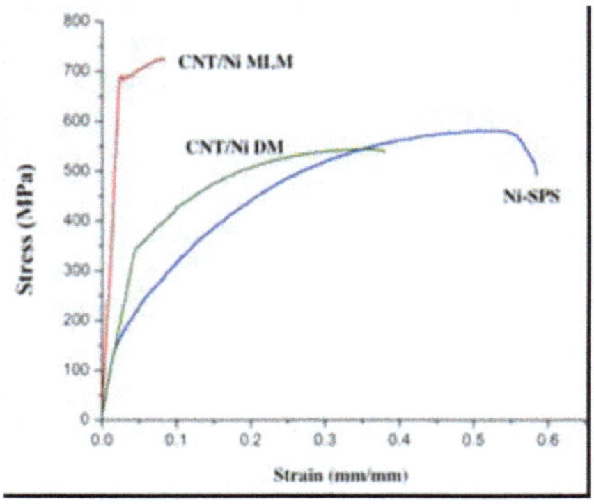

Figure 3: **Stress-strain curves of CNT/Ni (DM and MLM) composites and pure Ni obtained from tensile test.**

However, from the perspective of strength, the CNT/Ni (MLM) exhibits a yield strength that is almost the twice that of the CNT/Ni (DM) nanocomposite. The enormous strengthening in case of CNT/Ni (MLM) nanocomposites arises mainly due to the homogeneous distribution of individual CNTs within the nickel matrix [24]. The strong interfacial bonding between the CNTs and the Ni matrix is effective to improve mechanical properties of CNT/metal nanocomposites. Despite the fact that the CNTs are in the form of sub-micron bundles in case of the CNT/Ni (DM) nanocomposite, a significant improvement in yield strength is observed and can be attributed in part due to grain size refinement. The experimental observations indicate that the uniform dispersion of CNT bundles is helping in grain refinement as well as possibly in load transfer between the metal matrix and the nanotube bundles. Therefore, comparing and contrasting the mechanical properties of these nanocomposites, it is evident that while the CNT/Ni (MLM) composite exhibits the highest tensile yield strength, the CNT/Ni (DM) composite exhibits a higher tensile ductility. The role of any residual stresses resulting from the SPS process should also be considered; especially since the graphite punches being in contact with the sintered material during the cooling step can result in considerably high cooling rates experienced by the sintered material. Such cooling during SPS ensures that heat does not unnecessarily stay retained in the sintered sample. However, when the sintering temperature is significantly low (800° C) compared to the melting point of Ni (1455° C), or $T_{sintering} = 0.62\ T_{melting}$, the considerably high cooling rate is still unlikely to cause any significant residual stresses in the sintered composite.

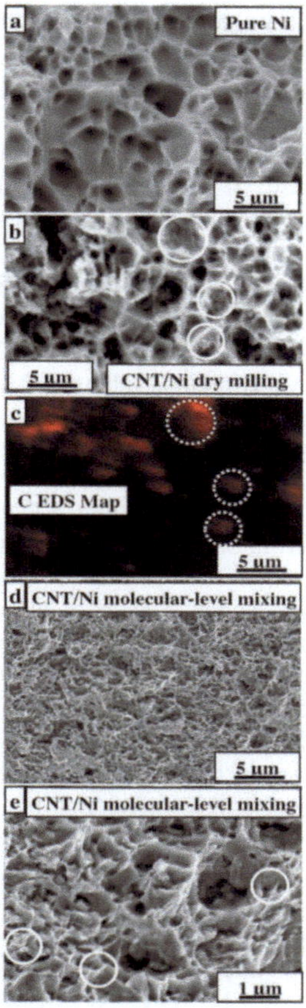

Figure 4: **SEM micrograph showing the fracture surfaces of the (a) pure Ni, (b) CNT/Ni DM, and (d) CNT/Ni MLM composites after tensile test (c) EDS C map corresponding to the fracture surface of CNT/Ni DM composite (e) SEM micrograph showing the presence of CNTs in the fracture surfaces of the CNT/Ni MLM composite.**

Additionally, it is worthwhile to compare the deformation behavior of CNT/Ni MLM processed nanocomposites with ultra-fine grained (UFG) nickel samples. Very high tensile yield strengths about 990MPa, have been reported in the literature for UFG nickel samples, processed via equal channel angular pressing (ECAP) plus cold-rolling [66]. These samples had grain sizes about 0.33 μm and exhibited an elongation to failure about 7%. SEM images of the fracture surface of these UFG nickel samples appear to exhibit a typical ductile failure with relatively large dimple sizes (larger than the grain size) [66]. Comparing the reported images of the fracture surface in case of UFG nickel with the fracture surface for the CNT/Ni MLM nanocomposite in the present study (Fig. 4(d)), the dimple sizes appear to quite similar and in both cases exhibit extended edges. However, the grain size in case of the CNT/Ni MLM nanocomposite is substantially larger about 15-25 μm, almost two orders of magnitude larger than that of the UFG nickel sample. Consequently, it can be concluded that the homogeneous distribution of CNT's in case of the MLM nanocomposite is primarily responsible for its very high yield strength (about 690 MPa).

Transmission electron microscopy analysis:

Fig. 5(a) shows a bright-field TEM image of the CNT/Ni (DM) nanocomposite showing a CNT bundle embedded in the Ni matrix, that is consistent with the results obtained from SEM observations. A high-resolution TEM image from the same nanocomposite is shown in Fig. 5(b). This high-resolution image shows a sharp interface between a nanotube and the nickel matrix with no significant evidence of interfacial mixing across the two phases. A selected area diffraction (SAD) pattern of the [011] zone axis of pure nickel is included as an inset in Fig. 5(b) A low magnification bright-field TEM image of the CNT/Ni (MLM) nanocomposite is shown in Fig. 5(c). This image clearly shows that the CNTs are homogeneously distributed within the nickel matrix. Fig. 5(d) shows a high-resolution TEM image of an individual CNT embedded in the nickel matrix. In this case too, the CNT/Ni interface is very sharp without any evident intermixing between the nanotube and the nickel. Unlike, CNT/Ni (ball milled) nanocomposites, in the molecular level mixed nanocomposite, individual CNTs are homogeneously dispersed within the nickel matrix, thus providing a large number of very strong nickel-CNT interfaces that leads to improvement in mechanical properties such as high yield strength and microhardness. The distribution of individual CNTs was found to be more homogeneous in case of the CNT/Ni (MLM) nanocomposite as compared to the CNT/Ni (DM) nanocomposite that largely exhibited a distribution of small bundles of CNTs. Nevertheless, these CNT bundles in case of the CNT/Ni (DM) nanocomposite resulted in a high degree of grain refinement in the nickel matrix. While the CNT/Ni (MLM) nanocomposite exhibits the best nanotube distribution as well as the highest strength level, the processing of the CNT/Ni molecular-level powders is a relatively slow process and the scalability of such a process can be challenging. In contrast, while the CNT/Ni (DM) nanocomposite does not exhibit the ideally desired distribution of nanotubes and consequently exhibits a lower strength level, the processing steps involved are relatively simple and in principle can be scaled up.

3.2 Graphene Nanoplatelet (GNP)/Nickel Nanocomposites:

A: Scanning Electron Microscopy (SEM) Analysis of GNP/Ni nanocomposites

Back-scattered SEM images of the SPS sintered pure nickel and GNP/Ni nanocomposites are shown in Fig. 6. Pure nickel (Fig. 6(a)) exhibits a uniform grain structure without any porosity with an average grain size ~ 12 μm. The full and rapid densification of powder compacts without any subtantial grain growth is one of the major advantages of SPS processing that is a result of sintering by joule heating and the spark plasma generated by the pulsed high

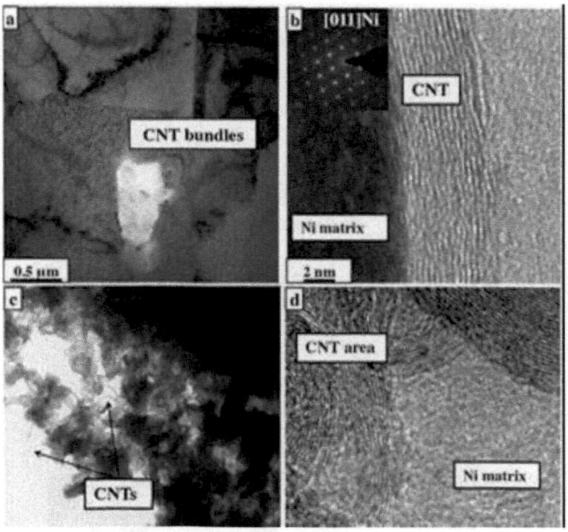

Figure 5:
- **(a) Bright-field TEM image showing CNT bundle in Ni matrix of the CNT/Ni (DM) nanocomposite**
- **(b) HRTEM image on the CNT/NI interface along with SAD pattern of pure Ni in inset of CNT/NI DM composite**
- **(c) Bright-field TEM images of highlighted CNT region, and**
- **(d) Embedded CNTs in the Ni matrix of CNT/NI MLM composite.**

electric current passing through the compact. Ni-1GNP nanocomposite shows a substantially refined grain structure (Fig. 6(b)), as compared with the pure Ni, even though both samples were prepared with the same initial size of Ni powder. GNPs appear to be quite uniformly distributed along the grain boundaries within the nickel matrix. The initial ball milling followed by spark plasma sintering helps to obtain a refined grain structure due to homogeneous distribution of GNPs within the nickel matrix. The addition of GNPs into the nickel matrix assists in grain refinement by possibly inhibiting grain growth and providing nucleation sites for new grains during recrystallization. Similar results were observed in Ni/CNT (DM) nanocomposites sintered using SPS [62]. Fig. 6(c) and (d) show SEM images of the Ni-2.5GNP and Ni-5GNP nanocomposites that also exhibit a refined grain structure, but additionally, coarse clusters of GNPs are observed within the nickel matrix. A pair of a backscatter SEM image and the corresponding EDS carbon map, from the Ni-2.5GNP sample, is shown in Figs. 6(e) and (f). Based on this map, the corresponding locations of GNP clusters have been marked with dotted circles in both Figs. 1(e) and (f). Based on these observations, it is evident that as the volume fraction of GNPs increases in these nanocomposites, there is a marginal refinement of the matrix nickel grains but this is accompanied by substantial clustering of the GNPs into coarser bundles within the matrix. Also, cross-sectional view of these composites didn't show any change in the shape of these GNP clusters. The main advantage of SPS sintered alloys and metal-matrix composites is refined grain structure can be retained without any substantial texturing effect.

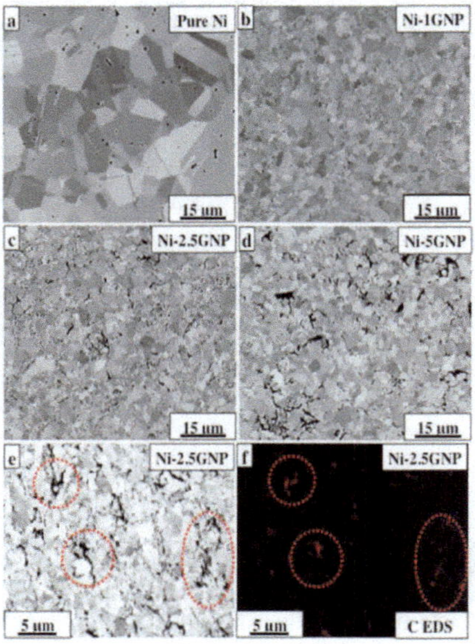

Figure 6. **Backscatter SEM images of**
(a) pure nickel,
(b) Ni-1GNP,
(c) Ni- 2.5GNP, and
(d) Ni-5GNP nanocomposites,
(e) SEM micrograph of Ni-2.5GNP nanocomposite, and
(f) corresponding EDS C map and circles are corresponding to GNP clusters.

B: X-ray Diffraction (XRD) Analysis of GNP/Ni nanocomposites

The x-ray diffraction (XRD) patterns for the SPS processed pure nickel and GNP/Ni nanocomposites are shown in Fig. 7. The XRD pattern for pure nickel shows peaks corresponding to the (111), (200), and (220) crystallographic planes of nickel. The addition of GNPs leads to the presence of a new peak which is referred to (0002) carbon peak corresponding to 2θ=26.5°. Presence of carbon peak in both the GNP/Ni nanocomposites confirms the presence of GNPs, without any other interfacial product/carbide formation. With increasing volume fraction of GNPs, the intensity of the GNP peak/ (0002) carbon peak, systematically increases. In all three XRD patterns from the GNP/Ni nanocomposites, the maximum intensity peak corresponds to the Ni(111) peak. The Ni(111)/Ni(002) intensity ratio for SPS processed pure Ni is 2.35; while that for Ni-1GNP, 2.43; for Ni-2.5GNP, 2.21; and finally for Ni-5GNP, 2.25.

These three ratio values are in good agreement with the standard ratio expected for randomly oriented Ni grains of 2.38 (based on the International Center for Diffraction Data (ICDD) files obtained from Joint Committee of PowderDiffraction Standards (JCPDS)). No significant change in texture results from the introduction of GNPs in the nickel matrix. Therefore, these XRD results substantiate the EBSD results indicating a random texture in these nanocomposites.

C. Tensile Properties of GNP/Ni nanocomposites

The yield strength and elongation (tensile) properties of SPS processed pure nickel, processed under identical conditions as those used for the nanocomposites, and GNP/Ni nanocomposites are listed in Table 2. The corresponding engineering stress-strain curves for these materials are shown in Fig. 8. The tensile yield strength of Ni-1GNP nanocomposite is 370- MPa, which is about 2.3 times higher than that of pure nickel (160 MPa), processed under the same SPS conditions. Also, Ni-1GNP nanocomposite exhibits about 40% elongation to failure/ ductility, which is well comparable with that of pure nickel (i.e., 50% elongation to failure). With increasing volume fraction of GNPs for Ni-2.5GNP nanocomposite exhibits tensile yield strength of 390 MPa, which is about 2.4 times higher than that of pure nickel with a compromise in ductility / elongation to failure (i.e. about 10%). Therefore, while increasing the volume fraction of GNPs from 1 to 2.5-vol% is accompanied with a marginal increment in tensile yield strength, there is a significant reduction in ductility (i.e., from 40% to 10%). This reduction in ductility can be attributed to the agglomeration of GNPs into coarse bundles within the nickel matrix. These bundles fail to effectively transfer load/stress during tensile deformation. Even more striking is that, the tensile yield strength of the Ni- 5GNP nanocomposite is 350 MPa, lower than that of both the Ni-1GNP and Ni-2.5GNP nanocomposites, due to more pronounced GNP clustering and coarser bundles within the nickel matrix. There are multiple previous reports discussing similar observations with respect to the deterioration in mechanical properties of CNT/metal composites due to the agglomeration of CNTs into bundles, that fail to effectively transfer load/ stress during tensile deformation [67-70]. Table 3 shows the grain size for the three GNP/Ni nanocomposites being investigated, the predicted yield strength for these grain sizes based on a Hall-Petch type relationship, and the experimentally measured yield strength of all these nanocomposites. The values of the Hall- Petch coefficient, σ_o and k, were obtained from an optimization study for pure nickel, SPS processed under different conditions. Varying the SPS processing parameters permitted a systematic variation in the grain size of the pure nickel samples from 5 to 45 μm. The corresponding tensile properties of these pure nickel samples were measured and their yield strength values were plotted against $d^{-1/2}$ (inverse square root of grain size), shown in Fig. 7 using a dotted line, to determine the σ_o and k values in the classical Hall-Petch relationship,

$$\sigma_y = \sigma_0 + kd^{-1/2}$$

Thus, values of σ_0= 77.7 MPa and k=434.5 MPa.$\mu m^{1/2}$ were determined from this optimization study on SPS processed pure nickel. This relationship was used to predict the effective contribution of grain refinement to the yield strength of the GNP/Ni nanocomposites, as shown in Table 3 under the column, σ_y(matrix). The experimentally measured yield strength values for the SPS processed pure Ni samples (filled diamonds) and the GNP/Ni nanocomposites (open circles) are also shown in Fig. 9 as discrete points. Based on both the yield strengths of the Ni-1GNP and Ni-2.5GNP nanocomposites, shown in both Table 3 as well as in Fig. 9, it is apparent that while the grain refinement in the nickel matrix contributes to the strength enhancement, this cannot be the sole factor for the high yield strength of these nanocomposites.

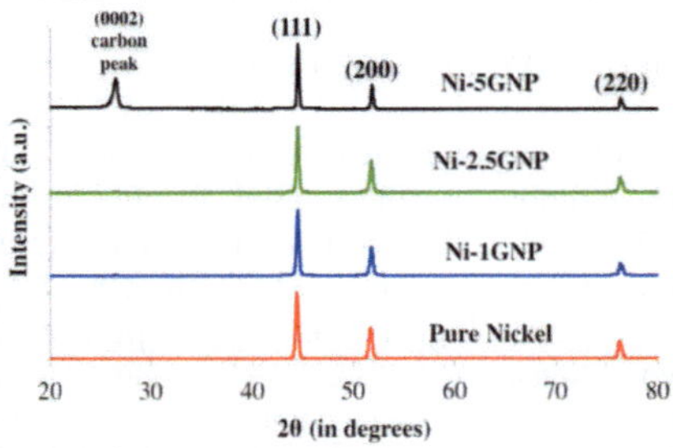

Fig. 7: XRD pattern obtained from pure nickel, Ni-1GNP, Ni-2.5GNP, and Ni-5GNP nanocomposites.

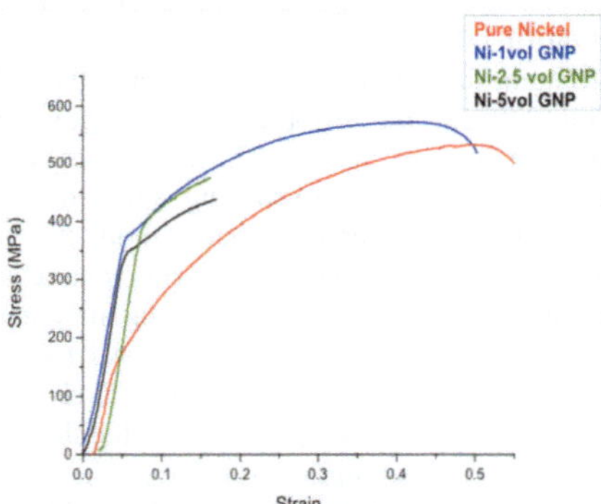

Fig. 8: Stress-strain curves of GNP/Ni nanocomposites and pure Ni obtained from tensile test.

Table 2: Microhardness, yield strength and elongation properties of pure nickel and GNP/Ni nanocomposites.

Sample	Hardness (HV)	Yield Strength (MPa)	Ductility/Elongation (%)
Ni (processed under the same conditions)	77.52±5	160±5	50
Ni-1GNP	165±10	370±3	43
Ni-2.5GNP	170±4	390±3.5	10
Ni-5GNP	150±3	350±6	10

Table 3: Grain size, predicted yield strength based on grain size (Hall-Petch relationship) and experimentally measured yield strength of GNP/Ni nanocomposites.

$$\sigma y = \sigma_o + kd^{-1/2}$$

$$\sigma_o = 77.7 \quad k=434.5$$

Sample	V_f(%)	Grain size d (μm)	σ_y(matrix)	Yield Strength σ_y (Expt)	⊗σy=σy(Expt)-
Ni-1GNP	1	4	295	370	75
Ni-2.5GNP	2.5	3	329	390	61
Ni-5GNP	5	2	385	350	-35

Thus, in both cases there is at least an additional 60-70 MPa enhancement in the yield strength beyond the value predicted by the Hall-Petch relationship, which can be attributed to the GNP reinforcement. Therefore, the Ni- 1GNP and Ni-2.5GNP nanocomposites exhibited a significant improvement in tensile yield strength due to the coupled effect of grain refinement as well as GNP reinforcement in the nickel matrix. Furthermore, the strengthening effect of the GNPs indicates that this reinforcement effectively transfers stress during tensile deformation. Typically, the effective yield strength of composites due to the presence of the reinforcement has often been modeled using a shear-lag model based on the load transfer theory [71-74] as follows:

$$\sigma_c = \sigma_m (1 + S_{eff}V_f)$$

where σ_c is the effective yield strength of the composite, σ_m is the yield strength of the matrix without the reinforcement, S_{eff} is the effective aspect ratio of the reinforcement, and V_f is the volume fraction of the reinforcement. This model was originally developed for composites reinforced with randomly oriented short fibers assuming perfect bonding between the reinforcement and the matrix [71-74] and predicts an a linear increase in the effective yield strength of the composite with increasing volume fraction of the reinforcement. While in the present study, the experimentally measured yield strength for the Ni-2.5GNP exceeds that of the Ni-1GNP nanocomposite, this increase is a combined effect of grain size refinement and reinforcement from the GNPs. Therefore, in order to de-couple the effect of strengthening due to grain refinement from strengthening due to composite reinforcement, the

$$\otimes\sigma_y(GNP) = \sigma_y(expt) - \sigma_y(matrix)$$

has been plotted as a function of volume % of GNPs, in Fig. 9(b). Here the σ_y(matrix) corresponds to the yield strength of SPS processed pure Ni with a similar grain size as that observed in the corresponding GNP/Ni nanocomposite.

This plot appears to depict a functional relationship where the $\otimes\sigma_y$(GNP) initially increases with increasing volume % of the GNPs, but after achieving a maximum value between 1 to 2 volume % GNPs, the $\otimes\sigma_y$(GNP) decreases with further increase in the GNP content. This decrease in yield strength for higher volume % of GNPs can be largely attributed to the agglomeration of the GNPs into bundles within the nickel matrix.

As noted earlier, the Ni-5GNP nanocomposite exhibited substantial deterioration in its tensile properties, including yield strength, which is a result of substantial agglomeration of the GNPs into bundles within the matrix. Despite the more refined nickel grain size in this nanocomposite (~2 μm) its yield strength is only 350 MPa, lower than that of both Ni-1GNP and Ni-2.5GNP. Comparing the ductility of the three GNP/Ni nanocomposites, it should be noted that the Ni-1GNP nanocomposite exhibits the largest elongation to failure ductility, ~40%, while the Ni- 2.5GNP and Ni-5GNP exhibit elongations of only ~10%. Interestingly, the stress-strain behavior of the Ni-1GNP nanocomposite largely resembles that of pure Ni, with significant strain hardening reaching an ultimate tensile strength (UTS) of ~570MPa. While both the Ni-2.5GNP and Ni-5GNP nanocomposites exhibit the early stages of strain hardening, they abruptly fail before achieving their ultimate tensile strengths. Such a well-behaved stress-strain curve, observed in case of the Ni-1GNP nanocomposite, coupled with its high yield and tensile strengths as well as excellent ductility, makes it a very promising material for engineering applications.

D: Fractography Analysis of GNP/Ni nanocomposites

Figs. 10(a-d) are SEM images (secondary electron images) showing the fracture surfaces of the pure nickel, Ni-1GNP, Ni-2.5GNP, and Ni-5GNP nanocomposites, respectively. From these images, it is evident that while pure nickel (Fig. 10(a)) and the Ni-1GNP (Fig. 10(b)) nanocomposite exhibit similar fracture surfaces, the Ni- 2.5GNP (Fig. 10(c)) and Ni-5GNP (Fig. 10(d)) nanocomposites exhibit somewhat different fracture surfaces. The Ni-1GNP nanocomposite fracture surface (Fig. 10(b)) exhibits the typical cup and cone type failure features, with dimples of the order of few microns in size. This is indicative of a very ductile deformation behavior in agreement with the greater than 40% elongation observed for these nanocomposites, similar to that of pure nickel. Contrastingly, the fracture surfaces of the Ni-2.5GNP and Ni-5GNP nanocomposites exhibit much smaller dimples as well as GNP clusters marked by arrows in Figs. 10(c) and (d). This is indicative of strain localization and eventual failure at a much lower level of strain, as clearly observed in their stress-strain behavior shown in Fig. 8. The significant enhancement of yield strength coupled with reasonable ductility, obtained in case of the Ni-1GNP nanocomposite, can be attributed to a coupled effect of grain refinement, homogeneous distribution of GNPs in the nickel matrix, and strong Ni-GNP interfacial bonding, leading to effective transfer of load across the metal/GNP interface during tensile deformation.

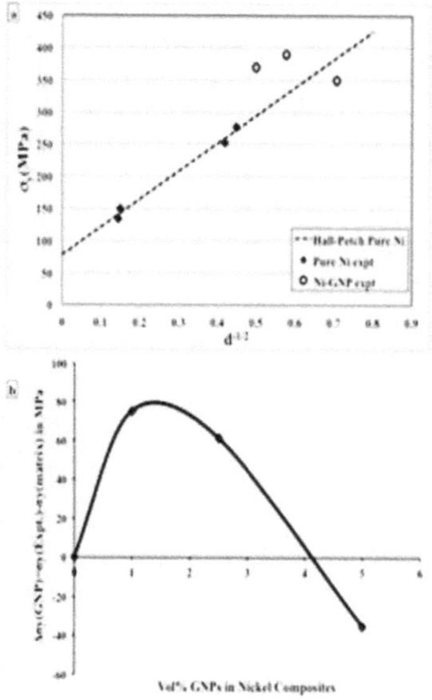

Figure 9: Stress-strain curves of GNP/Ni nanocomposites and pure Ni obtained from tensile test.

E: Transmission Electron Microscopy (TEM) Analysis of GNP/Ni nanocomposites

TEM results from the Ni-1GNP nanocomposite are shown in Fig. 11. While the GNPs in this nanocomposite are homogeneously distributed, most of them are located at the boundaries or triple junctions of nickel grains as shown in Fig. 11(a). An enlarged bright-field view of a single GNP embedded in the nickel matrix at a triple junction, is shown in Fig. 11(b). High-resolution TEM micrographs of the GNP/Ni interface in case of the Ni-1GNP nanocomposite, are shown in Figs. 11(c) and (d). The interface is relatively sharp, exhibits excellent bonding between the GNP and nickel grain, and is also contaminant-free. Furthermore, these high-resolution TEM images do not exhibit any interfacial reaction layer between the GNP and the Ni matrix in corroboration with XRD. Such a well-bonded GNP/Ni interface is expected to effectively transfer load between the matrix and the reinforcement, leading to a higher strength and ductility. Fig. 12 shows the TEM results from the Ni-2.5GNP nanocomposite. The lower magnification bright field TEM image shown in Fig. 12(a) shows both larger grains (order of microns) as well as nanocrystals of nickel, with an average diameter of 50nm, that were surrounded by GNPs, 50 nm wide and 300nm long. High-resolution TEM images of this microstructure are shown in Figs. 12(b)-(d). Figs. 12(b) and (c) clearly reveal the individual graphene sheets within platelets and their interface with the surrounding nickel grains. Additionally, a higher magnification view of a single GNP/Ni interface, shown in Fig. 12(d), revealed the well bonded and sharp nature of this interface. The lattice planes of nickel as well as graphene on either sides of the interface are clearly visible in this image and confirm the absence of any interfacial reaction product.

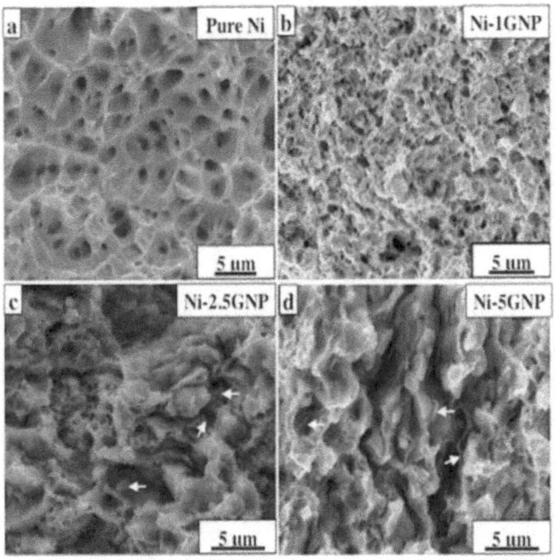

Figure 10: **SEM micrograph showing the fracture surfaces of the (a) pure Ni, (b) Ni-1GNP, (c) Ni-2.5GNP, and Ni-5GNP nanocomposites after tensile test; arrows are showing GNP clusters.**

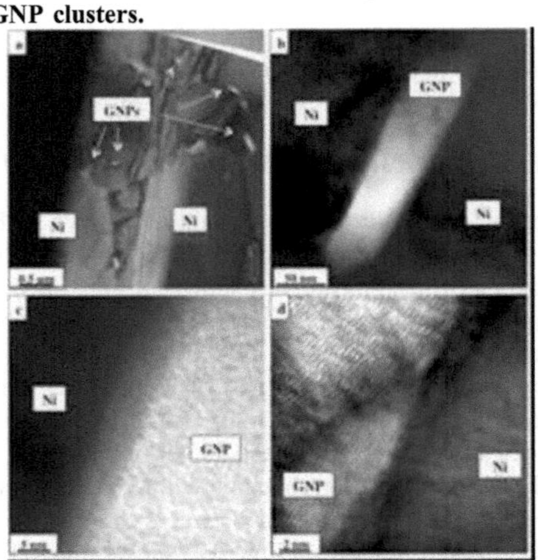

Figure 11: **(a) and (b) Bright-field TEM image showing GNPs along the boundary of nickel grains, (c) and (d) HRTEM image on the GNP/Ni interface of Ni-1GNP composite.**

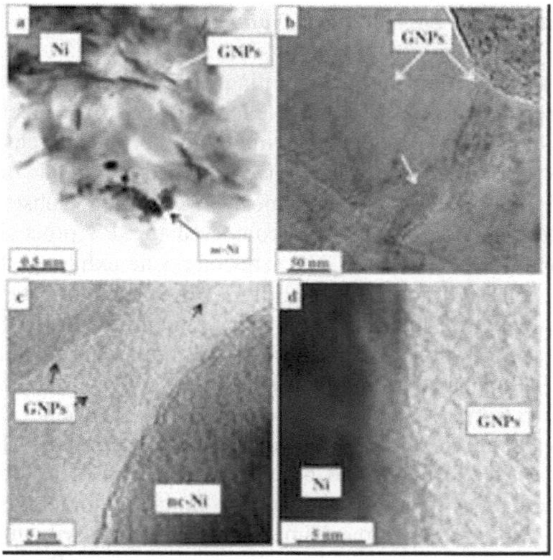

Figure 12: (a) and (b) **TEM micrographs showing distribution of GNPs and nanocrystalline Ni (nc-Ni) in the composite,**
(c) and (d) **HRTEM image showing the entangled configuration of GNPs in the matrix,**
(e) **HRTEM image of the interface between a GNP and a nc-Ni clearly showing a large difference between the interlunar spacing of Ni and graphite of Ni-2.5GNP composite.**

4. Summary and Conclusions

(i) Carbon Nanotube (CNT)/Nickel Nanocomposites:

Two types of CNT/Ni nanocomposites have been prepared using the spark plasma sintering (SPS) process, one using ball milled nickel and CNT precursors, referred to as CNT/Ni (DM) and the second one using molecular level mixed CNT/Ni composite powders as a precursor, referred to as CNT/Ni (MLM). While both types of nanocomposites exhibited a substantial enhancement in microhardness and tensile yield strength as compared to SPS processed pure Ni, the CNT/Ni (DM) nanocomposite exhibited much higher ductility while the CNT/Ni (MLM) nanocomposite exhibited much higher tensile yield strength. The improvement in mechanical properties of the CNT/Ni (DM) nanocomposite results from a combination of composite strengthening due to the nanotubes, coupled with grain refinement of the nickel matrix, both effects resulting from the homogeneous distribution of small bundles of CNTs. The microhardness and yield strength of CNT/Ni (MLM) nanocomposite was significantly higher as compared to both CNT/Ni (ball milled) as well as pure nickel. The high strength of CNT/Ni (MLM) nanocomposite is mainly achieved due to the homogeneous distribution of individual CNTs in the nickel matrix and high interfacial strength due to a well-bonded Ni/CNT interface, permitting excellent load transfer between the matrix and the nanotubes.

(ii) Graphene Nanoplatelet (GNP)/Nickel Nanocomposites:

A new class of nanocomposites, exhibiting an excellent combination of strength and ductility, has been prepared by spark plasma sintering (SPS) of a mechanically mixed (ball-milled) mixture of elemental nickel powder with graphite nanoplatelets (GNP). The volume fraction of the GNP reinforcement has been varied from 1 to 5 vol% in these nanocomposites. All the GNP/Ni nanocomposites exhibited substantial enhancement in microhardness and tensile yield strength as compared to SPS processed pure nickel, under similar processing conditions. The Ni- 1GNP nanocomposite exhibited the best combination of high yield strength (~350MPa) and high ductility (~40% elongation to failure), while the Ni-2.5GNP nanocomposite exhibited marginally higher yield strength (~370MPa) but substantially lower ductility (~10% elongation to failure). The yield strength enhancement in the GNP/Ni nanocomposites can be attributed to a synergistic effect of matrix grain coupled with the reinforcement effect resulting from a homogeneous distribution of GNPs with a strong interfacial bonding with the nickel matrix, which effectively transfers load across metal/GNP interface. While 1 vol% GNP appears to be an optimal level of reinforcement in the nickel matrix, higher levels of 2.5 or 5 vol% result in substantial agglomeration of the GNP's within the matrix leading to deterioration in mechanical properties primarily manifested by reduction in ductility. Furthermore, the substantial agglomeration of GNPs in case of the Ni-5GNP composite, results in less effective transfer of load from the metal matrix to the GNPs, leading to reduction in yield strength as compared to the Ni-1GNP and the Ni-2.5GNP nanocomposites. The yield strength of these nanocomposites has been modeled using a combined model incorporating the effects of composite reinforcement strengthening via load transfer with Hall-Petch type strengthening from grain refinement.

Acknowledgements

This work was financially supported by a grant (code No. FA2386-11-1-4008) from AOARD and partially supported by U. S. Air Force Research Laboratory (AFRL ISES Contract No. FA8650-08-C-5226) and a grant (code # 2010K000275) from 'Center for Nanostructured Materials Technology' under '21st Century Frontier R&D Programs' of the Ministry of Education, Science and Technology in Korea. The authors also gratefully acknowledge the Center for Advanced Research and Technology (CART) at the University of North Texas. The Korea Institute of Science and Technology (KIST) Institutional Program in the Republic of Korea has also provided partial support for this work.

References

1. P. Harris, "Carbon nanotube composites," ***International Material Reviews***, 49 (2004), 31 -43.

2. S. C. Tjong, "Recent progress in the development and properties of novel metal matrix nanocomposites reinforced with carbon nanotubes and graphene nanosheets," ***Materials Science and Engineering: R***, 74 (2013), 281-350.

3. S. Iijima, "Helical microtubules of graphitic carbon," ***Nature***, 354 (1991) 56-58.

4. S. Iijima and T. Ichlhashi, "Single-shell carbon nanotubes of 1-nm diameter," ***Nature***, 363 (1993) 603-605.

5. Y. Zhang, T. Ichihashi, E. Landree, F. Nihey, and S. Iijima, "Heterostructures of single-walled carbon nanotubes and carbide nanorods," ***Science***, 285 (1999) 1719-1722.

6. J. Liao, M. Tan, and I. Sridhar, "Spark plasma sintered multi-wall carbon nanotube reinforced aluminum matrix composites," ***Material & Design***, 31 (2010) S96-S100.

7. K. Chu, C. Jia, L. Jiang, and W. Li, "Improvement of interface and mechanical properties in carbon nanotube reinforced Cu-Cr matrix composites.," ***Material & Design***, 45 (2013) 407-411.

8. Q. Li, A. Vierecki, C. Rottmair, and R. Singer, "Improved processing of carbon nanotube/magnesium alloy composites," ***Composites Science and Technology***, 69 (1993) 1193-1199.

9. C. Guiderdoni, C. Estournes, A. Peigney, A. Weibel , V. Turq, and C. Laurent, "The preparation of double-walled carbon nanotube/Cu composites by spark plasma sintering, and their hardness and friction properties," ***Carbon***, 49 (2011) 4535-4543.

10. A. Esawi, E. Morsi, A. Sayed, A. Gawad, and P. Borah, "Fabrication and properties of dispersed carbon nanotube-aluminum composites," ***Materials Science and Engineering A***, 508 (2009) 167-173.

11. D. Lahiri, S. Bakshi, A. Keshri, Y. Liu, and A. Agarwal, "Dual strengthening mechanisms induced by carbon nanotubes in roll bonded aluminum composites," ***Materials Science and Engineering A***, 523 (2009) 263-270.

12. H. Kwon, M. Estili, K. Takagi, T. Miyazaki, and A. Kawassaki, "Combination of hot extrusion and spark plasma sintering for producing carbon nanotube reinforced aluminum matrix composites," ***Carbon***, 47 (2009) 570-577.

13. S. Uddin, T. Mahmud, C. Wolf, C. Glanz, I. Kolaric, C. Volkmer, H. Holler, U. Wienecke, S. Roth, and H. Frecht, "Effect of size and shape of metal particles to improve hardness and electrical properties of carbon nanotube reinforced copper and copper alloy composites," ***Composites Science and Technology***, 70 (2010) 2253-2257.

14. S. Cho, K. Kikuchi, T. Miyazaki, K. Takagi, A. Kawasaki, and T. Tsukada, Multiwalled carbon nanotubes as a contributing reinforcement phase for the improvement of thermal conductivity in copper matrix composites, ***Scripta Materialia***, 63 (2011) 375-378.

15. S. Bakshi, D. Lahiri, and A. Agarwal, "Carbon nanotube reinforced metal matrix composites- a review," ***International Material Reviews***, 55 (2010) 41-64.

16. A.R.P. Singh, J.Y. Hwang, T. Scharf, J. Tiley, and R. Banerjee, "Bulk nickel-carbon nanotube nanocomposites by laser deposition," ***Materials Science & Technology***, 26 (2010) 1393-1400.

17. C. Wei, F. Xue, and S. Jiehe, "Preparation of multi-walled carbon nanotube-reinforced TiNi matrix composites from elemental powders by spark plasma sintering," ***Rare Metals*** 31 (2012) 48-50.

18. R. George, K. Kashyap, R. Rahul, and S. Yamdagni, "Strengthening in carbon nanotube/aluminum (CNT/Al) composites," ***Scripta Materialia***, 53 (2005) 1159-1163.

19. K. Kim, S. Cha, S. Hong, and S. Hong, "Microstructures and tensile behavior of carbon nanotube reinforced Cu matrix nanocomposites," ***Materials Science and Engineering A***, 430 (2006) 27-33.

20. T. Scharf, A. Niera, J. Hwang, J. Tiley, and R. Banerjee, "Self-lubricating carbon nanotube reinforced nickel matrix composites," ***Journal of Applied Physics***, 106 (2009) 013508-1-7.

21. K. Kim, J. Eckert, S. Menzel, T. Gemming, and S. Hong, "Grain refinement assisted strengthening of carbon nanotube reinforced copper matrix nanocomposites," ***Applied Physics Letter***, 92 (2008) 121901-1-3.

22. S. Yamanaka, R. Gonda, A. Kawasaki, H. Sakamoto, Y. Mekuchi, M. Kuno, and T. Tsukada, "Fabrication and thermal properties of carbon nanotube/nickel composite by spark plasma sintering method," ***Materials Transactions***, 48(9) (2007) 2506-2512.

23. A. Esawi, K. Morsi, A. Sayed, M. Taher, and S. Lanka, "Effect of carbon nanotube (CNT) content on the mechanical properties of CNT-reinforced aluminium composites," ***Composites Science and Technology***, 70 (2010) 2237-2241.

24. S. Cha, K. Kim, S. Arshad, C. Mo, and S. Hong, "Extraordinary strengthening effect of carbon nanotubes in metal-matrix nanocomposites processed by molecular-level mixing," ***Advanced Materials***, 17 (2005) 1377-1381.

25. J.Y. Hwang, B. Lim, J. Tiley, R. Banerjee, and S. Hong, "Interface analysis of ultra-high strength carbon nanotube/ nickel composites processed by molecular level mixing," ***Carbon***, 57 (2013) 282-287.

26. S. Cha, K. Kim, H. Lee, C. Mo, and S. Hong, "Strengthening and toughening of carbon nanotube reinforced alumina nanocomposite fabricated by molecular level mixing process," ***Scripta Materialia***, 53 (2005) 793-797.

27. K. Kim, S. Cha, and S. Hong, "Hardness and wear resistance of carbon nanotube reinforced Cu matrix nanocomposites," ***Materials Science and Engineering*** *A*, 449 (2007) 46-50.

28. K. Kim, J. Eckert, G. Liu, J. Park, B. Kim, and S. Hong, "Influence of embedded-carbon nanotubes on the thermal properties of copper matrix nanocomposites processed by molecular-level mixing," **Scripta *Materialia***, 64(2) (2011) 181-184.

29. S. C. Tjong, "Recent progress in the development and properties of novel metal matrix nanocomposites reinforced with carbon nanotubes and graphene nanosheets," ***Materials Science and Engineering: R: Reports***, 74 (2013) 281-350.

30. M. A. Rafiee, J. Rafiee, Z. Wang, H. Song, Z.-Z. Yu, and N. Koratkar, "Enhanced Mechanical Properties of Nanocomposites at Low Graphene Content," *ACS Nano*, 3 (2009) 3884-3890.

31. D. Kuang, L. Xu, L. Liu, and W. Hu, "Graphene-nickel composites," ***Applied Surface Science***, 273 (2013) 484-490.

32. J. Wang, Z. Li, G. Fan, H. Pan, Z. Chen, and D. Zhang, "Reinforcement with graphene nanosheets in aluminum matrix composites," ***Scripta Materialia***, 66 (2012) 594-597.

33. L.-Y. Chen, H. Konishi, A. Fehrenbacher, C. Ma, J.-Q. Xu, H. Choi, "Novel nanoprocessing route for bulk graphene nanoplatelets reinforced metal matrix nanocomposites," ***Scripta Materialia***, 67 (2012) 29-32.

34. F. H. Latief and E.-S. M. Sherif, "Effects of sintering temperature and graphite addition on the mechanical properties of aluminum," ***Journal of Industrial and Engineering Chemistry***, 18 (2012) 2129-2134.

35. D. Prasai, J. C. Tuberquia, R. R. Harl, G. K. Jennings, and K. I. Bolotin, "Graphene: corrosion-inhibiting coating," ***ACS nano***, 6 (2012) 1102-1108.

36. S. Stankovich, D. A. Dikin, G. H. B. Dommett, K. M. Kohlhaas, E. J. Zimney, E. A. Stach, "Graphene-based composite materials," ***Nature***, 442 (2006) 282-286.

37. S. F. Bartolucci, J. Paras, M. A. Rafiee, J. Rafiee, S. Lee, D. Kapoor, "Graphene–aluminum nanocomposites," ***Materials Science and Engineering****: A*, 528 (2011) 7933-7937.

38. J. Hwang, T. Yoon, S. H. Jin, J. Lee, T. S. Kim, S. H. Hong, "Enhanced Mechanical Properties of Graphene/Copper Nanocomposites Using a Molecular- Level Mixing Process," ***Advanced Materials***, 25(46) (2013) 6724-6729.

39. N. T. Kirkland, T. Schiller, N. Medhekar, and N. Birbilis, "Exploring graphene as a corrosion protection barrier," ***Corrosion Science***, 56 (2012) 1-4.

40. K. Rajkumar and S. Aravindan, "Tribological behavior of microwave processed coppernanographite composites," ***Tribology International***, 57 (2012) 282-296.

41. F. H. Latief, E.-S. M. Sherif, A. A. Almajid, and H. Junaedi, "Fabrication of exfoliated graphite nanoplatelets-reinforced aluminum composites and evaluating their mechanical properties and corrosion behavior," *Journal of Analytical and Applied Pyrolysis*, 92 (2011) 485-492,.

42. K. Jagannadham, "Electrical conductivity of copper–graphene composite films synthesized by electrochemical deposition with exfoliated graphene platelets," ***Journal of Vacuum Science & Technology B: Microelectronics and Nanometer Structures***, 30 (2012) 03D109-03D109-9.

43. Ramanathan T, A. A. Abdala, Stankovich, D. A. Dikin, M. Herrera Alonso, R. D. Piner, "Functionalized graphene sheets for polymer nanocomposites," **Nature Nanotechnology**, 3(6) (2008) 327-331.

44. P. Song, Z. Cao, Y. Cai, L. Zhao, Z. Fang, and S. Fu, "Fabrication of exfoliated graphene-based polypropylene nanocomposites with enhanced mechanical and thermal properties," **Polymer**, 52 (2011) 4001-4010.

45. S. H. Song, K. H. Park, B. H. Kim, Y. W. Choi, G. H. Jun, D. J. Lee, et al., "Enhanced Thermal Conductivity of Epoxy–Graphene Composites by Using Non-Oxidized Graphene Flakes with Non-Covalent Functionalization," ***Advanced Materials***, 25 (2013) 732- 737.

46. D. Lee, S. H. Song, J. Hwang, S. H. Jin, K. H. Park, B. H. Kim, "Enhanced Mechanical Properties of Epoxy Nanocomposites by Mixing Noncovalently Functionalized Boron Nitride Nanoflakes," **Small**, 9 (2013) 2602-2610.

47. T. Yoon, W. C. Shin, T. Y. Kim, J. H. Mun, T.-S. Kim, and B. J. Cho, "Direct Measurement of Adhesion Energy of Monolayer Graphene As-Grown on Copper and Its Application to Renewable Transfer Process," **Nano Letters**, 12 (2012) 1448-1452.

48. S. C. Tjong, "Recent progress in the development and properties of novel metal matrix nanocomposites reinforced with carbon nanotubes and graphene nanosheets," ***Materials Science and Engineering: R: Reports***, 74 (2013) 281-350.

49. M. Tokita, "Mechanism of spark plasma sintering," in Proceeding of NEDO International Symposium on Functionally Graded Materials, 22 (1999) 1-13.

50. S. Somiya, Handbook of Advanced Ceramics: Materials, Applications, Processing, and Properties: Access Online via Elsevier, 2013.

51. Z. A. Munir, D. V. Quach, and M. Ohyanagi, "Electric current activation of sintering: a review of the pulsed electric current sintering process," ***Journal of the American Ceramic Society***, 94 (2011) 1-19.

52. Z A. Munir, U. Anselmi-Tamburini, and M. Ohyanagi, "The effect of electric field and pressure on the synthesis and consolidation of materials: A review of the spark plasma sintering method," ***Journal of Materials Science***, 41 (2006) 763-777.

53. R. Orru, R. Licheri, A. M. Locci, A. Cincotti, and G. Cao, "Consolidation/synthesis of materials by electric current activated/assisted sintering," ***Materials Science and Engineering: R: Reports***, 63 (2009) 127-287.

54. V. Mamedov, "Spark plasma sintering as advanced PM sintering method," ***Powder Metallurgy***, 45 (2002) 322-328.

55. X. Song, X. Liu, and J. Zhang, "Neck Formation and Self- Adjusting Mechanism of Neck Growth of Conducting Powders in Spark Plasma Sintering," ***Journal of the American Ceramic Society***, 89 (2006) 494-500.

56. M. Omori, "Sintering, consolidation, reaction and crystal growth by the spark plasma system (SPS)," ***Materials Science and Engineering**: A*, 287 (2000) 183-188.

57. W. Yucheng and F. Zhengyi, "Study of temperature field in spark plasma sintering," ***Materials Science and Engineering:*** *B*, 90 (2002) 34-37.

58. J. E. Garay, "Current-activated, pressure-assisted densification of materials," ***Annual Review of Materials Research***, 40 (2010) 445- 468.

59. M. Suárez, A. Fernández, J. L. Menéndez, R. Torrecillas, H. U. Kessel, J. Hennicke, "Challenges and Opportunities for Spark Plasma Sintering: A Key Technology for a New Generation of Materials," 2013.

60. T. Borkar and R. Banerjee. "Influence of spark plasma sintering (SPS) processing Parameters on microstructure and mechanical Properties of nickel." ***Materials Science and Engineering: A,*** 618(17) (2014) 176-181.

61. T. Borkar, S. Nag, Y. Ren, J. Tiley, and R. Banerjee, "Reactive Spark Plasma Sintering (SPS) of Nitride Reinforced Titanium Alloy Composites." ***Journal of Alloys and Compounds,*** 617 (2014) 933-945.

62. T. Borkar, J. Hwang, J. Y. Hwang, T. W. Scharf, J. Tiley, S. H. Hong, & R. Banerjee, "Strength versus ductility in carbon nanotube reinforced nickel matrix nanocomposites," ***Journal of Materials Research***, *29*(06) (2014) 761-769.

63. T. Borkar, H. Mohseni, J. Y. Hwang, T. W. Scharf, J. Tiley, S. H. Hong, & R. Banerjee, "Excellent strength-ductility combination in Nickel-Graphite nanoplatelet (GNP/Ni) nanocomposites," ***Carbon*** (2014).

64. T. Borkar and S. Harimkar, "Microstructure and wear behavior of pulse electrodeposited CNT/Ni composite coatings," ***Surface Engineering,*** 27(7) (2011) 524-530.

65. T. Borkar and S. Harimkar, "Effect of electrodeposition conditions and reinforcement content on microstructure and tribological properties of nickel composite coatings," ***Surface & Coatings Technology,*** 205(17) (2011) 4124-4134.

66. N. Krasilnikov, W. Lojkowski, Z. Pakiela, and R. Valiev, "Tensile strength and ductility of ultra-fine-grained nickel processed by severe plastic deformation," ***Materials Science and Engineering: A,*** 397 (2005) 330-337.

67. T. Kuzumaki, K. Miyazawa, H. Ichinose, K. Ito, "Processing of carbon nanotube reinforced aluminum composite," ***Journal of Materials Research***, 13 (1998) 2445-2449.

68. A. Esawi, M. Borady, "Carbon nanotube-reinforced aluminum strips," ***Composites Science and Technology***, 68 (2008) 486-492.

69. S. Dong, J. Tu, X. Zhang, "An investigation of the sliding wear behavior of Cu-matrix composites reinforced by carbon nanotubes," ***Materials Science and Engineering*** *A*, 313 (2001) 83-87.

70. X. Chen, J. Xia, J. Peng, W. Li, S. Xie, "Carbon-nanotube metal-matrix composites prepared by electroless plating," ***Composites Science and Technology***, 60 (2000) 301-306.

71. D. Nam, S. Cha, B. Lim, H. Park, D. Han, S. Hong, "Synergistic strengthening by load transfer mechanism and grain refinement CNT/Al-Cu composites," ***Carbon***, 50 (2012) 2417-2423.

72. K. Kim, J. Eckert, J. Menzel, S. Gemming, S. Hong, "Grain refinement assisted strengthening of carbon nanotube reinforced copper matrix nanocomposites," ***Applied Physics Letters***, 92(12) (2008) 121901-1-3.

73 Y. Jeong, S. Cha, K. Kim, K. Lee, C. Mo, S. Hong, "Synergistic strengthening effect of ultrafine grained metals reinforced with carbon nanotubes," ***Small***, 3 (2007) 840-844.

74. Y. Tang, X. Yang, R. Wang, M. Li, "Enhancement of the mechanical properties of graphene-copper composites with graphene-nickel hybrids," ***Materials Science and Engineering A***, 599 (2014) 247-254.

Advanced Composites for Aerospace, Marine, and Land Applications II
Edited by: T. Sano and T.S. Srivatsan
TMS (The Minerals, Metals & Materials Society), 2015

EFFECT OF IN-SITU TITANIUM BORIDE PARTICLE ADDITION AND FRICTION STIR PROCESSING ON WEAR BEHAVIOR OF ALUMINUM ALLOY 2219

V.Sampath[1], N.R.Rajasekaran[2]

[1] *Department of Metallurgical and Materials Engineering*
Indian Institute of Technology Madras
Chennai 600036, India
Email: vsampath@iitm.ac.in)

[2] *Department of Mechanical Engineering*
Dhanalakshmi College of Engineering
Chennai 601301, India

Keywords: Composites, Friction stir Processing, Wear, Microstructure

Abstract

There is paucity of data on the effect of *in-situ* TiB_2 particles on the different behavior of 2219 Al alloy. In the present work, therefore, composites with 2219 Al alloy matrix reinforced with *in-situ* TiB_2 particles were produced by flux-assisted synthesis. Different amounts (5 and 10 wt.%) of TiB_2 were incorporated. The base alloy and the composites were subjected to friction stir processing to reduce particle clustering. The wear behavior of the materials was studied. The base alloy showed considerable improvement in the wear behavior due to the *in-situ* particle addition which is attributed to the presence of finer particles with good interfacial bonding and high hardness. Uniform distribution of TiB_2 particles and good interface between the matrix and the particles led to reduction in the wear of friction stir processed composites by 30%. The results are analyzed and discussed in detail in the paper.

1. Introduction

Al alloy 2219 is widely used as a structural material in aircraft applications because of its high specific strength, high specific modulus and good weldability [1]. A number of researchers have reported that the strength of this alloy can be improved by modifying its grain structure and uniformly dispersing the second phase particles in its matrix. To improve the strength of this alloy, various secondary processing techniques, such as equichannel angular pressing (ECAP) [2] and cryo-rolling [3], and addition of grain refiners (such as Sc, Mg, Zr) [4], have been used to achieve it by modification of its grain structure. In general, the increase in strength of the alloy is by and large accompanied by a large decrease in ductility.

Reinforcement particles are generally added to alloy matrices through *ex-situ* processes. The drawbacks of *ex-situ* processes are poor wettability, formation of reaction products at the interface, particle size limitation, etc. In *in-situ* processes, the particle/reinforcement is produced within the melt itself. They exhibit therefore good interfacial bonding with the matrix. Since the particles are also much smaller in size (< 2μm), they tend to agglomerate during casting/fabrication. The property of a composite depends on a number of microstructural features including the distribution of the particles within the matrix.

In this work an attempt has been made to study the effect of TiB_2 particle reinforcement on this alloy using friction stir processing (FSP) to disperse the TiB_2 particles within the alloy. Friction stir processing (FSP) is based on the principle of friction stir welding. In this process, a specially designed non-consumable cylindrical tool with pin attached to its bottom is plunged into a metal plate and is rotated over the plate at a suitable speed. The heat generated at the tool-plate interface causes the plate to soften and allow the tool to travel smoothly. The intense plastic deformation and frictional heat generated causes the material to flow over the pin and form a defect-free zone. Nowadays metal matrices reinforced with particulates via *in-situ* addition are quite popular because of their isotropic properties, refinement of the grain structure, good wettability, uniform distribution and absence of reactions at the particle-matrix interface.

Clustering of particles is a common problem in *in-situ* composites as it leads to deterioration of mechanical properties. Various mechanisms have therefore been adopted to redistribute TiB_2 particles within the matrix of the *in-situ* composites, such as ECAP and extrusion [5, 6]. In all these processes the microstructural features, such as particle-rich bands contributing to the breakage of particles, have been reported.

A metal matrix composite, with 2219 Al alloy as the matrix and 5 and 10 wt.% of TiB_2 as the reinforcement introduced via an *in-situ* reaction, was prepared by a flux-assisted synthesis process. The formation of the brittle intermetallic compound Al_3Ti was completely suppressed by controlling the reaction temperature and time. The agglomerated TiB_2 particles were dispersed uniformly throughout the stir zone by friction stir processing (FSP). The process refines grain size and eliminates casting defects, leading to improvement in wear properties. Uniform distribution of TiB_2 particles, their fine size and good interfacial bonding with the matrix result in higher wear resistance of the friction stir processed composites as compared to that for the as-cast alloy. The wear surface analysis confirms the formation of a mechanically mixed layer (MML) on the surface of the composites. The fractographic examination of the friction stir processed composite reveals a ductile mode of failure. The effect of TiB_2 particle addition on the dry sliding wear behavior of cast composites and the friction stir processed composites are compared and the results are presented.

2. Experimental Procedure

Al alloy 2219 used in this work contains 6 wt. % Cu. The *in-situ* composites were produced through a flux-assisted synthesis process. Potassium titanium hexafluoride (K_2TiF_6) and potassium fluoborate (KBF_4) were used to produce TiB_2 particles. The Al alloy was melted in a furnace and the salts were added at 800° C. the reaction time was 1 h. The reactions involved in the in-situ process are [7]:

$$3K_2TiF_6 + 13\,Al \longrightarrow 3Al_3Ti + 3\,KAlF_4 + K_3AlF_6$$

$$KBF_4 + 3\,Al \longrightarrow AlB_2 + 2\,KAlF_4$$

$$AlB_2 + Al_3Ti \longrightarrow TiB_2 + 4Al$$

The cast alloy was subjected to FSP. FSP was used to distribute the TiB2 particles in the matrix. The friction stir processing parameters are 1000 rpm, traverse speed 60 mm/min and load 8 kN. Wear studies were carried out on a pin-on-disc equipment. The load was varied between 40 and 80 N at a sliding speed of 1 m/sec for a sliding distance 1000 m. The specimens for wear studies were prepared from the stir zone or nugget zone. The wear rate was calculated by volume loss divided by sliding distance.

3. Results and Discussion

3.1 Microstructural Characterization Of As-Cast Alloy And Composites:

3.1.1 XRD:

Figure 1(a) shows the XRD patterns of 2219 Al alloy and its composites. The diffraction peaks belonging to the phases α and $CuAl_2$ are presented for the alloy and its composites. In addition to these peaks those corresponding to TiB_2 were also observed in the composites. The presence of TiB_2 particles is further confirmed by dissolving the composite in 20% NaOH solution. Figure 1 (b) shows the XRD pattern of the extracted TiB_2 particles.

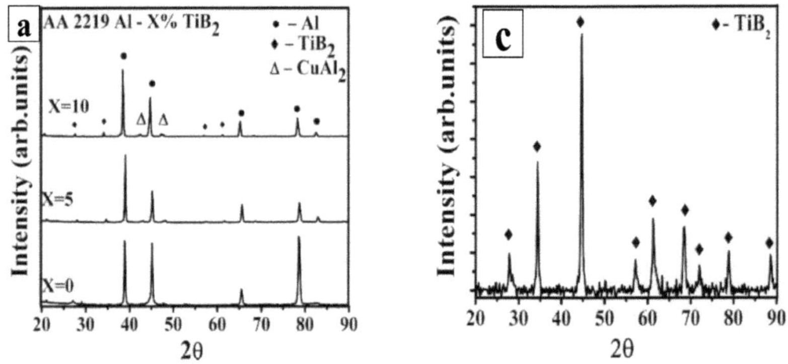

Figure 1. **XRD pattern of: (a) as-cast alloy and composites with 5 and 10 wt % TiB_2, respectively; (b) extracted TiB_2 particles.**

3.1.2 Optical Microscopy:

The optical micrographs of the as-cast alloy and those of the *in-situ* composites are shown in Figure 2. The microstructure of the as-cast alloy shows a dendritic structure. The grain size of the as-cast alloy is 115 μm. The presence TiB_2 particles transforms the dendritic structure into a more refined structure. The grain morphology changes from columnar to equiaxed. This is due to the grain refining action of TiB_2 particles. These particles act as heterogeneous nucleation sites in Al (alloy) matrix [8]. The particles refine the microstructure by retarding grain growth. The grain size of composites with 5 and 10 wt. % of TiB_2 are 36 μm and 25 μm, respectively.

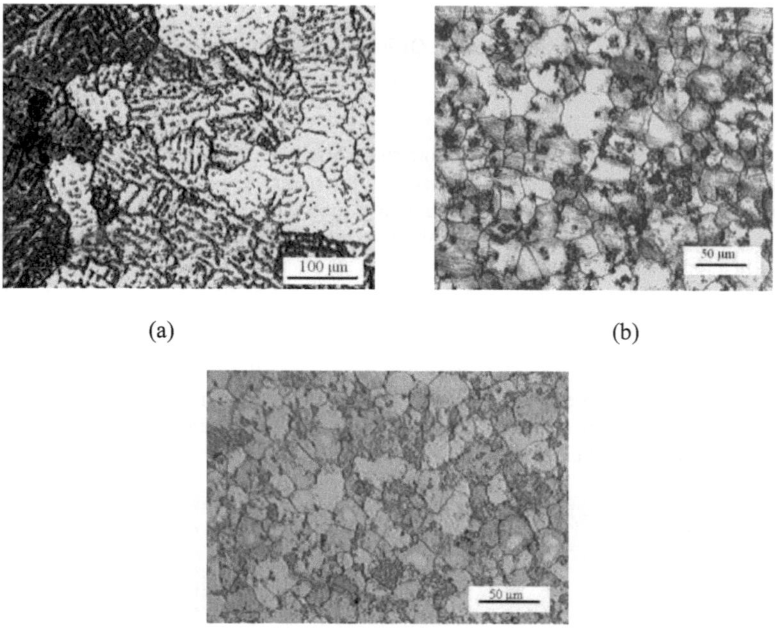

(a) (b)

(c)

Figure 2. **Optical micrograph of: (a) as-cast alloy; (b) composite with 5 wt.% TiB_2; (c) composite with 10 wt. %TiB_2**

3.1.3 SEM

The scanning electron micrographs of the as-cast alloy and composites with 5 and 10 wt % TiB_2, respectively, are shown in Figure 3. The microstructure of the as-cast alloy shows a coarse grain boundary network. The microstructures of the composites show distribution of TiB_2 particles. The particles were observed at the grain boundaries. The particles were pushed to the grain boundaries by the solidification front. The EDS analysis confirms the presence of TiB_2 in the composite.

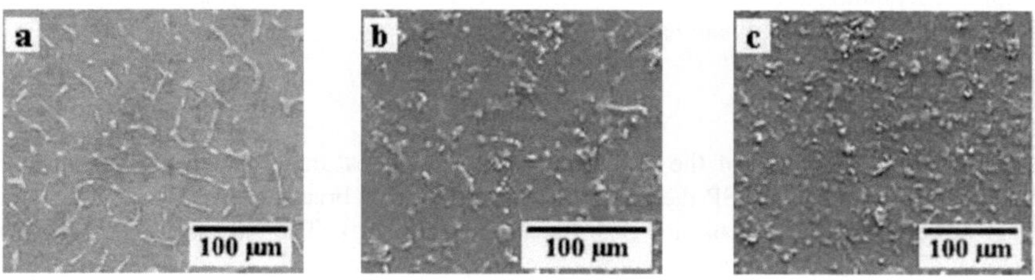

Figure 3. **SEM micrograph of: (a) as-cast alloy; (b) composite with 5 wt.% TiB_2 (c) composite with 10 wt.% TiB_2**

3.2 Microstructural Characterization Of Friction Stir Processed Alloy And Composites:

Friction Stir Processing:

In this process a rotating tool is plunged into the workpiece and along the line of interest. The friction between the shoulder and work piece results in localized heating, which increases the temperature of the work piece to a range where it is plastically deformed. During this process the workpiece is exposed to a combination of severe plastic deformation and high temperature, resulting in a modified structure.

3.2.1 Optical Microscopy

The optical micrographs of friction stir processed alloy and composites are shown in Figure 4. The microstructure of the as-cast alloy reveals elongated grains. The equiaxed grains were observed and broken dendrites were also found as black patches. The broken dendrites were uniformly distributed in the matrix.

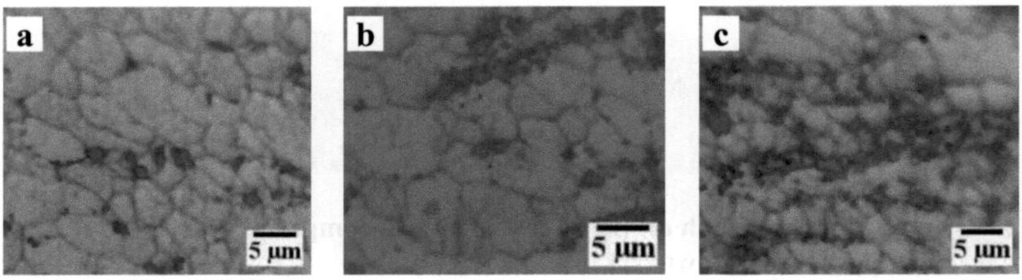

Figure 4. Optical micrograph of:
(a) FSPed alloy;
(b) FSPed composite with 5 wt.% TiB_2;
(c) FSPed composite with 10 wt. % TiB_2

Due to intense plastic deformation and frictional heat developed during FSP the grains in the stir zone were refined. It is well known that during FSP dynamic recrystallization occurs, leading to the generation of fine and equiaxed grains in the stir zone. The mechanisms proposed for the formation recrystallized grains at the stir zone are CDRX (continuous dynamic recrystallization) and DDRX (discontinuous dynamic recrystallization)

3.2.2 SEM:

The microstructures of the base-alloy and composites after friction stir processing are shown in Figure 5. During FSP the stirring action of the tool breaks the cast structure, removes the segregation of eutectic phase and distributes it in the matrix. The intense plastic deformation of material eliminates the shrinkage pores and casting defects as well.

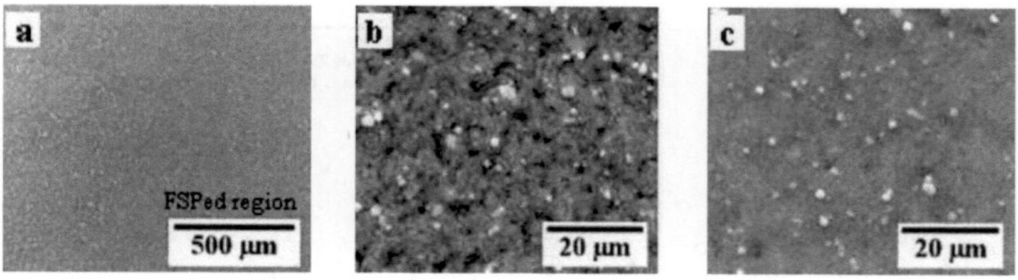

Figure 5. **SEM micrograph of:**
(a) FSPed base alloy;
(b) FSPed composite with 5 wt.% TiB_2;
(c) FSPed composite with 10 wt % TiB_2

The agglomerated TiB_2 particles were uniformly distributed in the matrix. The particles did not fracture during FSP as the particles were smaller (≤ 1.5 μm) and had a strong interfacial bonding with the matrix. The particles were mostly of a hexagonal morphology with straight edges and sharp corners. The particle size varied from 0.5-1.5 μm. The particle morphology and shape was retained after friction stir processing as shown in Figure 6. Particle fracturing was reported in *ex-situ* composites.

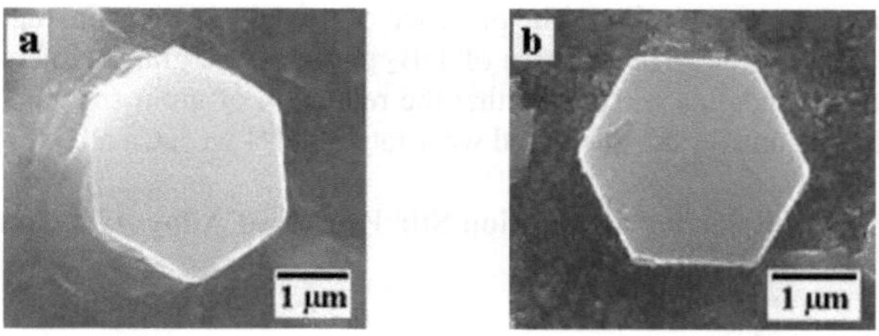

Figure 6: SEM micrograph of composite with 10 wt.% TiB_2: (a) before FSP ; (b) After FSP;

3.3 Dry Sliding Wear Behavior Of As-Cast Alloy And Composite With 5 And 10 Wt. % Tib_2:

3.3.1 Coefficient Of Friction:

The plots depicting the variation of coefficient of friction for the as-cast alloy and composites that were slid under a normal load of 40, 60 and 80 N, respectively, are shown in Figure 7. With increasing load, the coefficient of friction increases for the as-cast alloy, whereas the addition of TiB_2 particles reduces the coefficient of friction of the composites irrespective load. With the increase of TiB_2 particle content of composites, the coefficient of friction decreases. The incorporation of TiB_2 particles into the alloy decreases the friction coefficient irrespective of the load. In the case of the as-cast alloy, the coefficient of friction increases with increase of load and is attributed to its soft matrix.

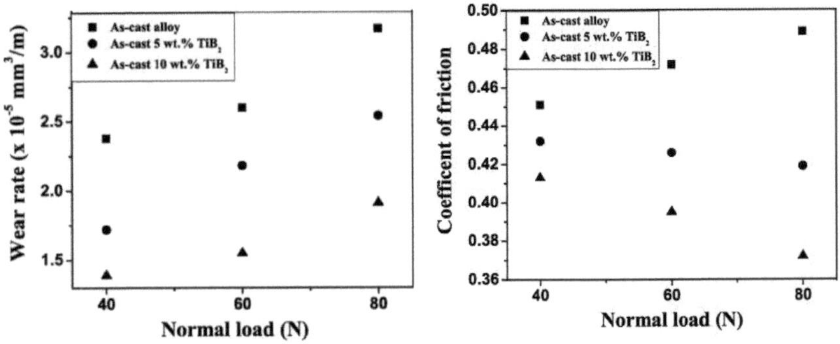

(a) (b)

Figure 7. Variation of: (a) wear rate with normal load; (b) coefficient of friction with normal load.

3.3.2 Wear Rate:

Figure 7(a) shows the variation of wear rate with normal load. The wear rate increases with increase of load for the as-cast alloy and composites. The wear rate of the composites is much lower than that of the as-cast alloy. The presence of particle restricts the metal flow during sliding and reduces the wear rate. The addition of TiB_2 particles and grain refinement reduce the wear rate significantly. Mandal *et. al* showed that the reduction of grain size and high hardness of composite were the reasons for the decreased wear rate of Al-4 wt.%Cu alloy [9].

3.4 Dry Sliding Wear Behavior Of Friction Stir Processed Alloy And Composites With 5 And 10 Wt.%Tib_2, Respectively:

3.4.1 Coefficient Of Friction:

The variation of coefficient of friction with sliding distance for friction stir processed alloy and composites for different loads is presented in Figure 8 (a). The coefficient of friction decreases with increase in particle content of the composites. *In-situ* TiB_2 particles have very good interfacial bonding with the matrix and their distribution reduces the coefficient of friction significantly. It was proved that the good interfacial bonding between the particles and matrix reduces the co-efficient of friction [10].

Tee *et a*l. also reported that the *in-situ* TiB_2 particle addition reduced the friction coefficient in Al and Al-Cu alloys [11]. The reduction in friction coefficient offered by *in-situ* reinforcement particles is higher compared to that for *ex-situ* reinforcement. The coefficient of friction for Al-TiB_2 and Al-SiC composites were 0.16-0.17 and 0.7, respectively [12]. The reduced friction coefficient of Al-TiB_2 composite is mainly due to the hard and fine *in-situ* TiB_2 particles, their uniform distribution in the matrix and good interfacial bonding with the matrix due to a cleaner interface.

Wood *et al.* also reported that the *in-situ* TiB_2 particles more drastically reduced the coefficient of friction compared to that by SiC. The lower coefficient of friction of friction stir processed *in-situ* composites is mainly due to its reduction in grain size and distribution of TiB_2 particles [13].

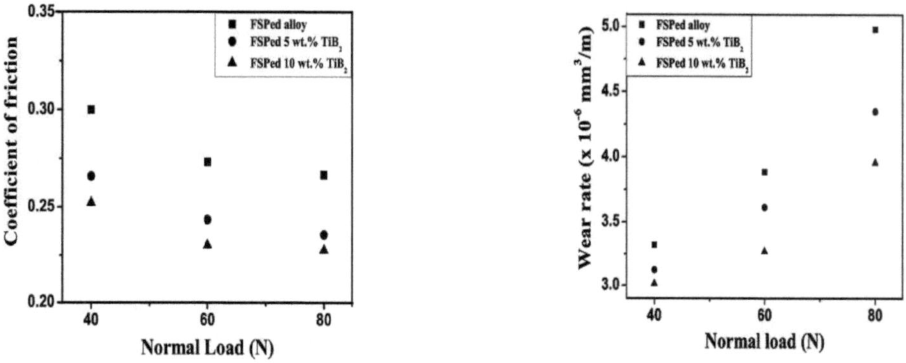

(a) (b)

Figure 8. **Variation of: (a) coefficient of friction with normal load; (b) wear rate with normal load.**

3.4.2 Wear Rate:

The wear rate of friction stir processed alloy and composites with normal load are shown in Figure 8(b). The wear rate increases with increase of load. But the increase in TiB_2 content reduces the wear rate. The friction stir processing reduces the grain size and eliminates the casting defects and dendritic structure in the cast alloy and composites. The agglomerated TiB_2 particles were well distributed in the matrix. Reduced grain size and uniform distribution of TiB_2 particle decrease the wear rate.

3.5 Worn Surface Analysis Of Friction Stir Processed Alloy And Composite:

The strong interfacial bonding between the particle and the matrix prevented the particle pullout from the Al alloy matrix. In friction stir processed composites, the surface is harder due to grain refinement and the presence of hard *in-situ* TiB_2 particles. Due to the high hardness of specimen more amount of material is removed from the counterface material and gets deposited over the surface. This is shown in Figure 9 (a) and (b).

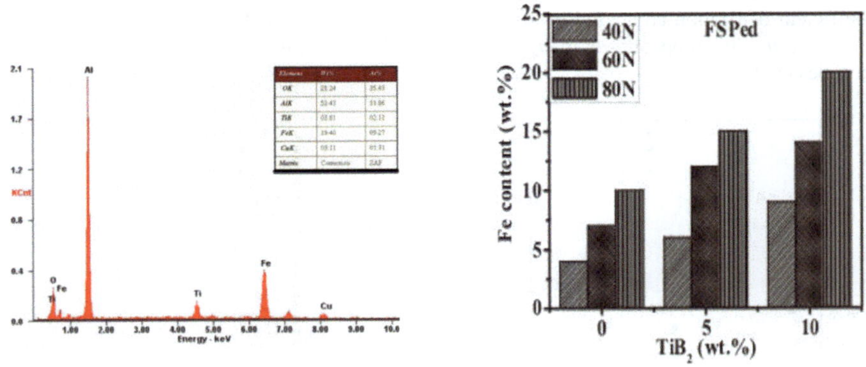

(a) (b)

Figure 9. (a) EDS analysis of worn surface of composite with 10 wt.% TiB_2;
(b) amount of Fe pickup vs. composition of material for different loads.

The wear mechanisms operating in the friction stir processed alloy and composites are assessed by EDS analysis. The surface oxidises easily due to frictional heat generated between the two sliding surfaces. The EDS analysis of the worn surface showed the presence of oxides of Al, Cu and Fe. A mechanically mixed layer (MML) consisting of iron oxide is formed on the counter face material. It acts as a lubricant film between the pin and counter face material thereby reducing the wear rate as well as coefficient of friction. The formation of MML has been reported by many researchers [14 and 15].

4. Conclusions

1. The agglomerated TiB_2 particles were completely removed after friction stir processing, leading to uniform distribution of discrete particles inside the matrix of the composites.

2. The wear properties of the alloy and composites were improved by the incorporation of *in-situ* TiB_2 particles and friction stir processing. Finer particles with good interfacial bonding and high hardness led to the reduction in wear rate of the composites.

3. Uniform distribution of TiB_2 particles and good interface between the matrix and particle reduced the wear rate of FSPed composites by 30% compared to that for the as-cast composites.

5. References

1. Srininivasa Rao P, Sivadasn K G, and Balasubramanian P K, ***Bull. Mater.Sci.*** **19** (1996) 549.

2. Mazurina I, Sakai T, Miura H, Sitdikov O and Kaibyshev R, ***Mater.Sci.Eng***. *A* **473** (2008) 297.

3. Shanumugasundaram T, Murty B S and Subramanya Sarma V, ***Scripta Materialia*** **54** (2006) 2013.

4. Nagaraju P, Srinivasa Rao K, Reddy G M, Kamaraj M, and Prasad Rao K, ***Mater.Sci.Eng.*** *A* **464** (2007) 192.

5. Ravi K R, Saravanan M, Pillai R M, Mandal A, Murty B S, Chakraborty M and Pai B C, ***J.Alloys and compounds*** **459** (2008) 239.

6. Watson I G, Forster M F, Lee P D, Dashwood R J, Hamilton R W and Chirazi A, ***Composites: Part A*** **36** (2005) 1177.

7. Lu L, Lai M O and Chen F L ***Acta. mater***. **45** (1996) 4297.

8. Murty B S, Kori S A and Chakraborty M, ***Inter. Mater. Review*** **47** (2002) 3.

9. Mandal, A, Chakraborty M, Murty B S ***Wear*** **262** (2007) 160.

10. Thakur, S K and B.K Dhindaw ***Wear*** **247** (2001) 191.

11. Tee, K , Lu L and Lai M, ***Wear*** **240** (2000) 59

12. Zhao, M, Wu G, Jiang L and Dou Z ***Composites: part A***, **37** (2006) 1916

13. Wood, J, Davies P, and Kellie J.L, **Mater.Sci. Tech**. **9**(10) 833

14. Venkatraman, B and Sunderrajan G, ***Wear*** **245** (1993) 22.

15. Mandal A, Maiti R, Chakraborty M and BS Murty, ***Mater. Sci.Eng.A*** **386** (2004) 296.

Advanced Composites for Aerospace, Marine, and Land Applications II
Edited by: T. Sano and T.S. Srivatsan
TMS (The Minerals, Metals & Materials Society), 2015

CHALLENGES IN FABRICATION OF SURFACE COMPOSITES BY FRICTION STIR PROCESSING ROUTE

Vipin Sharma[1], U. Prakash[1] and B.V. Manoj Kumar[1]

[1] *Department of Metallurgical and Materials Engineering;*
Indian Institute of Technology (Roorkee)
Roorkee, 247667, **India**

Keywords: FSP, Composite, Aluminum

Abstract

Friction stir processing is a newer technique for fabrication of surface composites. Surface composite fabrication by friction stir processing is simpler and offer many advantages as compared to conventional techniques. However, there are many challenges in composite fabrication due to asymmetric material flow and formation of defects. Reinforcement strategies, process parameters and reinforcement properties are key factors in fabrication of surface composites. In this study a newer reinforcement strategy is utilized to fabricate surface composites by pre-mixing the reinforcement particles with base alloy powders. The pre-mixing of ceramic reinforcement particles with metallic base alloy powder enhances the material flow and favours the uniform distribution of reinforcement particles.

1. Introduction

Surface composites are newer class of metal matrix composites intended for applications requiring harder surface and toughness of the bulk material. Surface composites are fabricated by variety of techniques but as only tailoring of surface is desired it needs proper control and careful fabrication. Friction stir processing is a promising technique for fabricating surface composites with fewer limitations.

FSP was introduced by Mishra et al. [1] as an adaptation of friction stir welding (FSW) for super plasticity in aluminum alloys. FSP also showed its potential in fabricating surface composites and first result in fabricating of surface composite is reported by Mishra et al. [2].FSP mechanism is simpler in which a rotating tool with co-axial probe is plunged in to the work-piece and then tool is moved in direction of interest to process the work-piece material. The tool and work-piece interaction generates heat which softens the material and this softened material stirred by the tool, finally deposited at the back of the tool [3]. In FSP, severe plastic deformation, stirring, forging and extrusion process occur simultaneously. The intense plastic deformation and elevated temperature causes dynamic recrystallization in the stir zone resulting in fine and equiaxed grains. Three distinct zones namely stir zone (SZ), thermo mechanically affected zone (TMAZ) and heat affected zone (HAZ) appears in microstructure of friction stir processed region [4]. The schematic illustration of FSP for fabrication surface composite and various zones is shown in Figure 1.

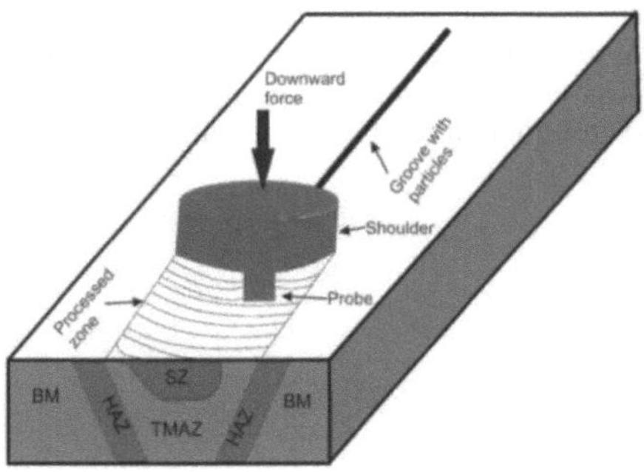

Figure 1. Schematic illustration of surface composite fabrication by FSP.

Initially surface composite was fabricated by applying ceramic particles slurry in volatile medium on the work-piece surface [2]. In the last decade various reinforcement strategies developed to effectively reinforce the ceramic particles in the metal matrix by FSP route. The groove method is commonly used for fabricating surface composites with better control in reinforcement volume percentage and thickness of surface composite. Drilled blind hole on the work-piece is also used in numerous studies of fabricating surface composites. Some studies used coating of ceramic particles on the work-piece and subsequent FSP for fabricating surface composites.

There are various issues in the fabrication of surface composites. Incorporation of ceramic particles in the matrix and there uniform distribution is an important issue. This issue becomes more critical when reinforcement particle size decreases. The uniform distribution of nano-particles in metal matrix composite is not fully achieved and nano-composites lacks in providing full strength. Secondly surface composite fabrication is also susceptible to many defects that generally not observed in friction stir processing of the material without particles. The pin driven flow in the stir zone becomes complicated due to presence of ceramic particles as the non-deformable ceramic particles reduce the amount of plasticized material and also obstruct the material flow. The pin stirs ceramic particles and their size reduction was also found in numerous studies. Generally the pin diameter and length is more than particle reinforcement zone but in the processed zone, 40-60% of volume is consisted by the ceramic particles.

In this study surface composite are fabricated by groove and drilled hole method with different reinforcement particles and alloys. The issues and challenges are discussed while a different approach of reinforcement hybridization is adopted to fabricate defect-free surface composites.

2. Experimental

In this work, 6.35 mm plates of commercial aluminum 2014 alloy (Cu-4.1 Si-0.64 Fe-0.2 Mn-0.56 Mg-0.6 in % and remainder Al) , Al-Si alloy (Si-11.8 Fe-0.4 Mn-0.4 Mg-0.2 Pb-0.1 in % and remainder Al) and reinforcement particles of TiC (2 μm) and SiC (10 μm) are used for fabricating surface composites by FSP route. The reinforcement particles are mixed with base alloy powder of 40 μm for four hours in bi-axis rotary mixer. The reinforcement particles are placed in the groove dimension of 1 mm width and 2 mm depth and blind hole of 2 mm diameter and depth drilled in the aluminum plate with 1 mm spacing. The volume fraction of the reinforcement particles in the stir zone area was calculated to be nearly 12 %. The vertical milling machine with fixture attachment to hold the plate was used for FSP. Schematic of FSP for fabrication of surface composite is shown in Figure 1. A non-consumable hot die steel tool (50 HRC) with a square pin of 4 mm length with 5 mm diagonal and concave shoulder of 18 mm diameter was used to process the plates filled with particles. The tool tilt angle of 2.5 ° is used and plate is allowed to cool at room temperature before each pass. The FSP of the plates was carried out at tool rotational and traverse speed of 1400 rpm and 40 mm/min, respectively. Microscopy was carried out using LEICA DMI 5000M optical microscope and Carl Zeiss SEM. Specimens for metallographic examination were cut normal to the processing direction and prepared using a standard grinding and polishing technique.

3. Result and Discussion

3.1 Al-Cu alloy with TiC reinforcement

The microstructure of AA2014/TiC surface composite fabricated at 1400 rpm and 40mm/min. with square tool are shown in Figure 2 (a-d). To avoid tool wear a fabrication strategy to reduce the ceramic particles and tool interaction is adopted in which tool is slightly offset from the groove containing reinforcement particles. Figure 2 (a) shows the micrograph of surface composite fabricated after four pass of FSP with tool offset of 2 mm from the groove. The lower portion of groove remains unfilled and material flows over the groove. Tool placed on the groove position shows no defect after four pass of FSP but reinforcement particle distribution is not uniform in the stir zone as shown in Figure 2 (b). The reinforcement particle distribution is having the banded feature with particle rich and particle free layers. Surface composite fabricated by placing tool probe on the groove exhibits agglomeration along material flow lines. Gandra et al. [5] found a triangular cavity in the fabrication of AA5083/SiC surface composites. They attributed the defects formation was due to the insufficient material flow influenced by the existence of SiC particles.

Improper process parameter selection results in various defects in friction stir processed zone. In FSW common types of defects pointed by researchers include worm hole, scalloping, ribbon flash, cavity, surface lack of fill, nugget collapse, surface galling etc. [6]. In surface composite fabrication by FSP route similar defects were found which were observed in FSW. The tendency towards defect formation is increased in fabrication of surface composites due to non-deformable ceramic particles and subsequent hindering offered by these particles in the material flow during FSP.

In surface composite, tool probe placed on the groove and tool direction is changed between each pass reduces agglomeration to some extent as shown in Figure 2 (c). However, many particle clusters exits in the stir zone. In adopting newer strategy of mixing ceramic particles with 10 weight percent of base alloy powder, the particle cluster reduces to great extent and uniform distribution of particles is achieved in the stir zone. This can be attributed to the presence of aluminum powder in the ceramic particles which enhances the material flow and wettability in the stir zone. However, reinforcement particles volume fraction decreases due to presence of aluminum powder. The reduction in volume fraction of reinforcement particles can be compensated by manipulating the groove dimensions.

Tool wear is a major issue in fabricating surface composite as tool probe stirred the ceramic particles. The ceramic particles also crushed by the high strain imposed and their size reduces during FSP. The tool wear can be minimized by using hard material like tungsten carbide or polycubic boron nitride tool but these tool material have low fracture toughness and high cost [7]. The tool wear is observed on tool probe after processing composites as shown in Figure 3 (a). Tool wear is more on specific location as demonstrated earlier in many studies. Liu et al. [8] in FSW of metal matrix composite by WC-Co tool pointed out that the wear of the pin was different at different locations of the pin, and the maximum wear was found at about one-third of pin length from the pin root. Prado et al. [9] reported that in FSW of AA 6061-20%SiC metal matrix composite (MMC) the wear of tool steel threaded pin tool resulted into a smooth self-optimized shape (without threads). They observed that the worn tool or self-optimized shape with no threads resulted in homogenous and integral welds without further significant tool wear. These observations suggest that tool wear can be greatly reduced even for MMCs when using the optimized tool shape. The surface composite fabricated by worn tool as in FSW worn tool influence on the joint is negligible. The surface composite is fabricated by four pass FSP and changing the tool direction after each pass. Surface composite shows agglomeration in the stir zone as shown in Figure 3 (b). The worn out tool is not effective in proper stirring of reinforcement particles and particles are trapped in the material flow layers.

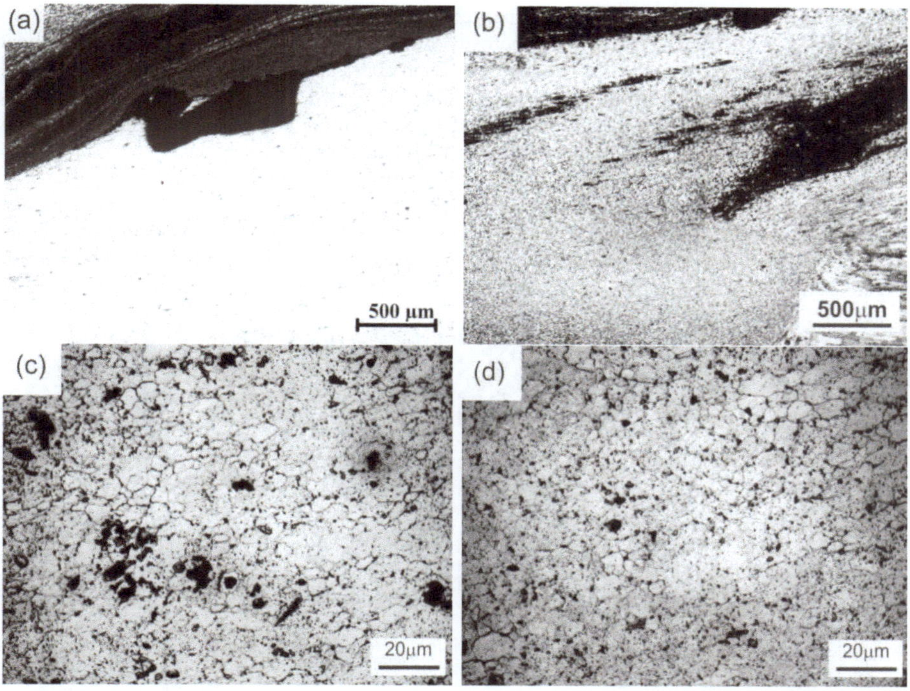

Figure 2. **Surface composites AA2014/TiC fabricated via FSP**
- **(a) tool probe offset by 2 mm from the groove**
- **(b) tool probe placed on the groove**
- **(c) tool direction is changed after passes**
- **(d) reinforcement particles mixed with alloy powders**

Figure 3. **Surface composite fabricated by the worn tool (a) worn tool (b) agglomeration of the reinforcement particles.**

3.2 Al-Si alloy with SiC reinforcement

The surface composite of Al-Si alloy reinforced with SiC by drilled hole and four passes of FSP is shown in Figure 4(a-d). Reinforcement particle distribution is not uniform even after four passes of FSP, the banded feature of particle rich and particle free layers is observed (Figure 4 a). The particle rich layers having porosity due to agglomeration and the weak bonded reinforcement particles as shown in Figure 4(b). Even after changing direction between each FSP pass the uniform distribution of reinforcement is not achieved, banded feature still exist in the stir zone (Figure 4 c). However, banding and agglomeration is reduced as compared to the surface composite fabricated without changing rotational direction.

The reinforcement particles mixed with 20% of aluminum alloy powder and fabricated by four passes of FSP with changing tool direction. The incorporation of aluminum powder in ceramic particles favors the material flow and distribution in the surface composite as shown in Figure 4 (d).

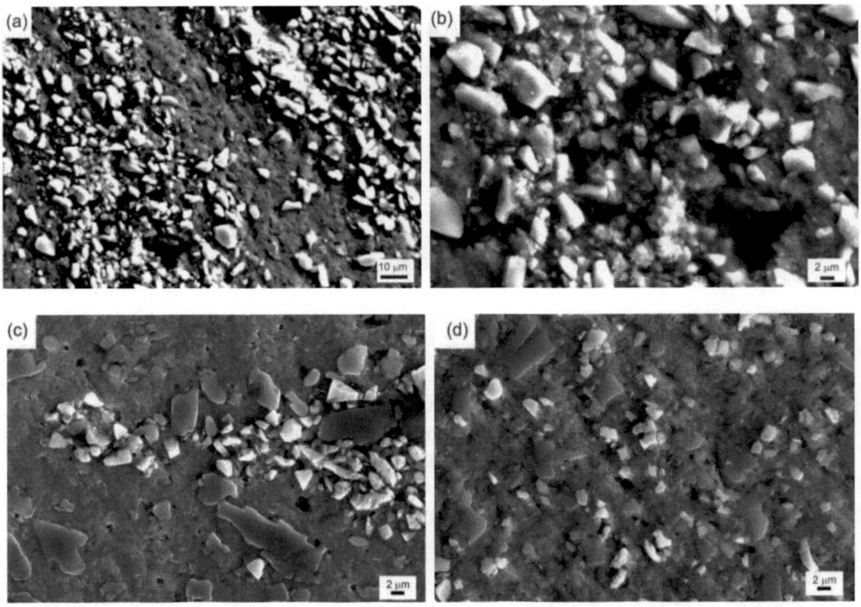

Figure 4. Surface composites Al-Si alloy/SiC fabricated via FSP
(a) four pass FSP
(b) agglomeration generated porosity in surface composite
(c) tool direction is changed after passes
(d) Composite fabricated by reinforcement particles mixed with alloy powders.

4. Conclusion

The surface composites are fabricated by using various reinforcement strategies. The defects in surface composite fabrication can be eliminated by proper selection of process parameters. Apart from process parameters a suitable reinforcement strategy is necessary to achieve the uniform distribution of reinforcement particles. In this study a newer strategy of mixing ceramic particles with base alloy powder is used. It was found that the pre-mixed reinforcement particles with the base alloy powder provide a uniform distribution. The pre-mixed particles are initially separated with alloy powders and during processing, the metallic particles softens and enhances the material flow in the stir zone.

Acknowledgement

One of the Author (VS) gratefully acknowledges the financial support of Department of Science & Technology, Government of India received under INSPIRE Fellowship Program.

References

[1] R.S. Mishra, M.W. Mahoney, S.X. McFadden, N.A. Mara, A.K. Mukherjee, "High strain rate superplasticity in a friction stir processed 7075 Al alloy," ***Scripta Materialia***, 42 (1999), 163-168.

[2] R.S. Mishra, Z.Y. Ma, I. Charit, "Friction stir processing: a novel technique for fabrication of surface composite," ***Materials Science and Engineering*** A, 341 (2003), 307-310.

[3] S. Mironov, Y.S. Sato, H. Kokawa, "Microstructural evolution during friction stir processing of pure iron," ***Acta Materialia***, 56(2008), 2602-2614.

[4] R.S. Mishra, Z.Y. Ma, "Friction stir welding and processing," ***Materials Science and Engineering: R: Reports***, 50(2005), 1-78.

[5] J. Gandra, R. Miranda, P. Vilaça, A. Velhinho, J.P. Teixeira, "Functionally graded materials produced by friction stir processing," ***Journal of Materials Processing*** Technology, 211(2011), 1659-1668.

[6] W.J. Arbegast, "A flow-partitioned deformation zone model for defect formation during friction stir welding," ***Scripta Materialia***, 58(2008), 372-376.

[7] R. Rai, A. De, H.K.D.H. Bhadeshia, T.DebRoy, "Review: friction stir welding tools," Science and Technology of Welding and Joining, 16(2011), 325-342.

[8] H.J. Liu, J.C. Feng, H. Fujii, K. Nogi, "Wear characteristics of a WC–Co tool in friction stir welding of AC4A+30vol%SiCp composite," ***International Journal of Machine Tools and Manufacture***, 45(2005), 1635-1639.

[9] R.A. Prado, L.E. Murr, K.F. Soto, J.C. McClure, "Self-optimization in tool wear for friction-stir welding of Al 6061+20% Al2O3 MMC," ***Materials Science and Engineering: A,*** 349(2003), 156-165.

METAL-MATRIX COMPOSITES

Advanced Composites for Aerospace, Marine, and Land Applications II
Edited by: T. Sano and T.S. Srivatsan
TMS (The Minerals, Metals & Materials Society), 2015

INFLUENCE OF REINFORCEMENT CONTENT ON TENSILE RESPONSE AND FRACTURE BEHAVIOR OF AN ALUMINUM ALLOY METAL MATRIX COMPOSITE

K. Manigandan[1], T. S. Srivatsan[2], Zhencheng Ren[1] and Jingyi Zhao[1]

Division of Materials Science and Engineering
Department of Mechanical Engineering
The University of Akron
Akron, Ohio 44325-3903, **USA**
1: Graduate student; 2: Professor

Abstract

In this paper, the results of a study that examined the intrinsic influence of ceramic particulate reinforcements on tensile deformation and fracture behavior of aluminum alloy 2014 is presented and discussed. The candidate aluminum alloy was reinforced with two volume fractions of the alumina (Al_2O_3) particulates. Influence of particulate reinforcement on microstructural development is highlighted. Samples of the aluminum alloy-based metal matrix composite were deformed in uniaxial tension. The influence of nature of loading and intrinsic microstructural effects on stress versus strain response, mechanical properties, deformation and final fracture behavior are neatly elaborated and discussed in light of intrinsic microstructural effects, deformation characteristics of the microstructural constituents and nature of loading.

Keywords: aluminum alloy, ceramic particle reinforcement, metal-matrix composite, microstructure, tensile response, fracture behavior, mechanisms.

1. Introduction

Discontinuously-reinforced aluminum (DRA) alloy-based metal matrix composites (MMCs) containing particulate, whisker or short fiber reinforcements in a 2XXX, 6XXX or 7XXX series aluminum alloy metal matrix have in the time period spanning the last three decades, i.e., since the 1980's, engendered considerable scientific and technological interest for selection and use in automobile, aerospace and a spectrum of other high performance markets [1- 12]. These composites are preferred primarily because they offer a number of advantages. A few of these are a 15-40 pct. increase in strength and a 30-100 pct. increase in stiffness when compared with the unreinforced alloy [9, 13-15]. Besides, the metal-based composite maintains receptiveness to processing and characterization techniques that are traditionally used for the conventional unreinforced counterpart. Increase in modulus of greater than one hundred percent have been reported for aluminum alloy that is reinforced with a volume fraction of 40 pct. of silicon carbide [12]. From a design engineering perspective that attractiveness stems from an improvement in elastic modulus, i.e., density compensated increase in elastic modulus. The moduli obtained are greater than the values obtained for typical titanium alloys and only marginally less than those of steels. Besides, the DRA MMCs based on particulate reinforcements are attractive primarily because they can be made to have properties that are isotropic in three dimensions, or in a plane, when compared one-on-one with the continuously-reinforced metal matrices. Also, conventional secondary fabrication methods and metallurgical processing can be used to produce a wide range of product forms, making them relatively inexpensive when compared one-on-one with metal-based composites that are reinforced with continuous filaments. The discontinuously-reinforced metal matrices based on whisker reinforcements offer the potential for enhanced properties, but often tend to suffer from whisker damage and breakage during secondary fabrication [16, 17].

The use of discontinuous ceramic particle reinforcements in aluminum alloy metal matrices has been shown to provide attributes such as improved abrasion resistance [18], enhanced strength at high temperatures [19], improved creep rupture properties [20], good micro-creep performance [21], improved corrosion resistance [22], and enhanced resistance to fatigue crack initiation when compared one-on-one with the unreinforced matrix alloy [23]. The noticeable disadvantages with these materials are that they often suffer from low tensile ductility (quantified by elongation or strain to failure), inadequate fracture toughness and inferior fracture resistance compared to the unreinforced metal matrix [24-32]. In recent years the emerging DRA MMCs have been synthesized using the techniques of ingot metallurgy, powder metallurgy and mechanical alloying. Each of these techniques results in an MMC that has different microstructure and thus different properties.

The objective of this short technical manuscript is to present experimental evidence of the influence of particulate volume fraction on the tensile response of a ceramic particulate reinforced aluminum alloy based MMC. The tensile properties and resultant fracture behavior were evaluated for two different volume fractions of the discontinuous ceramic particle reinforcement phase in the aluminum alloy metal matrix. The material produced had the same matrix composition, aging condition, precipitation characteristics and reinforcement particle characteristics to include type, size distribution, shape and interfacial characteristics. The quasi static fracture behavior of the two composites is discussed in light of concurrent and mutually interactive influences of intrinsic composite microstructural features, deformation characteristics of the metal matrix and the discontinuous particulate reinforcement phase.

2. Material

The DRA MMCs selected for this study were based on aluminum alloy 2014 reinforced with fine particulates of Al_2O_3. The ceramic particle–reinforced 2014 aluminum alloy MMCs were made using a proprietary casting technique by DURAL Aluminum Composites Corporation (DURALCAN USA, San Diego, CA). Two different fractions of the discontinuous Al_2O_3 particulate reinforcement phase (10 to volume percent and 15 volume percent) were chosen. The nominal chemical composition (in weight percent) of the matrix alloy is given in **Table 1**. The alloy has balanced levels of copper and magnesium, the primary strengthening agents. The iron and silicon elements in the alloy are impurities. The ingot-based Al_2O_3 particulate-reinforced MMC was made available as an extrusion billet. The billets were produced by extrusion of the cast ingot at 450°C through a standard shear-face die. The starting ingot was 178 mm in diameter and extruded at an extrusion ratio of 17:1 to produce rectangular blanks that measured 19 x 76 mm. The extruded 2014/Al_2O_3/ *xxp* (where *xx* denotes the volume percent of the discontinuous particulate reinforcement phase) material was solutionized at 500°C for 2 hours, cold water quenched, and artificially aged at 160°C for 2 hours to get the peak-elevated aged condition (T6). DURALCAN USA (California) provided the 2014/Al_2O_3/10*p* and 2014/Al_2O_3/15*p* MMCs, in the T6 condition.

Table 1: **Chemical Composition of matrix Al-Cu-Mg Alloy 2014 (in weight Percent)**

Element	**Cu**	**Mg**	**Mn**	**Si**	**Al**
Percent	4.4	0.50	0.80	0.80	Balance

3. Specimen Preparation

Blanks of size 150 x 20 x 20 mm were cut from the as-received 2014/Al_2O_3/10*p*-T6 and 2014/Al_2O_3/15*p*-T6 extrusions using a diamond-coated saw blade. Tension test specimens were precision machined from the blanks using a diamond-tipped cutting tool. The test specimens were machined with the stress axis parallel to the extrusion direction and conformed to specifications in ASTM E-8 [33]. The machined test specimens were smooth and cylindrical in the gage section, which measured 6.25 mm in diameter and 25 mm in length. To minimize the effects of any surface irregularities and finish, final surface preparation was achieved by mechanically polishing the entire gage section of the test specimens, prepared from the two MMCs, using progressively finer grades of silicon carbide–impregnated emery paper, to remove any and all circumferential scratches and surface machining marks.

4. Mechanical Testing

4.1 Initial Microstructure Characterization

Samples were cut from both the MMCs and prepared by standard metallographic procedures for observation in an optical microscope. The morphology of the reinforcing phase, i.e., the ceramic particulates (Al_2O_3), and their distribution through the aluminum alloy metal matrix and other observable intrinsic microstructural features were carefully examined in an optical microscope and photographed using a bright-field illumination technique.

4.2 Mechanical Testing

All mechanical tests were performed on a fully automated, closed-loop, servo hydraulic structural test machine (INSTRON: Model 8500 Plus) equipped with a 100 KN load cell. A precision alignment was given to the load train prior to the initiation of testing so as to eliminate and/or minimize grip-induced bending. Uniaxial tensile tests were carried out in the room temperature, laboratory air environment (T = 23°C, Relative Humidity = 55 pct.). The specimens were deformed at a constant strain rate of 10^{-4}/s. An actual 12.5 mm gage-length clip-on extensometer was attached to the test specimen (made from the MMC material) at the gage section with rubber bands. The stress and strain variation, parallel to the load line, and resultant mechanical properties of stiffness, strength and ductility was provided as a computer output of the test machine.

4.3 Failure-Damage Analysis

Fracture surfaces of the deformed and failed tensile test specimens of the two chosen MMCs were comprehensively examined in a JEOL scanning electron microscope (SEM) to determine the following:

(i) Macroscopic fracture mode, and

(ii) To characterize the fine-scale topography and microscopic mechanisms governing final fracture.

The distinction between macroscopic and microscopic fracture mechanisms is based entirely on the magnification level at which the observations are made. Samples for observation in the SEM were obtained from the deformed and failed MMC specimens by sectioning parallel to the fracture surface.

5. Results and Discussion

5.1 Initial Microstructure

The optical micrographs illustrating the microstructure of the 2014/Al_2O_3/*xxp*-T6 composites are shown in **Figures 1 and 2**. The Al_2O_3 particulate reinforcement phases, in the 2014-aluminum alloy metal matrix, were (a) non-uniform in size, (b) irregularly shaped, and (c) dispersed randomly through the aluminum alloy metal matrix. Some of the reinforcing Al_2O_3 particulates had sharp corners. At regular intervals, a clustering or agglomeration of the reinforcing Al_2O_3 particulates, of varying size and shape, was observed, resulting in local concentration of particulate-rich and particulate-depleted regions. An agglomerated site consisted of the smaller Al_2O_3 particulates, of varying sizes and shapes, intermingled selectively with a few larger Al_2O_3 particulates, which appeared either rectangular or oblong in shape. The degree of agglomeration or clustering of the reinforcing particulates was found to increase with an increase in the amount of the Al_2O_3 reinforcement phase in the 2014-aluminum alloy metal matrix. The non-uniform distribution of the reinforcing aluminum oxide (Al_2O_3) particulates along the three orthogonal directions (extrusion, long transverse, and short transverse) results in an anisotropic microstructure of the 2014/Al_2O_3/*xxp*-T6 composites. The alumina particles varied in size. However, no attempt was made, in this study, to determine the average particulate size and deviation for the two chosen MMC materials.

5.2 Tensile Properties

The tensile properties of the 2014/Al_2O_3/*xxp*-T6 MMCs, for two different volume fractions of ceramic particulate (Al_2O_3) reinforcement at ambient (27°C) temperature are summarized in **Table 2**. The results are the mean values based on duplicate tests. Factors contributing to the improved stiffness and strength of the 2014/Al_2O_3/*xxp*-T6 composites can be found elsewhere [34, 36]. The ultimate tensile strength (σ_{UTS}) of the composite microstructure was only marginally greater than the tensile yield strength (σ_{YS}), providing sufficient evidence that the work hardening rate past yielding is low. In fact, the ultimate tensile strength followed the same trend as the yield strength of the composite. The improvement in strength of the 2014/Al_2O_3/15*p*-T6 composite over the 2014/Al_2O_3/10*p*-T6 counterpart was only marginal at ambient test temperature (27°C). For both volume fractions of the particulate reinforcement phase in the aluminum alloy matrix, both the reduction in area and tensile ductility decreased with an increase in reinforcement content in the aluminum alloy metal matrix. The overall strain to failure (ε_f) of the 2014/Al_2O_3/*xxp* MMCs decreases with an increase in reinforcement content and is quite consistent with the observed decrease in both strength and stiffness with an increase in reinforcement content in the aluminum alloy metal matrix. Influence of Al_2O_3 particulate reinforcement content on engineering stress versus engineering strain response is shown in **Figure 3**. An increase in reinforcement content in the soft aluminum alloy metal matrix resulted in an observable decrease in ductility quantified by elongation or strain to failure.

Variation of true stress with true strain for the two composites is shown in **Figure 4**. This curve in the region of uniform plastic deformation follows the power law relationship

$$\sigma = K\,[\varepsilon_p]^n$$

Where n is the strain hardening exponent and K is the monotonic strength coefficient. An increase in reinforcement content in the soft aluminum alloy matrix does result in an increase in strain hardening exponent. For both composites the strain hardening values followed the trend shown by most metals and their alloy counterparts. The increase in strain hardening with increase reinforcement content was at best minimal.

Table 2: **A compilation of the room temperature tensile properties of 2014 Aluminum Alloy/Al_2O_3 Composites (T = 25°C).**

Material	Elastic Modulus		Yield Strength		UTS		Elongation GL=0.9" (%)	Reduction in Area (%)
	Ksi	GPa	ksi	MPa	ksi	MPa		
AA 2014/10P	12,444.0	85.8	81.20	560	82.70	570	4.20	**9.5**
AA 2014/ 15P	13,372.5	92.2	74.50	514	77.70	536	2.24	**6.6**

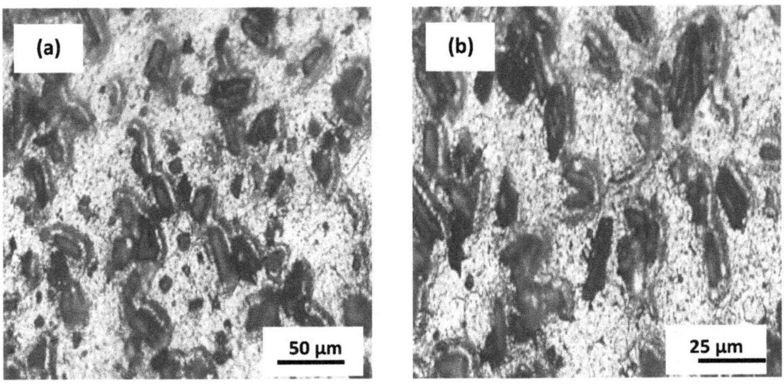

Figure 1. **Optical micrographs showing distribution of the reinforcing particulates in the 2014/Al_2O_3/10p composite**

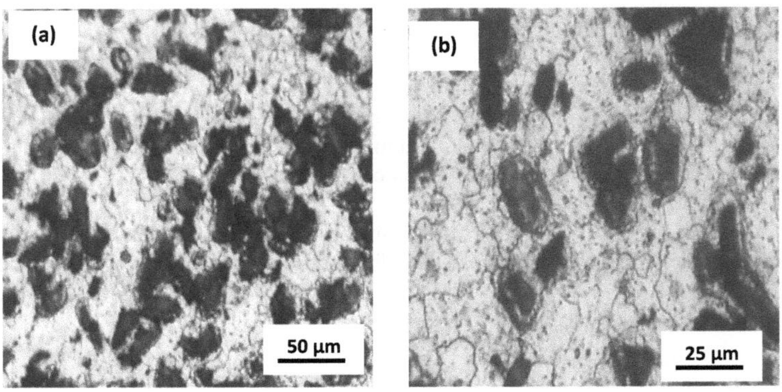

Figure 2. **Optical micrographs showing distribution of the reinforcing particulates in the 2014/Al_2O_3/15p composite.**

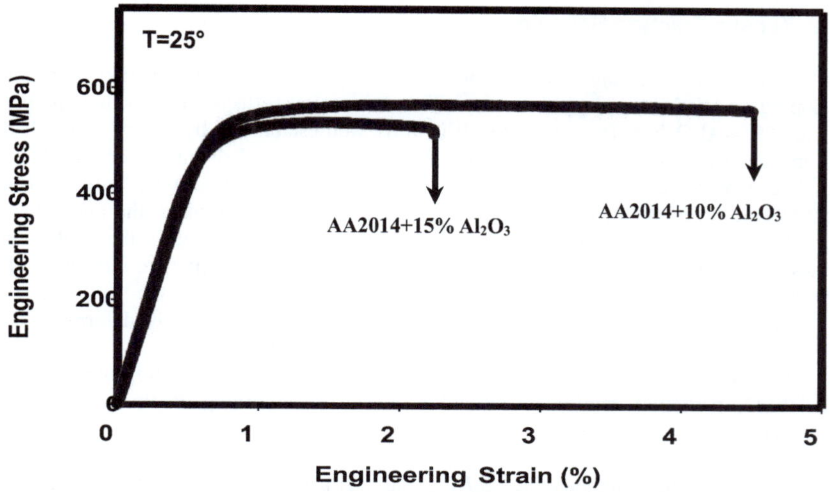

Figure 3. Influence of particulate Al_2O_3 reinforcement on engineering stress versus engineering strain response of the two composites.

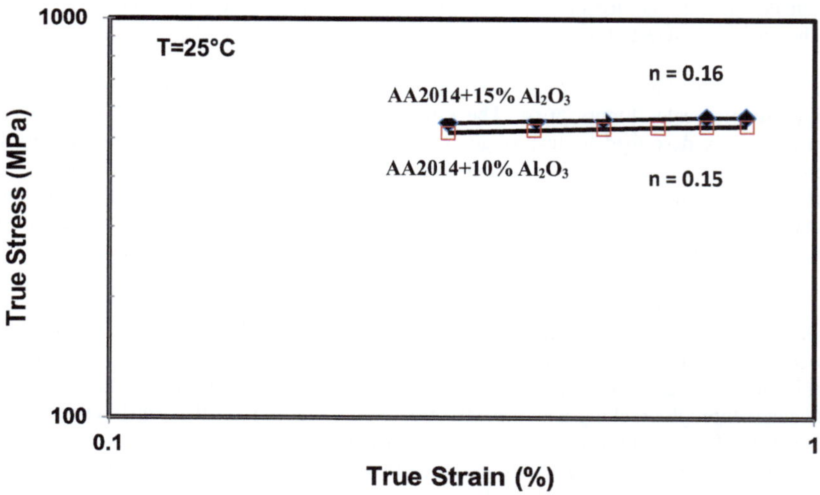

Figure 4. The monotonic stress - strain response of the chosen composites. Increase in reinforcement content had a marginal influence on strengthening or hardening at the fine microscopic level.

5.3 Tensile Fracture Behavior

The tensile fracture surfaces are helpful in providing information relevant to developing an understanding of intrinsic microstructural effects on fracture properties of the 2014/Al_2O_3/*xxp*-T6 composites. It is fairly well established by several researchers in their independent studies that the overall fracture behavior of unreinforced alloys is often dictated by events associated with void nucleation and growth, with nucleation essentially occurring at the coarse constituent particles dispersed through the microstructure [35, 36]. Furthermore, for the unreinforced aluminum alloy metal matrix, the nucleation of cavities and voids is promoted by the concurrent and mutually competitive influences of the following:

(1) Failure by cracking of the hard, brittle, and essentially elastically deforming second- phase inclusions and particles.

(2) Failure through separation at the interfaces between the elastically deforming Al_2O_3 particulate reinforcements and the plastically deforming aluminum alloy metal matrix.

At the ambient temperature (27°C), the 2014/Al_2O_3/*xxp*-T6 composites exhibited limited ductility, on a macroscopic scale, with fracture occurring essentially normal to the tensile stress axis. High-magnification examination of the tensile fracture surfaces revealed features reminiscent of locally ductile and brittle mechanisms. Representative fractographs of the tensile fracture surface of the two composites, i.e., 2014/ Al_2O_3/10*p* and 2014/Al_2O_3/15p are shown in **Figures 5** through **Figure 8.**

Tensile fracture of the 2014/Al_2O_3/10*p* composite was relatively rough when viewed on a microscopic scale (Figure 5(a)). The reinforcing Al_2O_3 particulates were surrounded by ductile regions described as tear ridges (Figure 5(b)), particulate failure through cracking (Figure 5(c)), and failure through separation at the matrix-reinforcement interfaces (Figure 5(d)). Isolated pockets of voids were intermingled with tear ridges and isolated regions of dimpled rupture (Figure 6(c)). Failure of the particulate reinforcement was evident through both cracking (Figure 6a) and separation at the interfaces with the matrix (Figure 6(b)).

Tensile fracture of the 2014/Al2O3/15*p* composite was also flat and normal to the far-field stress axis when viewed on a macroscopic scale (Figure 7a), but rough when viewed on a microscopic scale (Figure 7(b)). The matrix of the composite revealed combinations of (a) tear ridges (Figure 7c)) and (b) failure of the reinforcing alumina particulates through cracking and separation at the interfaces with the matrix (Figure 7 (d)). The matrix of the composite was covered with isolated microscopic voids, of varying size and shape (Figure 8a). Multiple microscopic cracks were observed in regions of particulate agglomeration, which contributed to the observed degradation in tensile ductility (Figure 8b).

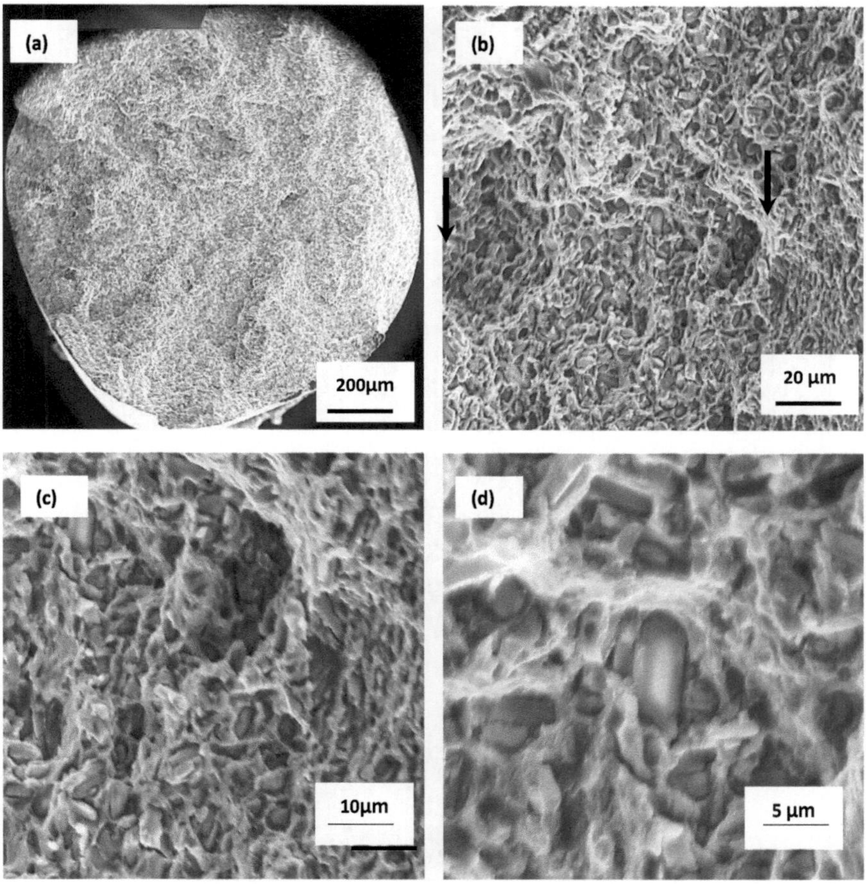

Figure 5. **Scanning electron micrograph of the tensile fracture surface of the 2014/Al_2O_3/10p composite deformed in tension showing:**
(a) Overall morphology of failure: rough on microscopic level
(b) Presence of tear ridges surrounding the reinforcing Al_2O_3 particulate reinforcements
(c) Failure of the reinforcing particulate through cracking
(d) Separation at the reinforcement – matrix interfaces

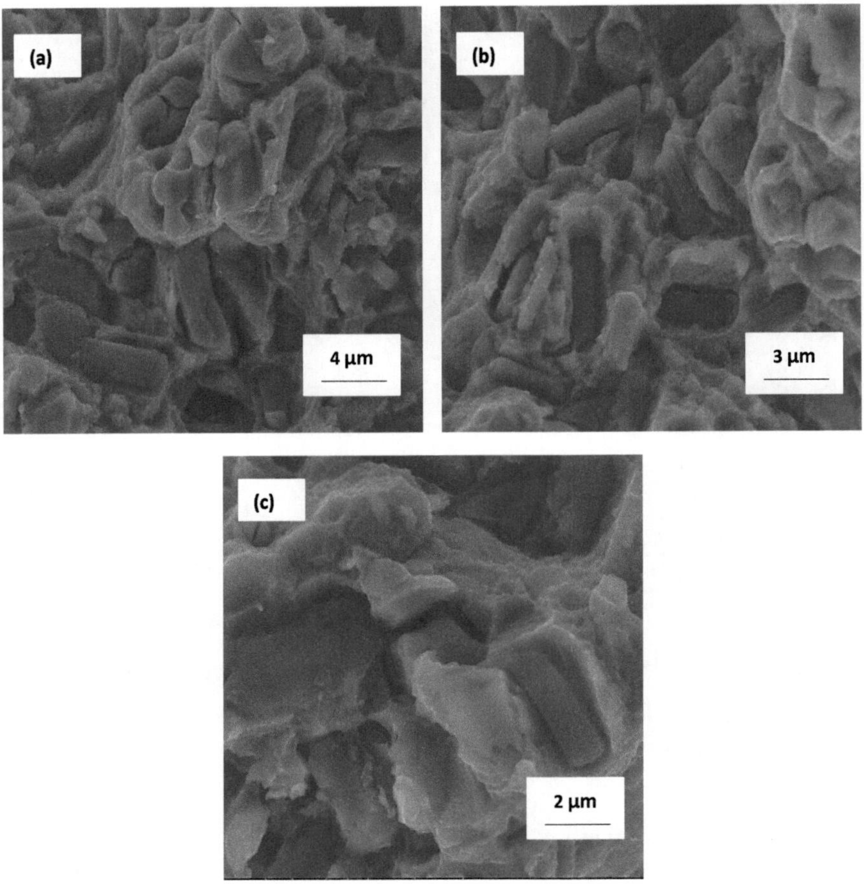

Figure 6. **Scanning electron micrograph of the tensile fracture surface of the 2014/Al_2O_3/10p composite deformed in tension showing:**

(a) Noticeable failure by cracking of the reinforcing particulates

(b) Clear and convincing evidence of separation at Al2O3 particulate -matrix interfaces.

(c) Isolated pockets of dimpled rupture

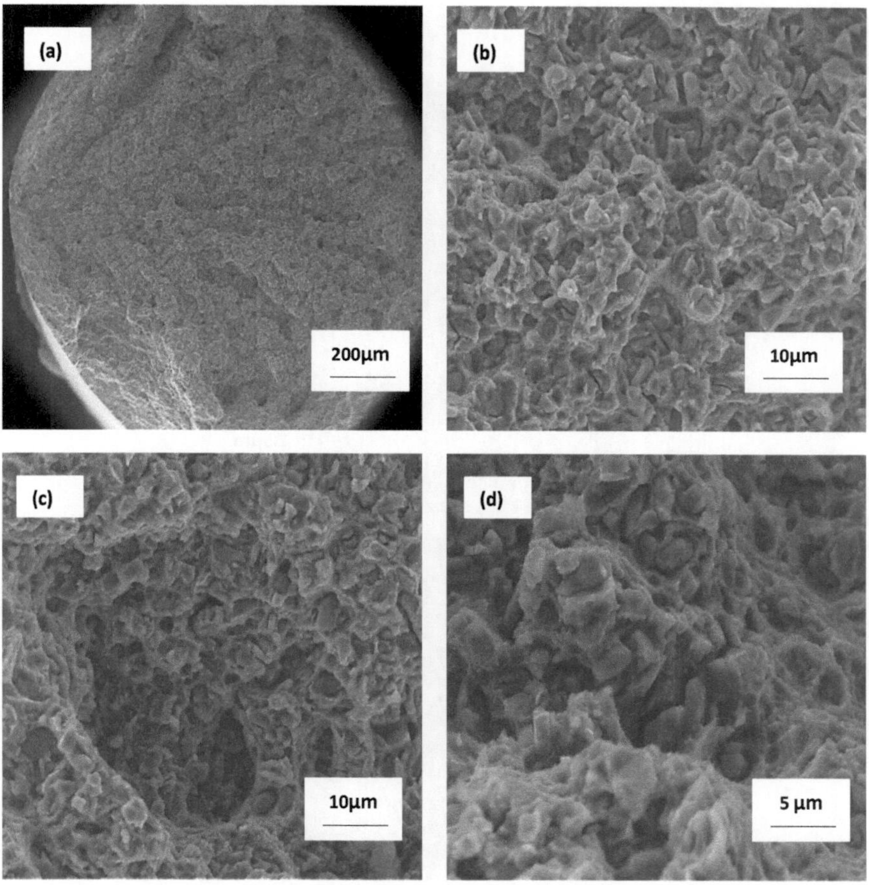

Figure 7. Scanning electron micrograph of the tensile fracture surface of the $2014/Al_2O_3/15p$ composite deformed in tension showing:

(a) Overall morphology of failure
(b) High magnification observation of (a) showing locally rough surface at fine microscopic level.
(c) High magnification observation of (b) showing tear ridges reminiscent of locally ductile failure
(d) Failure of the reinforcing Al_2O_3 particulates by both cracking and separation at the interfaces with the metal matrix.

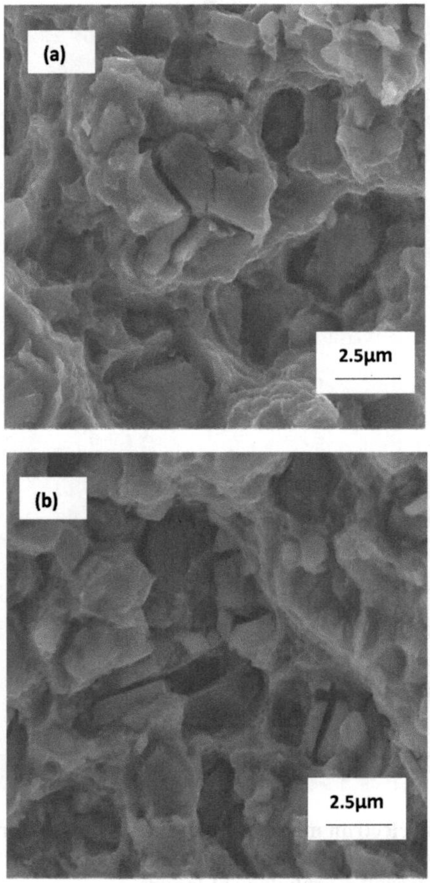

Figure 8. **Scanning electron micrograph of the tensile fracture surface of the 2014/Al_2O_3/15p composite deformed in tension showing:**

(a) Voids and shallow dimples scattered through the aluminum alloy metal matrix.

(b) Multiple microscopic cracks at regions of particulate agglomeration.

5.4 Microscopic Mechanisms Governing Fracture

The constraints in deformation caused by the presence of the hard, brittle, and essentially elastically deforming Al_2O_3 particles in the soft, ductile, and plastically deforming 2014 aluminum alloy metal matrix coupled with the resultant development of a triaxial stress state at the local level, in the metal matrix, aids in limiting the overall flow stress of the composite microstructure, while concurrently favoring the following events at the fine microscopic level:

(a) Void initiation and growth of voids in the matrix, and
(b) Failure by either cracking or separation at the reinforcement particulate-matrix interfaces.

Furthermore, as a consequence of the deformation constraints induced by the hard, brittle and elastically deforming Al_2O_3 particulate reinforcements, a higher applied stress is required to initiate plastic deformation in the adjoining matrix. This translates to higher elastic constant and higher yield strength of the composite material. Under the influence of a far-field tensile load, the microscopic voids appeared to have undergone limited growth confirming a possible contribution from particle constraint–induced triaxiality on failure of the composite matrix. Failure of the reinforcing Al_2O_3 particulate(s) is governed by competing influences of the following microscopic events: (1) local plastic constraints, (2) particle size, and (3) degree of agglomeration. The local plastic constraints are particularly important for the larger-sized particles and for particulate clusters during composite fracture [41].

A close examination of the fracture surfaces revealed the extent of damage to be highly localized at the Al_2O_3 reinforcing phase through cracked particles and failure at the interfaces of the reinforcing ceramic particle with the soft aluminum alloy metal matrix, with little evidence of void formation away from the cracked or fractured Al_2O_3 particle. This suggests that the plastic strain became localized during the early stages of deformation. The intrinsic brittleness of the reinforcing alumina particulates, coupled with the propensity for it to fracture due to localized deformation, results in both particle cracking and particle-matrix interface debonding (Figure 6 b and Figure 7 d) being the dominant damage mode. The higher yield strength, coupled with damage to the composite microstructure from the conjoint influence of cracking of the reinforcing Al_2O_3 particles and interfacial failure through decohesion, is responsible for the observed inferior ductility of the 2014/Al_2O_3/*xxp*-T6 MMCs. Also, the triaxial stresses generated during far-field tensile loading favor limited growth of the fine microscopic voids in the matrix of the composite. The limited growth of the voids during far-field tensile loading, coupled with lack of their coalescence as a dominant fracture mode for the two 2014/Al_2O_3/*xxp*- T6 MMCs studied, clearly indicates that the deformation properties of the 2014 matrix are significantly altered by the presence of the discontinuous Al_2O_3 particulate reinforcements. In Al_2O_3 particulate-rich regions of the matrix, fracture occurred early and the damage propagated rapidly among the reinforcing particle agglomerate. However, the reinforcement-lean regions, that is, the matrix, aid in retarding the progression and linkage of the damage.

Fracture of the brittle Al_2O_3 particles increases local damage and serve as a starting point for the formation of a long crack. This, in conjunction with failure of the surrounding matrix, results in the formation of fine microscopic voids at the particle-matrix interfaces. Very few of the fine microscopic voids coalesce and the halves of these voids are the shallow dimples observed surrounding the cracked particles. The lack of formation of ductile dimples in the matrix, as a dominant fracture mode, can be attributed to the constraints imposed on plastic flow caused by the presence of discontinuous Al_2O_3 particles, that is, the deformation incompatibility between the plastically deforming metal matrix and the elastically deforming Al_2O_3 reinforcement phase, and not due to limited ductility of the 2014 aluminum alloy metal matrix. With an increase in reinforcement content, i.e., Al_2O_3, microscopic fracture was found to be dominated by cracking of the reinforcing particulates on account of their intrinsic brittleness. With a progressive increase in strain, the larger-sized particles fracture first, at low values of applied strain, followed by the smaller-sized particles.

The quasi-static fracture surfaces revealed the fracture plane of cracked particles to be essentially normal to the far-field stress, suggesting importance of normal (tensile) stress in inducing particle fracture. The overall "damage" resulting from quasi-static deformation of the 2014/Al_2O_3/*xxp* MMC can be associated with failure of the discontinuous particulate reinforcement either through cracking or separation at the interface of the particulate with the soft and ductile aluminum alloy metal matrix. An observation of the quasi-static fracture surfaces of the two composites revealed that an excess of 50 pct. of the reinforcing particles had fractured during deformation. This indicates that not all of the particles were loaded to their fracture stress, suggesting the non-uniform distribution of the particles in the metal matrix and their influence in governing deformation. This situation is exacerbated by the mismatch strain and concomitant high compression stresses in the composite matrix due to differences in the thermal expansion coefficient between the two key constituents of the composite: (i) the reinforcing Al_2O_3 particulates, and (ii) the 2014 aluminum alloy matrix [37, 38] For the reinforcing particles to fracture, the applied far-field tensile stress will have to first overcome the internal compressive stress present in the particle.

For this particulate-reinforced 2014 metal matrix, the majority of damage is associated with particulate clusters and is in the form of cracked particles and voids, which have formed around the particles. The broken Al2O3 particles elevate the local stress, which raises the probability of failure of the matrix and other particles in the immediate vicinity. Final fracture is achieved by damage propagation through the matrix between particle-rich regions. Few of the voids generated by particle cracking did not grow extensively in the tensile stress direction, which is generally the case in ductile fracture of unreinforced aluminum alloy [38]. The lack of extensive void growth in this Al_2O_3 particulate–reinforced 2014 aluminum alloy metal matrix suggests that the fracture strain is critically controlled by both the void nucleation strain and associated linkage strain.

6. Conclusions

The influence of volume fraction of discontinuous particulate reinforcement on tensile fracture of 2014 aluminum alloy–based MMCs provide the following key observations.

1. The as-received microstructure of the 2014/Al_2O_3-T6 composites revealed a non-uniform dispersion of the reinforcing particulates, of non-uniform size, along the different directions of the extruded composite plate. At regular intervals, an agglomeration or clustering of the Al_2O_3 particulates, of varying size, was observed.

2. The non-uniform size and dispersion of the reinforcing phase caused the particles to crack at low values of applied stress. The cracks initiated both at and near the particle- matrix interfaces and in regions of particle agglomeration.

3. The ultimate tensile strength (σ_{UTS}) of the composite microstructure was only marginally greater than the tensile yield strength (σ_{YS}), providing sufficient evidence that the work hardening rate past yielding is low. The ultimate tensile strength followed the same trend as the yield strength of the composite. The improvement in strength of the 2014/Al_2O_3/15*p*-T6 composite over the 2014/Al_2O_3/10*p*-T6 counterpart was only marginal at ambient test temperature (27°C).

4. For both volume fractions of the particulate reinforcement phase in the aluminum alloy matrix, both the reduction in area and tensile ductility decreased with an increase in reinforcement content in the aluminum alloy metal matrix.

5. The quasi-static fracture surfaces revealed limited ductility or brittle appearance on a macroscopic scale but, at the microscopic level, features reminiscent of locally ductile and brittle rupture mechanisms. Tensile fracture of both the composites examined was relatively rough when viewed on a microscopic scale. The reinforcing Al_2O_3 particulates were surrounded by ductile regions described as tear ridges, particulate failure through cracking, and failure through separation at the matrix-reinforcement interfaces. Isolated pockets of voids were intermingled with tear ridges and isolated regions of dimpled rupture. Failure of the reinforcing Al_2O_3 particulates was evident through both cracking and separation at the interfaces with the soft metal matrix.

References

1. D. F. Hasson and C. R. Crowe: in 'Strength of Metals and Alloys, ICSMA 7' Pergamon Press, Oxford, 1985, pp.1515-1520
2. S. G. Fishman: 'Role of Interfaces on Material Damping' ASM, Materials Park, OH, 1985, pp. 33-41.
3. P. K. Liaw, H. G. Greggi, and W. A. Logsdon: ***Materials Science Letters***, 1987, 22, 1613
4. J. F. Mandell, K.C. Hong, and D. H. Grande: ***Ceramic Engineering Science***, Proc. 1987, 8, 937
5. D. H. Grande, J. F. Mandell, and K. C. J. Hong, ***Materials Science***, 1988, 23, 311
6. P. Niskanen, and W. R. Mohn:. ***Advances in Materials Processing***, 1988, 133, 39
7. M. S. Zedalis, J. D. Bryant, P. S. Gilman, and S. K. Das: ***Journal of Metals***, 1991, p. 29
8. Nair, S.V., Tien, J.K. and Bates, R.C. ***International Meals. Reviews***. 1985, 30, 285
9. T. S. Srivatsan, T. S. *Sudarshan and E. J. Lavernia:* ***Progress in Materials Science, An International Review Journal,*** Vol. 39 (4/5), 1995, pp. 317-325.
10. S. G. Fishman: Metal Matrix Composite Requirement: More Reliable Mechanical Properties Data, MMCIAC, Current Highlights, Vol. 1, May 1981.

11. W.C. Harrigan in Metal Matrix Composites (editors: R.JK. Everett and R. J. Arsenault) Academic Press, London, pp. 1-15, 1991.
12. R. J. Arsenault: ***Materials Science and Engineering***, Vol. 64, 1984, pp. 171-181.
13. W. R. East: ***Materials Engineering***, March 1988, p. 33.
14 R. DeMeis: ***Aerospace America***, March 1989, p. 26
15 T. C. Willis: ***Metals and Materials***, August 1988, p. 485
16 W. H. Hunt, Jr., C. R. Cook, and R. R. Sawtell: "Cost Effective High Performance Powder Metallurgy Aluminum Matrix Composites for Automotive Applications,' SAE Technical Paper Series 910834, February 1991, Warrendale, PA
17 W. H. Hunt, Jr., Paper presented at the International Conference on PM Aerospace Materials, November 1991, Lausanne, Switzerland
18. S. Norose, T. Sasdafa, and M. Okabe: in 'Proceedings 28th Japan Congress on Materials Research', 1985, pp. 231-241.
19 W.L. Phillips: in Proceedings of Conf. on Composite Materials' (Editor: B. Noton), Metallurgical Society of AIME, New York, 1978, pp. 567-577
20. T. G. Nieh: ***Metallurgical Transactions***, 1984, 15A, p. 139

21. G. Gould: in 'Proc. Third Int. Conf. on Isostatic Pressing', London, 1986, vol. 1, p. 10.
22. D. F. Hasson, C. R. Crowe, J. S. Ahearn, and D. S. Cooke: 'Failure Mechanisms in High Performance Materials', Cambridge University Press, Cambridge, 1985, pp. 147- 156
23. Y. Sugimura, and S. Suresh: ***Metallurgical Transactions***, 1992, 23A, p. 2231
24. P.K. Liaw abd W.A. Logsdon: ***Engineering Fracture Mechanics***, Vol. 24, 1986, pp. 737-745.
25. J. K. Shang and R. O. Ritchie: ***Metallurgical Transactions***, Vol. 20A, 1989, pp. 897- 908.
26. J. K. Shang and R. O. Ritchie: ***Acta Metallurgica***, 1989, Vol. 37, pp. 2267-2278.

27. J. J. Lewandowski, C. Liu and W. H. Hunt, Jr.: in Interfacial Phenomenon in Composites: Processing, Characterization and Mechanical Properties, TMS, AIME, Warrendale, PA, 1988.
28. J. J. Lewamdowski, C. Liu and W. H. Hunt, Jr.: ***Materials Science and Engineering***, Vol. 107A, 1989, pp. 49-55.
29. J. J. Lewandowski, C. Liu and W. H. hunt, Jr.: in Powder Metallurgy Composites (editors: P. Kumar, K. Vedula and A.M. Ritter), TMS-AIME, Warrendale, PA, 1989.
30. M. Manoharan and J. J. Lewandowski: ***Scripta Metallurgica***, Vol. 23, 1989, pp. 301- 305.
31. M. Manoharan and J. J. Lewandowski: ***Acta Metallurgica***, Vol. 38 (3), 1990, pp. 489- 496.
32. D. L. Davidson: ***Engineering Fracture Mechanics***, Vol. 33, 1989, pp. 965-977.
33. T.S. Srivatsan: ***Journal of Materials Science***, 1996, vol. 31, pp. 1375-88.
34. T.S. Srivatsan: "The Cyclic Fatigue and Fracture Behavior of Al2O3 Particulate- Reinforced 2014 Aluminum Alloy Metal-Matrix Compos2267-
35. F.A. McClintock: Ductility, ASM International, Metals Park, Cleveland, OH, 1968, pp. 256-61.
36. R.H. Van Stone, T.B. Cox, J.R. Low, Jr., and J.A. Psioda: ***International Metals Reviews***, 1975, vol. 30, pp. 157-71.
37. R.J. Arsenault, L. Wang, and C.R. Feng: ***Acta Metallurgica Materialia***., 1991.
38. A.S. Argon, J. Im, and R. Safoglu: ***Metallurgical Transactions*** A, 1975, Vol. 6A, pp. 825-837.
39. T.S. Srivatsan: ***International Journal of Fatigue***, 1995, vol. 17 (3), pp. 183-99.
40. D.J. Lloyd: Acta Metallurgica, 1991, vol. 39 (1), pp. 59-70.
41. E. Hochreiter, M. Panzenbock, and F. Jeglitsch: ***International Journal of Fatigue***, 1993, vol. 15 (6), p. 493.

Advanced Composites for Aerospace, Marine, and Land Applications II
Edited by: T. Sano and T.S. Srivatsan
TMS (The Minerals, Metals & Materials Society), 2015

MECHANICAL PROPERTIES OF STEEL ENCAPSULATED METAL MATRIX COMPOSITES

Sean Fudger [1,2], Eric Klier [2], Prashant Karandikar [3], Brandon McWilliams [2], Chaoying Ni [1]

[1]Department of Materials Science and Engineering, University of Delaware, 201 Du Pont Hall, Newark, DE 19716-1501, USA
[2]Army Research Laboratory Building 4600, Aberdeen Proving Ground, MD 21005, USA
[3]M Cubed Technologies, Inc., 1 Tralee Industrial park, Newark, DE 19711, USA

Keywords: Encapsulation, Metal Matrix Composite, Residual Stress

Abstract

This research evaluates a coefficient of thermal expansion (CTE) mismatch induced residual compressive stress approach as a means of improving the ductility of metal matrix composites (MMCs). MMCs are frequently incorporated into advanced material systems due to their tailorable material properties. However, they often have insufficient strength and ductility for many structural applications. By combining MMCs with high strength steels in a hybridized, macro composite materials system that exploits the CTE mismatch, materials systems with improved strength, damage tolerance, and structural efficiency can be obtained. Macro hybridized systems consisting of steel encapsulated light metal MMCs were produced with the goal of creating a system which takes advantage of the high strength, modulus, and damage tolerance of steels and high specific stiffness and low density of MMCs while mitigating the high density of steels and the poor ductility of MMCs. Aluminum and magnesium based particulate reinforced MMCs combine many of the desirable characteristic of metals and ceramics, particularly the unique ability to tailor their CTE. This work aims to compare the performance of macro hybridized material systems consisting of aluminum or magnesium MMCs reinforced with Al_2O_3, SiC, or B_4C particles and encapsulated by A36 steel, 304 stainless steel, or cold worked Nitronic® 50 stainless steels.

Introduction

Metal Matrix Composites (MMCs) have a wide range of applicability for advanced material systems due to the tailorability of their mechanical and physical properties [1,2]. Moreover, particulate reinforced aluminum or magnesium based MMCs with a high volume fraction of the ceramic phase provide for very high specific stiffness, low CTE, and low density seen in ceramics while maintaining the good strength, and processing advantages of cast light metals. These mechanical and physical properties can be fine-tuned by adjusting the size, distribution, or volume fraction of the reinforcement material, utilizing different reinforcement materials, matrix alloy selection, or adjusting the processing parameters used in creating the MMCs [3-5].

This study examines a coefficient of thermal expansion (CTE) mismatch induced residual compressive stress to provide for a macro composite materials system with increased specific stiffness, increased strength, increased damage tolerance, reduced density and allow for greater structural efficiency in structural applications. By controlling the CTE mismatch between two materials a residual stress can be introduced upon cooling which will result in increased strain to failure for the material in compression. A Hydrostatic pressure of 300 MPa (43.5 ksi) has been shown to increase the reduction of area from 10% to 80% and quadruple failure strain in a 6061 aluminum alloy reinforced with 15 vol% Al_2O_3 particulates [6,7]. The imposed hydrostatic pressure caused a shift in the microscopic failure mechanism. Without the induced hydrostatic pressure, typical failure observed was due to ductile failure in the matrix due to nucleation and growth of voids, void formation at the reinforcement-matrix interface, or ductile failure in the matrix in regions adjacent to the reinforcement. Following a superimposed hydrostatic pressure, the ductility significantly increases because the failure mechanism shifted to fracture of the brittle Al_2O_3 reinforcement phase which then resulted in void nucleation, growth, and coalescence in the matrix [8-14].

By carefully selecting an encapsulating material with a substantially higher CTE than the MMC the aforementioned residual compressive stress is introduced. This work utilized a low CTE mild steel (A36), a high CTE austenitic stainless steel (304), and a high strength/high CTE austenitic stainless steel (Nitronic® 50, cold worked 25%) encapsulating an MMC core consisting of an aluminum or magnesium alloy matrix discontinuously reinforced with a high volume fraction of Al_2O_3, SiC, or B_4C ceramic particles. More literature exists on MMCs reinforced with SiC and Al_2O_3 as compared to B_4C due to fabrication complications and higher costs [15,16]. However, B_4C has a lower density and higher hardness along with better chemical and thermal stability compared to SiC and Al_2O_3 and deserves a thorough investigation [17-19].

Experimental Procedure

Steel/MMC macro hybridized tensile bars and shear push-out specimens were produced by M-Cubed Technologies Inc. Newark DE. Split-Hopkinson Pressure bars will be produced at a future date to measure the high strain rate response of the materials. For the macro hybridized tensile bars, steel (A36, 304, and Nitronic ® 50) bar with a diameter of greater than 0.5" dia. was bored to produce a tube with an inside diameter of 0.188". The steel tubes were then processed to fill the internal void with various compositions of particulate MMCs, Al/SiC, Al/Al_2O_3, and Mg/B_4C using a cast or infiltration approach [20-23]. Finally, the gauge section of the steel/MMC bars were turned to a diameter of 0.300", 0.400" or 0.500" creating cylindrical dog-bone tensile samples as per ASTM B557-10. The various MMC cores therefore remained constant at 0.188" but with 3 different thicknesses of steel to provide for 3 different residual stress conditions. A similar process was used for the production of the shear push-out specimens. However, in this case, three different diameter macro hybridized rods were produced, 0.626", 0.874", and 1.0". In all cases, the MMC core remained constant at 0.364" diameter. The rods were then sliced to a thickness of 0.236" as shown in Figure 1b below [24].

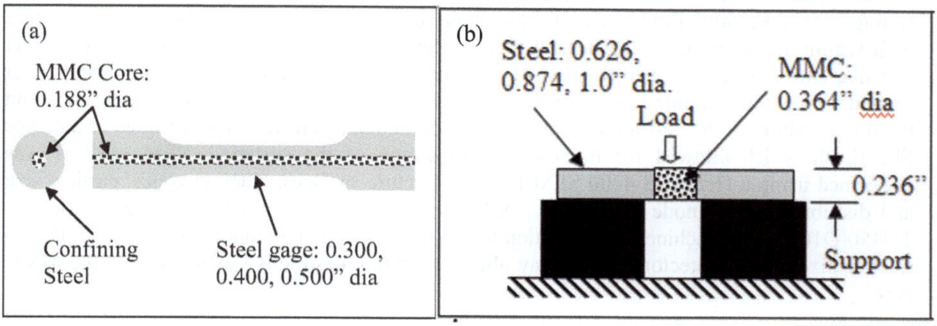

Figure 1. Design of composite tensile and shear test samples.

Two different aluminum alloys were used for the matrix material of the particulate reinforced aluminum MMCs, Al-10Si and Al-4Mg. The matrix alloys were chosen specifically for the reinforcement type in order to avoid certain thermodynamic instabilities. Since SiC tends to form Al_4C_3 when it reacts with aluminum, Al-10Si was chosen to avoid formation of Al_4C_3 and subsequent degradation of the reinforcement particles which can drastically diminish mechanical properties [5]. An Al-4Mg alloy was selected for use with the Al_2O_3 reinforcement as this alloy provides better ductility than Al-10Si. Table I lists the hybridized systems produced for this study. Aluminum matrix composites were produced with SiC and Al_2O_3 particles as reinforcement at 30 and 55 vol%. Similarly, a magnesium matrix composite was produced with 45 vol% B_4C particles as reinforcement in a Mg AZ91E alloy. The initial macro hybridized systems produced utilized either A36 steel or 304 stainless steel to encapsulate the Al/Al_2O_3 or Al/SiC MMC core. The second iteration utilized an Al/SiC or Mg/B_4C MMC core encapsulated by Nitronic® 50 stainless steel. Based on in-house thermal exposure and hardness studies, a Nitronic 50 alloy with 25% cold work was selected as it provides for the optimum combination of strength and ductility in the steel following the thermal processing of the macro hybridized material system. The corresponding Nitronic bar stock was procured from High Performance Alloys, Inc. Windfall, IN.

Table I. Description of the hybridized systems studied

Steel Encapsulant	**Matrix Material**	**Reinforcement**	**Reinforcement percentage**
A36	Al-10Si	SiC	30 vol%
A36	Al-10Si	SiC	55 vol%
A36	Al-4Mg	Al_2O_3	30 vol%
A36	Al-4Mg	Al_2O_3	46 vol%
304SS	Al-10Si	SiC	30 vol%
304SS	Al-10Si	SiC	55 vol%
304SS	Al-4Mg	Al_2O_3	30 vol%
304SS	Al-4Mg	Al_2O_3	55 vol%
Nitronic® 50	Al-10Si	SiC	55 vol%
Nitronic® 50	Mg AZ91E	B_4C	45 vol%

Extensive characterization was performed on the tensile specimens in the form of optical microscopy, scanning electron microscopy (SEM), and X-Ray analysis. Optical microscopy was performed on the fracture surface of the samples to check for any imperfections, casting defects, or other variations at the surface. Subsequently, cross sections transverse to the direction of testing, were taken at multiple locations thoughout the gage section to investigate microstructure both within and away from the necked region. Specimens were cross sectioned using a wet abrasive cut-off saw with a feed speed of 0.05 mm/s. The resulting cross-sections were first ground with silicon carbide paper to 800 grit. Subsequently, they were polished using 3, 1, and 0.3 micron alumina slurries in a Buhler VibroMet2 vibratory polisher. This resulted in a mirror like finish which allowed for microscopic inspection of the surface. SEM analysis was performed using a Hitachi S-4700 SEM to study fracture surfaces, reaction zones, particle size and distribution, and mode of fracture. X-Ray imaging was performed utilizing a Yxlon Y. TU450-D10 X-Ray machine in conjunction with a GE Inspection Technologies DXR-250P (200 micron pixel pitch) detector. The X-ray allowed for non-destructive evaluation of the MMC core.

Results and discussion

A 3D finite element modeling framework was implemented in the commercial software (ABAQUS/Explicit) to predict the residual stresses due to thermal expansion mismatch and predict the effect of hydrostatic confining pressure on the tensile damage behavior in steel encapsulated MMC tensile bar specimens. The framework involves a two step process in which a thermo-mechanical simulation is first conducted to determine the residual stresses present in the hybrid structure after cool down from MMC processing conditions. The stress state from this simulation is imported into a simulation of a tension test as the initial stress state. The density, elastic modulus, and coefficient of thermal expansion of the MMC were assumed to be 2.96 g/cc, 200 GPa, and 11.8 x $10^{-6}/^{o}C$ respectively. The density, elastic modulus, and coefficient of thermal expansion of the steel were assumed to be 7.85 g/cc, 200 GPa, and 17.3 x $10^{-6}/^{o}C$ respectively. In the tensile loading simulations the MMC was modeled as elastic to failure with a hydrostatic pressure dependent failure criterion using a user subroutine. In tension the MMC was assumed to have an ultimate strength (σ_u^t) of 300 MPa. In compression the MMC was assumed to have an ultimate strength (σ_u^c) which was dependent upon the confining stress state. In the absence of experimental data for the pressure dependent strength of the current MMC system, it was necessary to assume this behavior for this preliminary work which was designed to test the implementation of the pressure dependence in the finite element framework. The compressive strength was assumed to follow the following hydrostatic pressure (*P*) dependent relation:

$$\sigma_u^c(P) = 300 + 3\text{P (MPa)} \tag{1}$$

The steel was modeled as elastic-plastic using Mises plasticity with an assumed initial yield strength of 1027 MPa with isotropic hardening to an ultimate strength of 1138 MPa at a true equivalent plastic failure strain of 0.19. The thermo-mechanical simulation did not consider heat transfer and assumed a uniform temperature profile which decreased linearly from an assumed stress free temperature of 400°C to room temperature (25°C). The interface between the two materials was assumed to be perfectly bonded.

The predicted residual stress distributions from thermo-mechanical simulations considering 0.500 inch (12.7 mm) and 0.300 inch (7.62 mm) thickness of steel encapsulation are shown in Figure 2a. The 0.500" specimen had a uniform compressive pressure of 65 MPa in the gage section while the 0.300" specimen had a uniform compressive pressure of 50 MPa. The quantitative accuracy of the residual stresses is not expected to be very realistic due to the assumptions of material properties and temperature profile used in the simulations. However, the modeling results are critical in that they indicate that the confining pressure is uniform within the gage section of the specimen and increases with the thickness of the encapsulating steel used and that the experimental design to test for the effect of confining pressure on the response of the MMC is valid. The predicted damage patterns from the tensile simulations considering hydrostatic pressure dependent strength are shown in Figure 2b and qualitatively predict the experimentally observed shear lag fracture patterns.

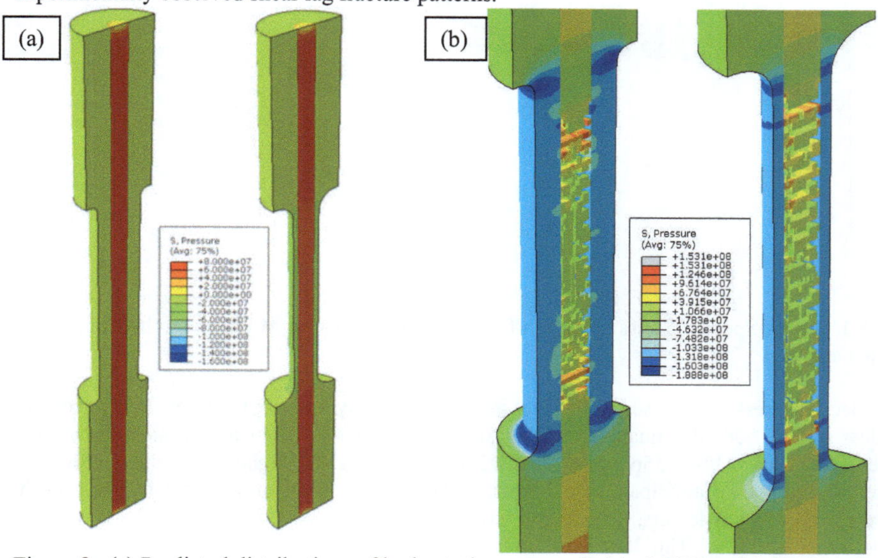

Figure 2. (a) Predicted distributions of hydrostatic pressure (Pa) in 0.500" and 0.300" steel encapsulated MMC tensile bar specimens showing uniform confining pressure in the gage section. (b) Predicted damage from tensile test simulations of 0.500" and 0.300" steel encapsulated MMC tensile bar specimens. Sign convention is negative for tension and positive for compression.

To realistically compare the various steel/MMC macro hybridized systems, the specific properties were calculated. For stress, the specific stress is the stress divided by the density of the material system in question. Specific properties accout for density differences between the steels, MMC system and volume fraction reinforcement, and the volume of MMC in each a given hybridized system and allows for a "normalized" comparison.

The specific stress versus strain obtained from the experimental tension tests for 3 MMC systems encapsulated with A36 steel (0.500" dia. guage section) is shown in Figure 3. Specific stress was calculated by normalizing the experimental stress by sample density in order to effectively compare the monolithic steel and the hybridized steel/MMC specimens. A36 has a CTE similar to the Al/SiC-55p and Al/Al2O3-46p, and lower than the Al/SiC-30p. Therefore, one would anticipate there would be no residual compressive stress in the MMC and this is confirmed by FEA. The A36-30p and A36-55p curves represent a solid A36 tensile bar subjected to the same thermal cycle as the macro hybridized system during fabrication. As all the specific stress strain curves in Figure 3 show, the macro hybridized system has a substantially lower specific stress than for the A36 steel alone. In this case, there appears to be no advantage to the hybridized system. However, the A36/MMC tensile bars do exhibit a substantial increase in the strain to failure versus the A36 steel.

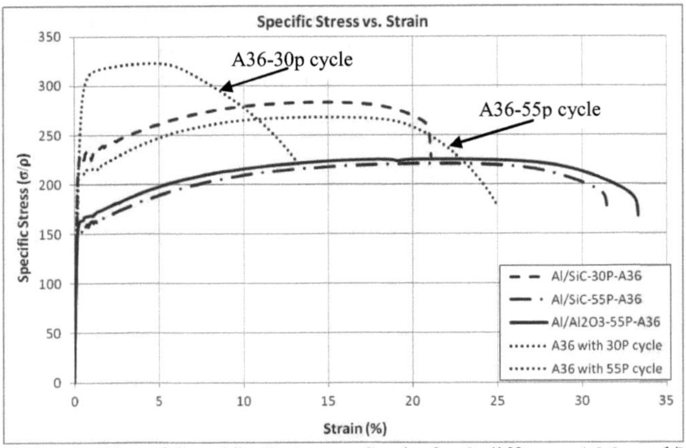

Figure 3. Plot of Specific Stress vs. Strain for 3 different A36 steel/MMC macro hybridized cast tensile bars.

The specific stress versus strain for the same 3 MMC systems but encapsulated with 304 stainless steel (0.500" dia. guage section) is shown in Figure 4. 304 stainless steel has a much higher CTE than Al/SiC-55p and Al/Al2O3-46p, and slightly higher than the Al/SiC-30p. Therefore, one would anticipate that following the thermal excursion required to fabricate the macro hybridized tensile bars, there would be a residual compressive stress in the MMC and this is confirmed by FEA. The 304SS-55p curve represents a solid 304 stainless steel tensile bar subjected to the same thermal cycle as the macro hybridized system during fabrication. In this case, all the specific stress strain curves in Figure 4 show the macro hybridized systems actually have a higher specific stress, particularly yield stress, than the 304 stainless steel alone. The strain to failure of the hybridized systems is similar to the 304 SS. In this case, the macro hybridized systems do in fact offer a specific yield strength performance advantage over the monolithic steel.

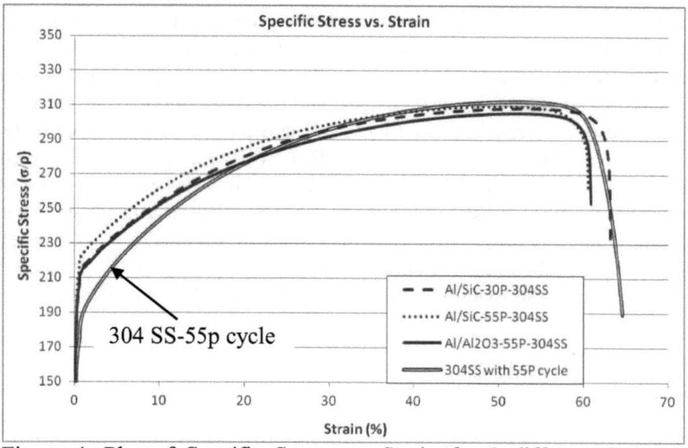

Figure 4. Plot of Specific Stress vs. Strain for 3 different 304 stainless steel/MMC macro hybridized cast tensile bars.

Despite the property enhancement in the 304 SS/MMC systems, it was recognized that this is not a very high strength system due to the relativley low strength of the annealed 304 SS utilized. Therefore a cold worked Nitronic 50® stainless steel was utilized to increase the yield strength of the system.

Figure 5 presents specific stress vs. strain data for Al/SiC 55 vol%, 0.500" diameter gauge section, macro hybridized systems with three different encapsulating steels. As shown in Figure 5, the Nitronic® 50/Al-SiC hybridized system has a drastically higher specific stress as compared to the 304 stainless and A36 steels yet retains in excess of 20% strain to failure despite the strain to failure of the Al/SiC-55p MMC of less than 0.2%. The 304 stainless steel has about a 40% higher specific stress than the A36 steel and nearly double the strain to failure.

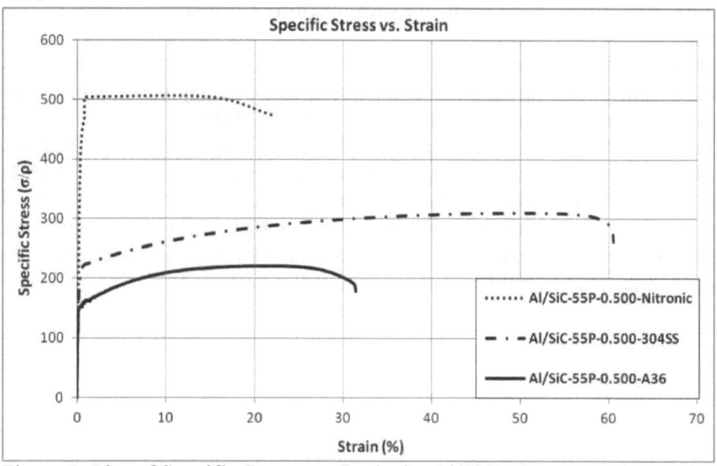

Figure 5. Plot of Specific Stress vs. Strain for Al/SiC-55p encapsulated with 3 different steels.

To evaluate the relative effectiveness of the ratio of steel to MMC, the specific yield stress was calculated for each of the macro hybridized materials systems and ploted vs. the %MMC. The %MMC refers to the ratio between the area of MMC core to the total area of the hybrid system (MMC + steel). The MMC core diameter remains constant and the outer diameter of the steel varies from 0.300" to 0.400" to 0.500" resulting in 14.1, 22.1, and 39.3% MMC respectively. The specific yield stress of the steel alone is indicated at 0% MMC while the specific yield stress of the MMC is indicated at 100%. The dashed line between these points represents the rule of mixtures (ROM) prediction for specific yield stress at a given %MMC. The results are shown in Figure 6 and clearly indicate what was observed above. For the low CTE A36 steel, and all %MMC combinations, the specific yield stress is below the ROM. However, for the high CTE 304 SS, and all %MMC combinations, the specific yield stress is above the ROM prediction.

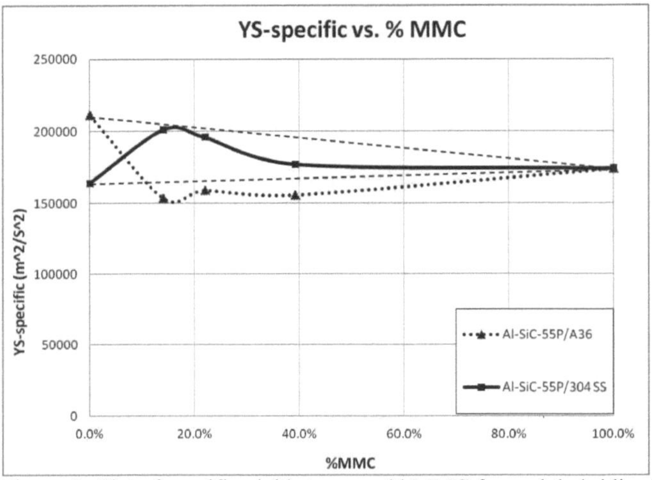

Figure 6. Plot of specific yield stress vs. % MMC for each hybridized system.

Another important parameter for many structural systems is the damage tolerance, or the specific energy, which is the area under the stress strain curve. The specific energy was calculated for each of the macro hybridized materials systesms and ploted vs. the %MMC. The specific energy of the steel alone is indicated at 0% MMC while the specific energy of the MMC is indicated at 100%. The dashed line between these points represents the rule of mixtures (ROM) prediction for specific energy at a given %MMC The results are shown in Figure 7. Due to the extreme strain to failure of the annealed 304 SS, the specific energy is very high. More importantly, and unlike the specific YS ROM results, both the A36 and 304 SS macro hybridized systems exhibit specific energy greater than the ROM would predict at all %MMCs evaluated.

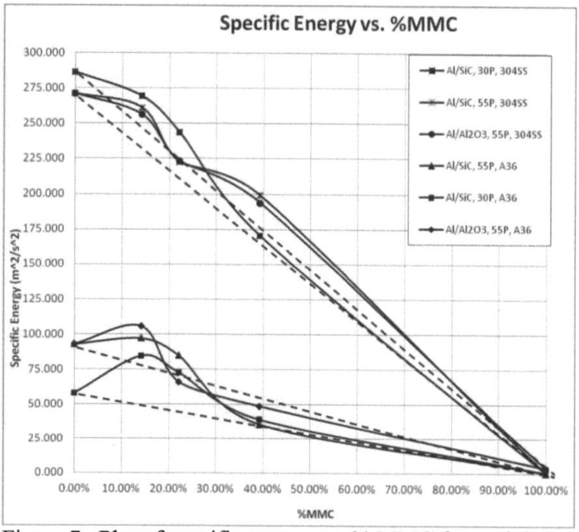

Figure 7. Plot of specific energy vs. % MMC for each hybridized system.

Finally, and perhaps most importantly, Figure 8 presents specific energy for the macro hybridized materials evaluated in this work compared to several common structural materials. The A36 mild steel has a specific energy similar to 5083 aluminum armor plate. The 304 SS has substantially higher specific energy, similar to 4340 steel and much greater than Ti-6Al-4V (annealed). However, this material has a realativley low yield stress. Finally, the hybridized system based on Nitronic® 50 (CW) with either an Al-SiC and Mg-B_4C MMC, which has greatly increased yield strength, exhibits specific energy similar to Ti-6Al-4V (annealed). This is significant as such a system utilizes lower cost raw materials and lower cost manufacturing processes than titanium.

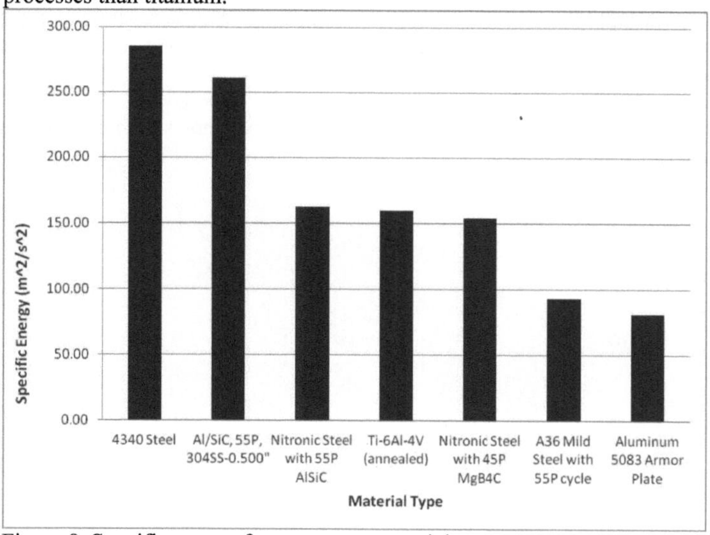

Figure 8. Specific energy for numerous materials.

The interface between the MMC and the constraining steel layer in a macro hybridized steel/MMC system is likely to be a critical parameter for a real world application of such a system. This interface was evaluated by a shear push-out test. Figure 9 shows the shear data for the 0.626" diameter steel sample for each of the hybridized systems encapsulated with the A36 steel and 304 stainless steels.

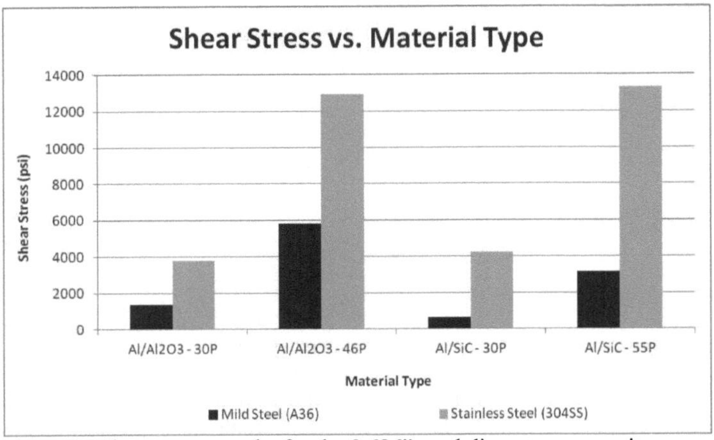

Figure 9. Shear stress results for the 0.626" steel diameter composite.

The nature of the material systems investigated herein means that this shear push-out test is measuring not only the bond strength between the steel and the MMC, but the residual stress due to the CTE mismatch. The 304 stainless steel shows a much higher shear stress presumably due to the much higher residual compression on the MMC as compared to the A36 steel. It is also apparent that MMCs with a higher reinforcement content result in greater shear stress as these MMCs have a lower CTE and lead to a larger CTE mismatch with the steel and subsequently a greater residual compressive stress in the MMC.

Figure 10 is a photomicrograph of Al/Al_2O_3 55 vol% MMC and shows a homogeneous distribution of reinforcement which is typical of the composites.

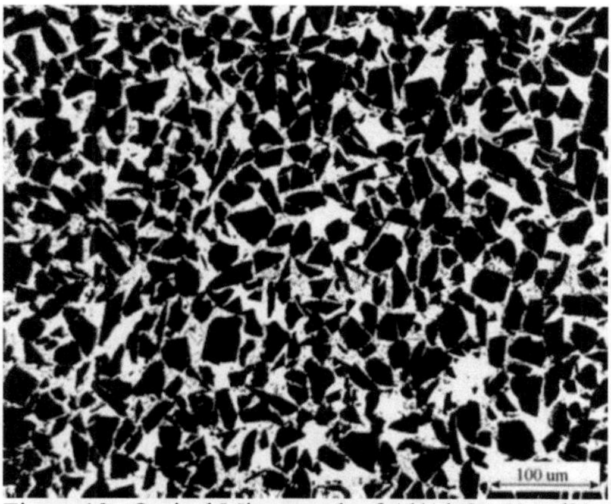

Figure 10. Optical Micrograph of Al/Al_2O_3 55 vol%.

Scanning electron microscopy was performed via secondary electron imaging and backscattered electron imaging. The fracture surfaces were further investigated using SEM checking for any abnormalities. A small number of the $Al\text{-}4Mg/Al_2O_3$ tensile bars exhibited microshrinkage porosity near the MMC-steel interface, which acted as a stress riser and ultimately resulted in fracture at this location. This result was not unexpected as Al-4Mg is inherently difficult to cast. Figure 11 shows an SEM image of a typical fracture surface.

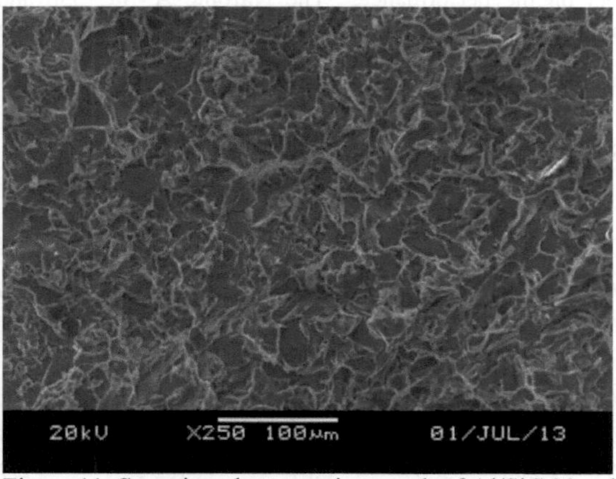

Figure 11. Scanning electron micrograph of Al/SiC 30 vol%.

The majority of the fracture surfaces confirm a uniform distribution of reinforcement particles with only minor agglomeration indicating a successful casting method at each reinforcement volume percent.

The reaction zone at the MMC-steel interface, both at the fracture surface and subsequent cross sections, was thoroughly investigated on each sample. It is apparent that the materials in contact along with the CTE mismatch affect the thickness of the reaction layer. Figure 12a shows a thin reaction layer observed at the interface of the A36/Al-SiC-55p system where Figure 12b shows a much thicker, brittle reaction layer at the interface in a 304 SS/Al-SiC-55p system.

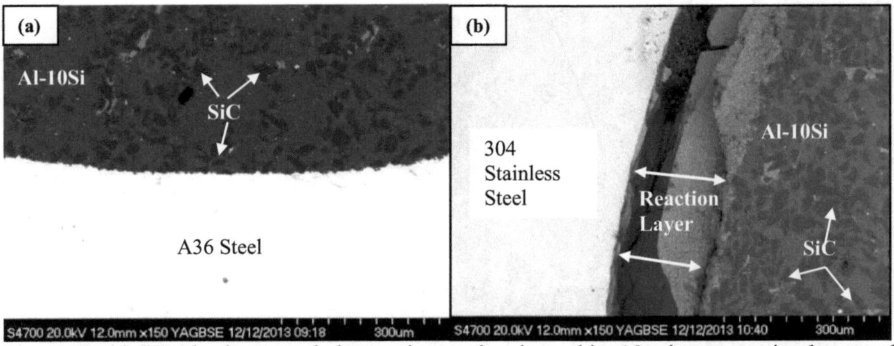

Figure 12. (a) SEM backscattered electron image showing a thin, 10 micron, reaction layer at the interface in an A36/Al-SiC-55p system. (b) SEM backscattered electron image showing a thicker, 180 micron, reaction layer at the interface in a 304 SS/Al-SiC-55p system.

The reaction layers tend to be thickest in systems with the largest CTE mismatch which may indicate the intimate contact provided throughout the processing may allow for an extended reaction time and thus a thicker reaction layer. Although thick reaction layers are observed on many of the systems, excluding the magnesium systems, and consist of brittle intermetallics, there seems to be no significant effect on the performance. The Nitronic Al-SiC 55p and Nitronic Mg-B_4C demonstrate very similar performance results with drastically different reaction layer thicknesses.

Table II. Summary of material properties

Material	Matrix	Vp	ρ (g/cc)	E (GPa)	Elongation (%)	Hardness	CTE ppm/K
A36	N/A	N/A	7.8	200	23	178B	11.7
304 SS (annealed)	N/A	N/A	8.03	200	40	201B	16.6
Nitronic 50	N/A	N/A	7.88	200	17	336B	16.1
Al_2O_3 CAP-3	N/A	N/A	3.9	370	0.1	1292	5.1
Hot Pressed SiC	N/A	N/A	3.22	453	0.1	1905	3
Sintered SiC Hexoloy	N/A	N/A	3.13	410	0.09	1924	3
SiC (RBSC)	N/A	N/A	3.03	380	0.07	1332	2.9
Al/Al_2O_3	Al-4Mg	46	3.3	160	0.51	-	14.1
Al/SiC	Al-10Si	30	2.78	120	0.18	-	15.6
Al/SiC	Al-10Si	55	2.96	202	0.09	-	11.8
Mg/B_4C	AZ91E	45	2.22	165	0.07	206B	-

ρ – density; E – Young's modulus; Hardness for metals Brinell, for ceramics - Knoop 2kg, Vp – particle volume fraction, CTE – coefficient of thermal expansion (20-100°C). Sources: Spectra: Honeywell; CAP-3: Ceralloy, Ekasic-T: Ceradyne; Hexoloy: Saint Gobain; SiC-N: Cercom (CoorsTek); RBSC, RBBC: M Cubed Technologies (MCT). Other material properties were obtained directly from manufacturer's websites or datasheets.

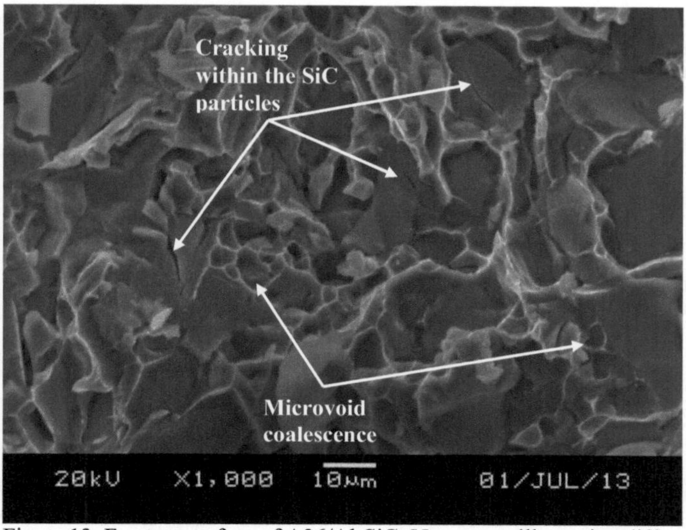

Figure 13. Fracture surface of A36/Al-SiC-55p system illustrating different failure modes.

Another focus of fracture surface investigation is identification of the mode of failure. Ductile microvoid coalescence was observed within each matrix material as expected from aluminum and magnesium alloys. Figure 13 shows cracking of the reinforcement particles within the matrix of the low residual stress A36/Al-SiC-55p system. For this system, we don't expect to see a lot of residual stress yet we see failure first occurring in the particles and expect good bonding between the matrix and reinforcement resulting in increased strain to failure. For the high residual stress systems, the bonding within the MMC is less important as it is dominated by the residual compressive stress and thus the residual stress is delaying microvoid formation and coalescence in the matrix of the MMC and the strain to failure is increased.

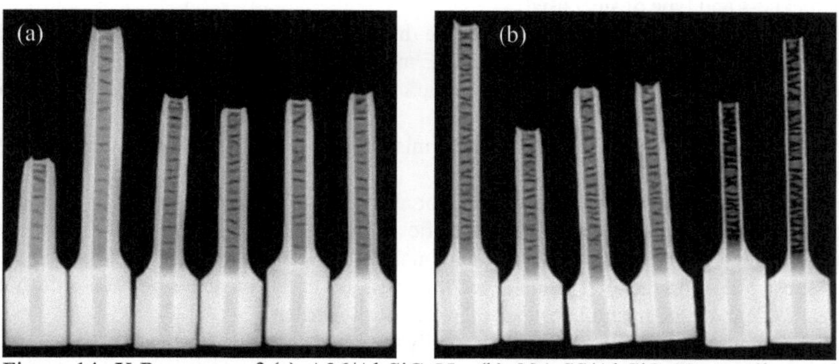

Figure 14. X-Ray scan of (a) A36/Al-SiC-55p (b) 304 SS/Al-SiC-55p tensile specimens post fracture.

Table III. Summary of MMC cracks in each system.

Average Number of Cracks by Diameter		
	A36 Steel	304 Stainless
0.500"	17.7	31.4
0.400"	13.8	35
0.300"	11.7	29
Average Number of Cracks by Reinforcement Type		
	A36 Steel	304 Stainless
Al/Al_2O_3	10.8	35.6
Al/SiC	16.2	30.3
Average Number of Cracks by Reinforcement Percentage		
	A36 Steel	304 Stainless
30p	11.0	24.2
55p	16.1	36.0
Average Number of Cracks by Surrounding Metal		
	A36 Steel	304 Stainless
	14.4	31.8

The x-ray images shown in Figure 14 show that the failure of the MMC core is characterized by a series of transverse cracks and relatively uniform segments of MMC. These cracks are due to a transfer of localized stresses explained by the shear lag effect. Analysis of the number of cracks and the resultant length of each MMC segment, as shown in Table III, indicate that there are roughly double the number of cracks in the 304 SS system as compared to the A36 steel. Similarly, the 55 vol% reinforcement MMC resulted in 50% more cracks than its 30 vol% counterpart. In both cases, the system that presumably provides for the most residual compressive stress displays more cracks and shorter MMC segments. Thus, each element of the MMC can carry more load due to the residual compressive stress. Despite the very poor ductility of the MMC, the macro hybridized system displays excellent ductility.

Conclusions

1. The FEA model predicts significantly different residual compressive stress in relation to the thickness and type of steel used.
2. Improved specific strength is observed in the high CTE mismatch system vs. steel alone whereas this trend is reduced for the low CTE mismatch system.
3. Greatly improved specific stress in Nitronic® systems while retaining good strain to failure.
4. Better than ROM specific YS of high CTE mismatch system, worse than ROM for low CTE mismatch system.
5. Better than ROM specific energy observed for all systems.
6. Macro hybridized systems can offer specific energy greater than or similar to more expensive high performance conventional monolithic metals, specifically Ti6Al-4-V.
7. The substantial interfacial reaction layer thickness observed on the Al systems do not appear to degrade the macro hybridized tensile bar properties.
8. The FEA model agrees with the experimentally observed shear lag fracture patterns.

Acknowledgements

This work was funded by a cooperative Agreement No. W911NF-11-2-0040 between M Cubed Technologies Inc. and the US Army Research Laboratory (ARL), APG, MD.

This project was supported in part by an appointment to the Research Participation Program at the US ARMY Research Laboratory administered by the Oak Ridge Institute for Science and Education (ORISE).

The Department of Materials Science and Engineering, University of Delaware.

References

1. A. Mazahery and M. Ostadshabani, "Investigation on mechanical properties of nano-Al_2O_3 reinforced aluminum matrix composites", Journal of Composite Materials, 45 (2011), pp. 2579-2585.
2. V. Kumar GB, Rao CSP, Selvaraj N, Bhagyashekar MS. "Studies on Al6061-SiC and Al 7075-Al_2O_3 metal matrix composites. J Miner Mater Charact Eng 2010;99(1):43-55.
3. McDaneals DL. "Analysis of stress, strain, fracture and ductility behavior of Al matrix composite containing discontinuous SiC reinforcement". Metall Trans A 1985; 16A: 1105–1115.
4. Ibrahim IA, Mohammed FA and Lavernia EJ. "Particulate reinforced metal matrix composites: a Review". J Mater Sci 1991; 26: 1137–1156.
5. Lloyd DJ. "Particle reinforced aluminum and magnesium matrix composites". Int Mater Rev 1994; 39: 1–21.
6. D. Liu and J. J. Lewandowski, "The effects of superimposed hydrostatic pressure on deformation and fracture part I: particulate reinforced 6061 composites", Metallurgical Transactions A 24 [3] (1993) 601-608.
7. D. Liu and J. J. Lewandowski, "The effects of superimposed hydrostatic pressure on deformation and fracture part II: particulate reinforced 6061 composites", Metallurgical Transactions A 24 [3] (1993) 609-614.
8. A.L. Grow, S. Patankar, T. Osman, R.W. Margevicius, and J.J. Lewandowski: Presented at *TMS-AIME Meeting*, San Diego, CA, March 1–4, 1992.
9. S. Patankar, R.W. Margevicius, and J.J. Lewandowski: Case Western Reserve University, Cleveland, OH, unpublished research, 1992.
10. D.S. Liu, M. Manoharan, and J.J. Lewandowski: *Scripta Metall.*, 1989, vol. 23, pp. 253–56.
11. D.S. Liu, M. Manoharan, and J.J. Lewandowski: *J. Mater. Sci. Lett.*, 1989, vol. 8, pp. 1447–48.
12. J.J. Lewandowski, D.S. Liu, and C. Liu: *Scripta Metall.*, 1991, vol. 25, pp. 21–26.
13. F. Zok, J.D. Embury, M.F. Ashby, and O. Richmond:*9th Risø Int. Symp. on Metallurgy and Materials Science*, S.I. Anderson, H. Lilholt, and O.B. Pederson, eds., Risø National Laboratory, Roskilde, Denmark, 1989, pp. 517–27.
14. A.K. Vasudevan, O. Richmond, F. Zok, and J.D. Embury: *Mater. Sci. Eng.*, 1989, vol. A107, pp. 63–69.
15. Khan KB, Kutty TRG, Surappa MK. "Hot hardness and indentation creep study on Al–5% Mg alloy matrix–B4C particle reinforced composites". Mater Sci Eng A 2006; 427:76–82.
16. Zhang H, Ramesh KT, Chin ESC. "High strain rate response of aluminum 6092/B_4C composites". Mater Sci Eng A 2004;384:26–34.
17. Aizenshtein M, Froumin N, Shapiro-Tsoref E, Dariel MP, Frage N. "Wetting and interface phenomena in the B4C/(Cu–B–Si) system". Scripta Mater 2005;53:1231–5.

18. Jung J, Kang S. "Advances in manufacturing boron carbide–aluminum Composites". J Am Ceram Soc 2004;87(1):47–54.
19. A. Baradeswaran and A. Elaya Perumal, "Influence of B_4C on the tribological and mechanical properties of Al 7075-B_4C composites". Composites: Part B 54 (2013) 146-152.
20 P. G. Karandikar et al., "Al/Al_2O_3 MMCs and macrocomposites for armor applications," Ceramic Engineering and Science Proceedings 34 (5) (2013), 63-74.
21. P. G. Karandikar et al., "Net shape Al/B_4C metal matrix composites (MMCs) for high specific stiffness and neutron absorption applications," TMS 2013 Annual Meeting Supplemental Proceedings, John Wiley & Sons, Inc., Hoboken, NJ, (2013), 683-689.
22. M. Aghajanian, et al., "Processing of Hybrid structures consisting of Al-based metal matrix composites (MMCs) with metallic reinforcement of steel or titanium", TMS2013 Annual Meeting Supplemental Proceedings, John Wiley & Sons, Inc., Hoboken, NJ, (2013), 629-638.
23. B. Givens, et al., "Effect of particle loading on the properties of Al/SiC metal matrix composites," in Aluminum Alloys: Fabrication, Characterization and Applications II, Yin et al. editors, TMS, Warrendale, PA (2009) 197-202.
24. P. Karandikar et al. "Metal Encapsulated MMC Macrocomposites" Materials Science & Technology 2014 Conference Proceedings.

Advanced Composites for Aerospace, Marine, and Land Applications II
Edited by: T. Sano and T.S. Srivatsan
TMS (The Minerals, Metals & Materials Society), 2015

THE EVOLUTION OF SOLID POWDERS IN LIQUID ALUMINUM AT LOW TEMPERATURE AND THE EFFECTS OF ULTRASOUND ON IT

Zhiwei Liu and **Qingyou Han**

School of Engineering Technology
Purdue University
401 North Grant Street
West Lafayette, IN 47906, USA

Keywords: Al_3Ti particles; Ultrasound; Aluminum composites

Abstract

This research investigated the evolution of solid Ti powders in liquid Al at low melting temperature of 730 °C. In situ Al_3Ti particles were synthesized due to the reactive diffusion between solid Ti powders and liquid Al. These Al_3Ti particles were blocky in morphology and smaller than 5 μm in size. It was found that the formation of Al_3Ti particles underwent the processes of nucleation, growth and rupture. Accordingly, a reaction-peeling model was suggested to illustrate the formation mechanism of small blocky Al_3Ti particles. In addition, the effects of high-intensity ultrasound on the diffusion reaction between solid Ti powders and liquid Al were studied. Ultrasonic fields in the Al melt could generate instantaneous high pressure and interparticle collisions, both of which could accelerate the formation rate of Al_3Ti particles significantly. An ultrasound assisted in situ method was proposed for producing in situ Al_3Ti/Al composites in this research.

Introduction

In situ Al_3Ti/Al-alloy composites have attracted considerable attention recently due to the excellent properties of Al_3Ti phase, such as its high melting point, low density, high Young's modulus, and good wet ability with aluminum [1, 2]. Particularly, the Al_3Ti phase also has an excellent resistance to oxidation and corrosion in the fluoride atmosphere above the melting point of aluminum [3]. Several manufacturing techniques have been developed in the fabrication of in situ Al_3Ti/Al-alloy composites, such as powder metallurgy [4], the centrifugal method [5], mechanical alloying [6] and the in situ casting technique [7].

In situ casting technique has been regarded as a prospective method for commercial production of Al matrix composites due to its characteristics of low cost and high efficiency. In the conventional methods, in situ Al_3Ti particles were generally synthesized by adding some salts containing Ti element, such as K_2TiF6, or the mixture of Ti and Al powders into the Al melt. Chen et al. [8] fabricated in situ Al_3Ti/Al composites by adding K_2TiF6 into the molten Al at 1100 °C. The Al_3Ti phase in their research was needle-like in morphology with the length exceeding 200 μm and width of 10 μm. Jie et al. [9] prepared Al-Al_3Ti composites by adding the mixture of Ti and Al powders into the Al liquid at 900 to 1000 °C. Most of Al_3Ti particles produced in their research were rod-like in morphology and longer than 50 μm. Based on the previous work of researchers, the synthesizing temperatures using in the conventional methods are usually higher than 900 °C. However, high temperature of Al-alloys melt can result in some issues such as the burning loss of alloying elements, high manufacturing cost, gas pick-up and oxidation of Al melt. It also should be indicated that needle-like or long rod-like Al_3Ti particles can influence the mechanical properties of the composites severely. Thereby, lowering the fabricating temperature and decreasing the Al_3Ti particle size are expected in the practical production of in situ Al_3Ti/Al composites.

Ultrasound, as a facile method, has been extensively applied in organic synthesis and fabrication of materials [10-12]. Ultrasonic fields in a liquid can give rise to nonlinear effects such as acoustic cavitation and acoustic streaming. Acoustic cavitation, or the formation of small cavities in the liquid, occurs as a result of the tensile stress produced by an acoustic wave in the liquid phase. These cavities continue to grow by inertia until they collapse under the action of compressing stresses during the compression half-period, producing high intensity shock waves in the fluid. In general, acoustic cavitation undergoes the formation, growth, and implosive collapse of bubbles coupled to the ultrasonic field [13].The above process of bubbles can occur simultaneously at millions of locations in a reactor within a short span time (a few microseconds). The collapse of cavitation bubbles can result in extremely intense local heating (around 5000 K) and high pressure (around 10^5 kPa), as well as super rapid heating and cooling rates (higher than 10^{10} K/s) [14-16]. These above extreme conditions produced in ultrasonic fields can promote the reactions which need high temperature, high pressure, or long reaction time. Particularly, for a suspension (liquid containing some solid particles), the high velocity interparticle collisions can be generated due to the effect of shock wave. Accordingly, the morphology and instant temperature of particle surface both can be changed significantly [17-19].

In this research, solid Ti powder was used as the only additive. The melting temperature of Al was designed at 730 °C, which was around 200 °C lower than the temperatures used in the conventional methods. The synthesis of Al_3Ti particles via the direct reaction between solid Ti powders and liquid Al at low melting temperature was studied. Since a suspension of the Al melt containing some Ti powders can be formed, the influence of ultrasound on the evolution of solid Ti particles in the Al melt was also investigated in order to clarify the effects of ultrasound in the reaction. Two groups of experiments with and without ultrasonic treatment were conducted, respectively.

Experimental

Materials and Experimental Process

Pure Al ingot (99.5 % commercial purity) and solid Ti powders (99.7 % commercial purity, average size of 20 μm) were used as the reactants. A 200 g Al ingot was melted in a graphite crucible in an electrical resistant furnace. When the temperature of molten Al was stable at 730 °C, solid Ti powders were added into the molten Al. In the mean time, high-intensity ultrasound was introduced in the melt by immersing a Nb radiator, as shown in Fig 1.The amount of solid Ti powders added into the Al melt corresponded to the composition of Al-3 wt.%Ti. During the reaction process, 4 ultrasonically treated samples were taken out from the melt after 1 min, 3 min, 5 min and 10 min. For simplicity, the four samples were denoted as US1, US3, US5 and US10, respectively. The power of ultrasonic generator was 1.5 kW, with the frequency of 20 kHz. The other 4 control samples without ultrasonic treatment were taken out from the melt after the same time intervals for making a comparison, which were denoted as CS1, CS3, CS5 and CS10, respectively.

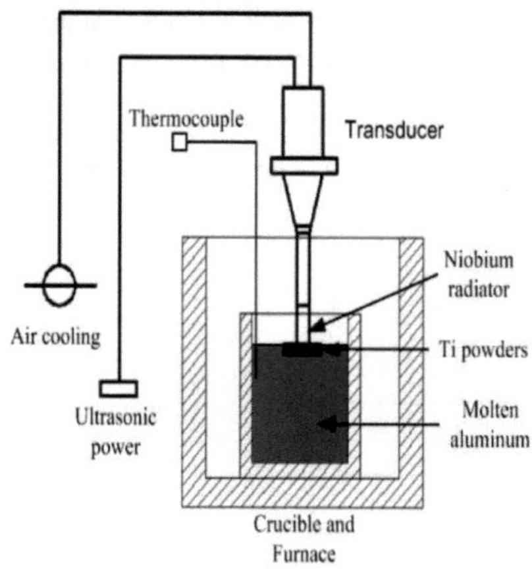

Figure 1. Schematic diagram of ultrasonic vibration assisted diffusion reaction between solid Ti powders and liquid Al.

Characterization

The phases in the samples were examined by X-ray diffraction (XRD, Bruker D8, Germany) by using Cu Kα radiation at 40 kV and 40 mA and a scan rate of 0.015 °/s. The microstructural features of the samples were observed by scanning electron microscopy (SEM s4800) equipped with Energy dispersive spectroscopy (EDS).

Results and Discussion

After adding solid Ti powders into molten Al at 730 °C, some sparks were observed on the surface of the Al melt, and the temperature of the melt increased a few degrees (almost in the range of 730-750 °C), indicating that exothermal reactions occurred between solid Ti powders and liquid Al. The related reaction equation was shown as follows.

$$3Al + Ti \rightarrow Al_3Ti \quad (1)$$

XRD Analysis

Figure 2 shows the XRD diffraction patterns of control samples (Figure 2a) and ultrasonically treated samples (Figure 2b), respectively. It is clear that Al_3Ti was the only newly formed phase in the reaction in both groups of samples. For control samples, the amount of in situ formed Al_3Ti phase increased as the reaction time prolonged according to the diffraction intensity of Al_3Ti phase (the intensity increased from CS1 to CS10). For ultrasonically treated samples, the amount of in situ formed Al_3Ti also increased with increasing reaction time. It was found that the diffraction intensities of Al3Ti phase in US5 and US10 samples were almost the same, indicating that solid Ti powders reacted with liquid Al completely to form Al_3Ti in 5 min. The degree of reaction between solid Ti powders and liquid Al in both groups of experiments could be observed directly through the following SEM test. In addition, comparing the two groups of samples with the same reaction time, one important feature was found that the amount of Al_3Ti phase in the ultrasonically treated sample was obviously higher than that in the control sample, showing that high-intensity ultrasound could promote the formation of Al_3Ti phase in the reaction.

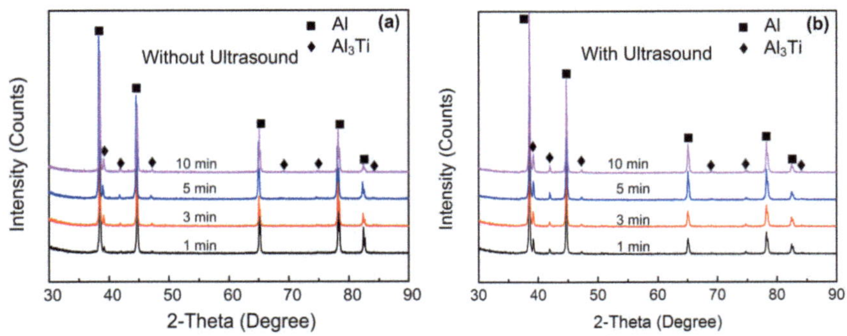

Figure 2. XRD patterns of samples with different reaction times:
(a) Control samples, and
(b) Ultrasonically treated samples.

SEM Analysis of Control Samples

Figure 3 shows the typical microstructures of control samples with different reaction times. The evolution of solid Ti powders in liquid Al was shown clearly by observing samples from CS1 to CS10. It is obvious that the solid Ti powders were unable to react with liquid Al completely in

10 min, for some Ti particles covered with reaction layers which were identified as Al_3Ti phase by EDS were found in the matrix after solidification. Also, the thickness of Al_3Ti reaction layers increased as the reaction time prolonged, which could reach around 5 μm, 10 μm, 15 μm and 25 μm in CS1, CS3, CS5 and CS10 samples, respectively. In addition, in situ formed Al_3Ti particles aggregated together around these Ti powders. As the reaction time increased, the amount of Al_3Ti particles increased, as shown in Figure 3b, c and d. This trend was in good agreement with the XRD results in Figure 2a.

Furthermore, based on the results of SEM, some important features about Al_3Ti reaction layers were also found in control samples. One was that the outside part of reaction layer was composed of by some loose Al_3Ti particles with weak bonds. Second one was that the size of Al_3Ti particles in the reaction layer became larger from inside to outside of the layer. Third one was that some inter-Al_3Ti particle spacings existed in the reaction layer. The above three features about reaction layers were shown clearly in Figure 3b, c and d, respectively. Also, it is clear that in situ formed Al_3Ti particles located in the matrix were blocky in morphology and were smaller than 5 μm in size in the four control samples. An important finding from control samples was that lower melting temperature can lead to the formation of blocky Al_3Ti particles with smaller size. The related formation mechanism of Al_3Ti particles will be discussed in detailed later in this paper.

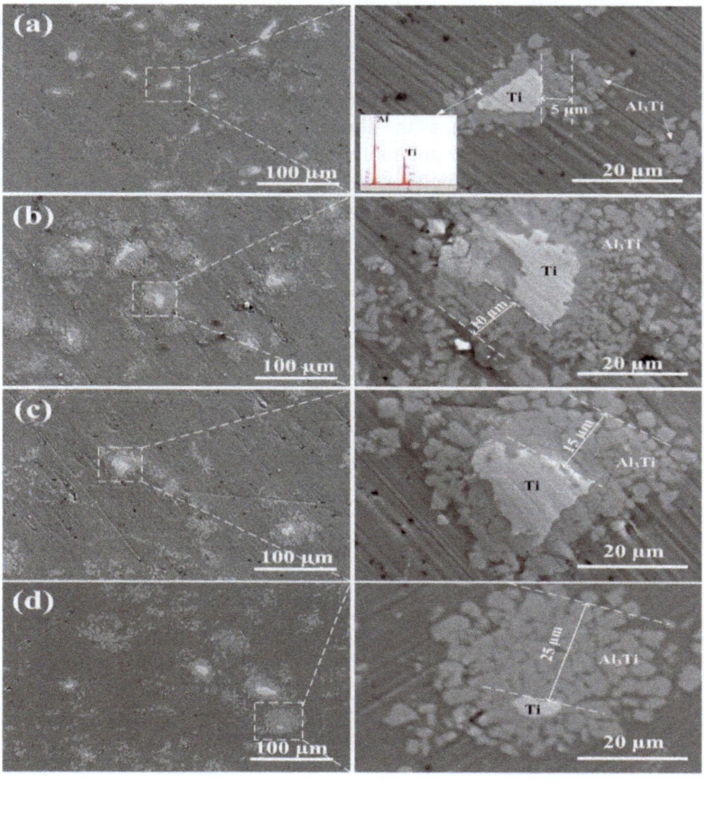

Figure 3. **Typical microstructures of the control samples with different reaction times and higher magnification of areas marked in images: (A) 1 min, (b) 3 min, (c) 5min, and (d) 10 min.**

3.3 SEM analysis of Ultrasonically Treated Samples

Similarly, Figure 4 shows the typical microstructures of ultrasonically treated samples with different reaction times. The effect of high-intensity ultrasound on the evolution of solid Ti in liquid Al was also shown clearly. After 1 min of ultrasonic treatment, a few of solid Ti powders covered by Al_3Ti layers existed in the matrix, and the thickness of the reactive layer was around 5 μm, as shown in Figure 4a. As the reaction time increased to 3 min, most of the solid Ti powders reacted with liquid Al completely, a very few of unreacted large-sized Ti powders covered with Al_3Ti layers with a thickness of around 5 μm were found in the Al matrix, as shown in Figure 4b. Moreover, it should be noted that the Al_3Ti layers in ultrasonically treated samples were gear-like in morphology (as shown in Figure 4a and b), suggesting that some outside Al_3Ti particles in reaction layer were peeled off from the reaction layer in the ultrasonic field.

After 5 min, all solid Ti powders reacted completely with liquid Al (Figure 4c), since no unreacted Ti particles were found in the Al matrix, indicating that high-intensity ultrasound was able to promote greatly the reaction between solid Ti powders and liquid Al. After 10 min, only in situ formed Al_3Ti particles existed in US10 sample (Figure 4d). This trend of the amount of in situ formed Al_3Ti phase in ultrasonically treated samples was also in good accordance with the XRD results in Figure 2b. In addition, as the ultrasonic processing time increased, in situ formed Al_3Ti particles were distributed more and more uniformly in the Al matrix due to the effect of ultrasonic stirring. Most of Al_3Ti particles synthesized in the four ultrasonically treated samples were blocky in shape, having the size smaller than 5 μm.

Overall, the reaction time between solid Ti powders and liquid Al could be shortened significantly by using high-intensity ultrasound comparing the two groups of samples. The related effects of ultrasound on the formation of Al_3Ti particles will be discussed in detail in the following section.

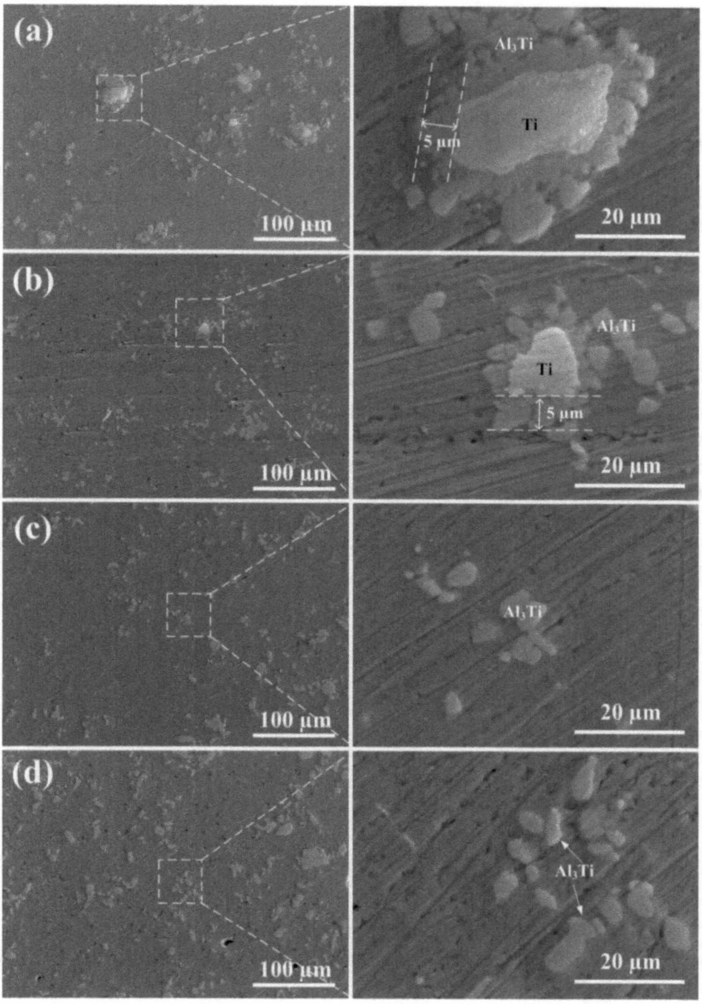

Figure 4. Typical microstructures of the ultrasonically treated samples with different reaction times and higher magnification of areas marked in images: (a) 1 min, (b) 3 min, (c) 5min, and (d) 10 min.

Formation Mechanism of Small Blocky Al_3Ti Particles

After solid Ti powders are added into the Al melt, a reactive diffusion can occur at the Ti/Al interface as soon as the solid Ti particle contacts the liquid Al. In reactive diffusion, the reaction rate for the formation of new phase can be evaluated by diffusion coefficients for the species. It is well known that the impurity diffusion coefficient in solids is usually described by the Arrhenius equation [20]:

$$D = D_0 \exp(-E_A / RT) \tag{2}$$

where D_0 is a temperature-independent pre-exponential (m^2/s), E_A the activation energy for diffusion (J/mol), R the gas constant, and its value 8.31 J/mol K; T the absolute temperature (K). Based on this empirical equation, the diffusion rates for Ti atoms to Al and Al atoms to Ti in Ti (solid)-Al (liquid) system can be approximately evaluated, and then the main diffusion phase can be defined. In the temperature range of 700-850 °C, the Arrhenius equation for the diffusion coefficients for Al in solid Ti ($D_{Al/Ti}$) [21] and for Ti in liquid Al($D_{Ti/Al}$) [22] can be expressed as the following equations, respectively.

$$D_{Al/Ti} = 9.58 \times 10^{-9} m^2/s \bullet \exp\left[\frac{-114600 J/mol}{8.31 J/molK \cdot T}\right] \tag{3}$$

$$D_{Ti/Al} = 4.29 \times 10^{-7} m^2/s \bullet \exp\left[\frac{-36300 J/mol}{8.31 J/molK \cdot T}\right] \tag{4}$$

Since some heat was released in the reaction, the temperature of the Al melt containing solid Ti powders was in the range of 730-750 °C, and the values of $D_{Al/Ti}$ and $D_{Ti/Al}$ are in the range of 1.0×10^{-14} to 1.35×10^{-14} m^2/s and 5.5×10^{-9} to 6.0×10^{-9} m^2/s, respectively, as shown in Figure 5 This result illustrates that solid Ti is the main diffusing species in the Al(l)-Ti(s) system, for the values of $D_{Ti/Al}$ are much greater than those of $D_{Al/Ti}$ by around five orders of magnitude. Thereby, the diffusion of Ti atoms into liquid Al (rich Al environment) leaded to the formation of Al_3Ti phase.

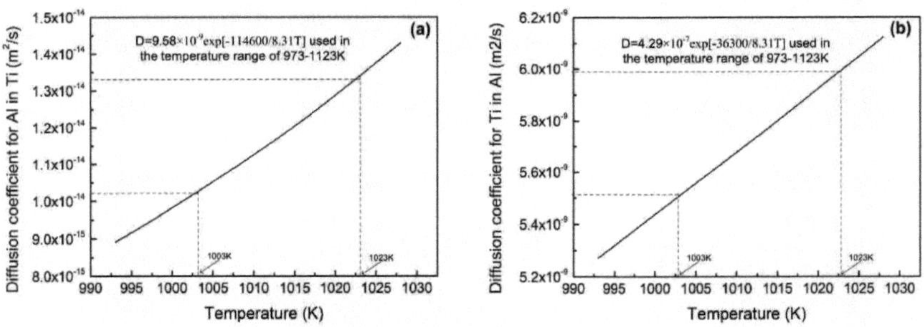

Figure 5. (a) **Diffusion coefficients for Al in Ti and**
(a) **diffusion coefficients for Ti in Al in the temperature range of 973-1123** K.

In the initial stage of diffusion of Al(l)-Ti(s) system, solid Ti powders totally exposure to liquid Al, and plenty of Ti atoms can cross Ti/Al interface into liquid Al in a very short time. According to the Al-Ti phase diagram[23] (as shown in Figure 6), It is clear that the solubility of Ti in liquid Al was rather low at 730 °C, which is less than 0.15 wt. %. As a result, the initial diffusion of Ti atoms across the Ti/Al interface into the liquid Al can produce a saturated solution adjacent to the interface. After that, the nucleation of Al_3Ti phase occurs, and these Al_3Ti nuclei attach to the surface of solid Ti particle.

As the reaction time increases, further diffusion of Ti atoms results in the growth of Al_3Ti particles on the solid Ti surface. Due to the spherical morphology of Ti particle, Al_3Ti particles grow along the radical direction rather than the planar direction. For Al_3Ti phase, it usually has the priority growth at <110> direction. As a result, the Al_3Ti reaction layer on the surface of Ti particle is not a compact layer and some inter-Al_3Ti particle spacings exist in the reaction layer due to the above growth characteristics of in situ Al_3Ti particles.

As the reaction time further prolongs, more Ti atoms are consumed to form Al_3Ti phase, and Al_3Ti particles grow larger. Thereby, for the reactive diffusion between solid Ti and liquid Al, it is a process of the decrease of solid Ti volume, but also a process of the increase of Al_3Ti volume. One important feature is that the volume of the Al_3Ti phase formed in the Al-Ti interface was greater than the consumption volume of Ti, which can be easily understood due to the lower density of Al_3Ti comparing with Ti. Also, Machowiak et al. [24] also reported the same finding in their research. This volume change easily gives rise to stresses in the reaction layer. During the growth of Al_3Ti phase, tensile stresses in the reaction layer develop parallel to the Ti surface. An easy rupture of Al_3Ti particles from reaction layer takes place when the tensile stresses become large enough. The existence of spacings between each Al_3Ti particles and the brittleness of Al_3Ti are attributed to the peeling process.

After the Al_3Ti particles peel from the reaction layer, the remaining Al_3Ti phase starts to grow following the above steps. The latter peeling particles push the earlier peeling particles away to outside. Thereby, a loose reaction layer composed of loose Al_3Ti particles can be formed.

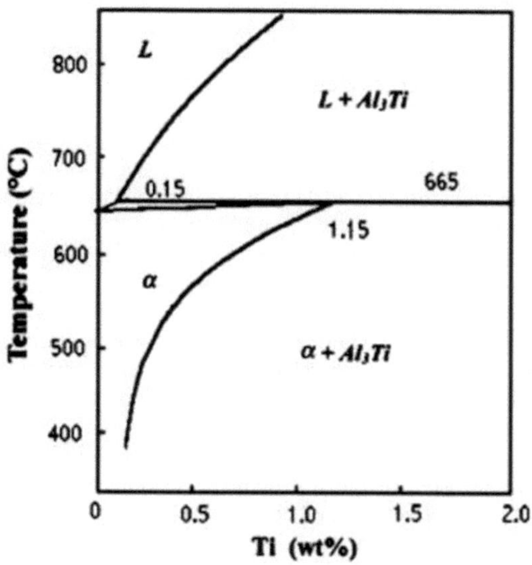

Figure 6. Al-Ti phase diagram (Al-rich corner).

Effects of Ultrasound on the Synthesis of In Situ Al_3Ti Particles

It is obvious that high-intensity ultrasound was able to promote the synthesis of Al_3Ti particles from the direct reaction between solid Ti powders and liquid Al. Based on the preceding analysis, the formation mechanism of in situ Al_3Ti particles at low melting temperature was explained by the reaction-peeling model.

When high-intensity ultrasound was introduced in the liquid-solid (powders) system, ultrasonic cavitation could take place in the Al melt due to the existence of liquid phase. A certain amount of small solid particles were included in the Al melt, which could be regarded as a suspension. Accordingly, some unique effects were generated in the suspension which was irradiated with ultrasound fields.

One was the instantaneously high pressure in the Al melt, the value of which could reach around 500 atmospheres. Once the high pressure was add on the Ti particle, the Al_3Ti particles on its surface would be peeled away more easily. On the other hand, due to the existence of shock waves, the ultrasonic irradiation of liquid-solid (powder) suspensions produced a very important effect: high velocity interparticle collisions. The shock waves could cause small particles to collide into one another with great force and the impact velocities range from 100-500 m/s for particles ~ 10 μm [17]. The interparticle collisions could introduce remarkable changes in surface morphology, composition and reactivity. In our research, a solid particle could collide with the other solid particle with high speed due to effect of ultrasound. Based on the momentum theorem, since the collision time was rather short, the force generated between the two particles would be extremely large. Thereby, the Al_3Ti reaction layer on the Ti surface could be destroyed effectively, leading to a thinner Al_3Ti reaction layer on the Ti surface. In addition, due to the viscosity of the Al melt, the high speed movement of Ti particles wrapped with Al_3Ti particles in the Al melt could produce shear force between the solid particles and liquid Al, which also contributed to the peeling of Al_3Ti particles. The accelerated peeling of Al_3Ti from the reaction layer in ultrasonic field can decrease the thickness of reaction layer, by which the contact of Ti atoms and Al atoms became much easier, and the reaction time was shortened remarkably.

Conclusions

This research studied the evolution of solid Ti powders in the Al melt at low melting temperature of 730 °C, in which the formation mechanism of blocky Al_3Ti particles was discussed, and the effects of high-intensity ultrasound on the evolution was also investigated. The following conclusions were drawn.

1. Small blocky Al_3Ti particles with the size smaller than 5 μm could be synthesized by adding solid Ti powders into the Al melt at 730 °C. A reaction-peeling model was proposed to explain the formation mechanism of Al_3Ti particles.

2. High-intensity ultrasound was able to shorten the reaction time significantly. Ultrasonic fields in the Al melt gave rise to instantaneous extreme high pressure, and this strong shear force could play effect on the peeling of Al_3Ti particles. On the other hand, high velocity interparticle collisions resulting from the collapse of ultrasonic cavitation bubbles also played an important effect on the peeling of Al_3Ti particles.

References

[1] J.M. Wu, S.L. Zheng, and Z.Z. Li, "Thermal Stability and its Effects on the Mechanical Properties of Rapidly Solidified Al-Ti alloys," *Materials Science and Engineering*, A289 (2000), 246-254.

[2] Q. Zhang, B.L. Xiao, W.G. Wang, and Z.Y. Ma, "Reactive Mechanism and Mechanical Properties of In Situ Composites Fabricated from an Al-TiO_2 System by Friction Stir Processing,"*Acta Materialia*, 60 (2012), 7090-7013.

[3] R.N. Wright, B.H. Rabin, and W.H. McFerran, "Combustion synthesis of cubic Al_3Ti alloys,"*Journal of Materials Research*, 7 (1992), 2733-2738.

[4] V.A. Chianeh, H.R.M. Hosseini, M. Nofar, "Micro structural features and mechanical properties of Al-Al_3Ti composite fabricated by in-situ powder metallurgy route," *Journal of Alloys and Compounds*, 473 (2009), 127-132.

[5] Y. Watanabe, N. Yamanaka, and Y. Fukui, "Wear Behavior of Al-Al_3Ti Composite Manufactured by a Centrifugal Method," *Metallurgical and Materials Transactions*, A30 (1999), 3253-3261.

[6] S.H. Wang, P.W. Kao, and C.P. Chang, "The Strengthening Effect of Al_3Ti in Ultrafine Grained Al-Al_3Ti Alloys," *Scripta Materilia*, 40 (1999), 289-295.

[7] X.M. Wang, A. Jha, and R. Brydson, "In Situ Fabrication of Al3Ti Particle Reinforced Aluminium Alloy Metal-matrix Composites," *Materials Science and Engineering*, A364 (2004), 339-345.

[8] T.J. Chen, J. Li, H. Yuan, "Casting Fabrication of In Situ Al_3Ti-Al Composites and Their Wear Behaviors," *China Foundry*, 6 (2009), 319-327.

[9] W.Q. Jie, E.G. Kandalova, R.J. Zhang, V.I. Nikitin, "Al_3Ti/Al Composites Prepared by SHS," *Rare Metal Materials and Engineering*, 29 (2000), 145-148.

[10] K.S. Suslick, *Ultrasound: Its Chemical, Physical, and Biological Effects* (New York, NY: Wiley-VCH, 1998).

[11] T.J. Mason and J.P. Lorimer, *Applied Sonochemistry: The uses of Power Ultrasound in Chemistry and Processing*, (Weinheim: Wiley-VCH, 2002).

[12] H.X. Xu, B.W. Zeiger, and K.S. Suslick, "Sonochemical Synthesis of Nanomaterials," *Chemical Society Reviews*, 42 (2013), 2555-2567.

[13] B.W. Zeiger, K.S. Suslick, "Sonofragmentation of Molecular Crystals," *Journal of American. Chemical Society*, 133 (2011), 14530-14533.

[14] W.B. McNamara III, Y. Didenko, and K.S. Suslick, "Sonoluminescence Temperatures during Multi-bebble Cavitation," *Nature*, 401 (1999), 772-775.

[15] K.S. Suslick, G.J. Price, "Application of Ultrasound to Materials Chemistry," *Annual Review of Materials Science*, 29 (1999), 295-326.

[16] S. Merouani, O. Hamdaoui, Y. Rezgui, and M. Guemini, "Theoretical Estimation of the Temperature and Pressure within Collapsing Acoustical Bubbles," *Ultrasonics Sonochemistry*, 21 (2014), 53-59.

[17] S. J Doktycz, and K. S. Suslick, "Interparticle Collisions Driven by Ultrasound", *Science,* 247 (1990), 1067-1069.

[18] K. S. Suslick, "Sonochemistry," *Science,* 247 (1990), 1439-1445.

[19] K. S. Suslick, and G. J. Price, "Applications of ultrasound to materials chemistry *Annual Review of Materials Science*,29 (1999), 295-326.

[20] Z.W. Liu, M. Rakita, Q.Y. Han, and J.G. Li, "A Developed Method for Fabricating In-Situ TiC_p/Mg Composites by Using Quick Preheating Treatment and Ultrasonic Vibration", *Metallurgical and Materials Transactions*, A43 (2012), 2116-2124.

[21] M. Thuillard, L.T. Tran, M.A. Nicolet, "Al_3Ti Formation by Diffusion of Aluminum through Titanium", *Thin Solid Films*, 166 (1988), 21-27.

[22] Y. Du, Y.A. Chang, B.Y. Huang, W.P. Gong, Z.P. Jin, H.H. Xu, Z.H. Yuan, Y. Liu, Y.H. He, and F.Y. Xie, "Diffusion Coefficients of Some Solutes in fcc and Liquid Al: Critical Evaluation and Correlation" *Materials Science and Engineering*, A363 (2003), 140-51.

[23] G.K. Sigworth, "The Grain Refining of Aluminum and Phase Relationships in the Al-Ti-B System", *Metallurgical Transactions*, A15 (1984), 277-282.

[24] J. Mackowiak, L.L. Shreir, "The Nature and Growth of Interaction Layers Formed during the Reaction between Solid Titanium and Liquid Aluminum," *Journal of the Less Common Metals*, 1 (1959), 456-466.

Advanced Composites for Aerospace, Marine, and Land Applications II
Edited by: T. Sano and T.S. Srivatsan
TMS (The Minerals, Metals & Materials Society), 2015

METAL MATRIX COMPOSITES DIRECTIONALLY SOLIDIFIED

Alicia Esther Ares [1,2] and **Carlos Enrique Schvezov** [1,2]

[1] *Materials Institute of Misiones*
IMAM (CONICET-UNaM)
University of Misiones
1552 Azara Street, Posadas
Misiones, 3300 **Argentina.**

[2] Member of CIC
National Research Council (**CONICET**)
Argentina.

Keywords: Metal Matrix Composites, Zinc-Aluminum Matrix, Properties, Characterization.

Abstract

The present work is focus on studying the dendritic solidification of metal matrix composites, MMCs, (using zinc-aluminum, ZA, alloys as matrix and the addition of SiC and Al_2O_3 particles). The compounds were obtained by as-cast solidification, under continuous stirring and in a second stage were directionally solidified in order to obtain different dendritic growth (columnar, equiaxed and columnar-to-equiaxed transition (CET)). The results in MMCs were compared with those obtained in directional solidification of ZA alloys, primarily with regard to structural parameters. The size and evolution of microstructure, according to the size of the MMCs particles and the variation of the thermal parameters was analyzing. In general it was found that the size of the microstructure (secondary dendritic spacing) decreases with the increase of particles in the matrix. When cooling rate increases, particle size decreases, and a higher cooling rate causes finer and more homogeneous dendrites Also, the segregation which was found in the matrix of the composites was significantly less than in the case of ZA alloys.

1. Introduction

Zinc-based alloys have been found suitable for a number of engineering applications in which they perform better than aluminium alloys, cast iron and bronzes, there has been a growing interest in the use of such alloys as matrix for the production of MMCs [1–4].
The nature of the interface between the matrix and the additional material (particles) has an important role in defining the characteristics of the composite, particularly the coherence between the matrix and SiC particles and in the preparation of composites [5-7].
Solidification properties are described by measuring secondary dendrite arm spacing (SDAS), which is the most important metallurgical factor that affects the mechanical properties of the materials. A decrease in the SDAS value increases the ultimate tensile strength and elongation. SDAS is controlled by the solidification time [8-17].

The presence of SiC particles in the ZA alloys increases the value of the elastic modulus and the hardness. Adding ceramic particles improves the strength and stiffness, but reduces the ductility. The wear resistance of the alloy increases with the volume fraction of the reinforcement [17]. In previous research of our group, the as-cast and directional solidification of MMCs was analyzed [18-22].

In the present research, because there is no sufficient experimental data about the columnar-to-equiaxed transition phenomenon on metal matrix composites (MMCs) directionally solidified in the literature, in the present research the CET was analyzed in ZA-27 alloys reinforced with SiC particles.

2. Experimental Procedure

The casting alloys (Zn-27wt.%Al) were prepared from Zn 99.998 wt.% and aluminum 99.999 wt.%, and composites by using the same alloys as matrix and adding SiC and Al_2O_3 particles in different quantities in volume percent. The alloys and composites were directionally solidified and K-type thermocouples, protected with ceramic shields, were longitudinally arranged using to measure thermal profiles during the growth [18-22]. The samples were melted in ceramic molds. The solidification time was approximately 100 s. After solidification, the samples were cut and polished with emery paper up to 1000 grit and then using 1 μm diamond paste. Samples were etched for approximately 10-20 s, at room temperature (20-25 °C) in a solution of chromic acid (i. e., 50 g Cr_2O_3; 4 g Na_2SO_4 in 100 ml of water [23]). The position of the CET was determined by observation under Arcano® and Olympus® optical microscopes. The microstructure was analized using SEM and an image analyzer. The distributions of elements in the microstructure were determined using EDXA. The compositions of the samples which are considered in this report are shown in Table I.

The secondary dendrite spacings were measured in the cross sections of the other half of the samples near the thermocouple positions using an image analyzer (Neophot). Each section was mounted, polished and etched and the spacing was determined by the number of intercepts along a straight line.

Table I. Samples composition.

#	Description
1	ZA27+7wt pctSi+2.5 wt pct Cu
2	ZA27+3.5wt pctSi+1 wt pct Cu
3	ZA27 + 5 vol pct SiC
4	ZA27 + 8 vol pct SiC
5	ZA27 + 3.5 wt pct Si+1wt pct Cu + 16 vol pct SiC
6	ZA27 + 3.5 wt pct Si+1wt pct Cu + 16 vol pct Al_2O_3

3. Results and Discussion

The solidification microstructure of the ZA27 alloy without an alloying element presented a dendritic structure consisting of primary dendrites rich in aluminum, evidenced by the EDXA shown in Figures 1 and Figure 2. The $\alpha+\eta$ eutectoid is formed from α dendrites and β peritectic through a transformation at 275°C, following the Al-Zn phase diagram (Figure 3) and formed during the final stage of solidification. This $\alpha+\eta$ eutectoid has a typical platelike form of α and η sheets. The X-ray analysis shows that the α and η phases consist of an Al (α) phase with an fcc structure and lattice parameter a = (4.0474 ± 0.0002) Å and a zinc (η) phase with an hcp structure and lattice parameters a = (2.6655 ± 0.0001) Å and c = (4.9494 ± 0.0003) Å . In addition it is observed a standard and finer eutectic structure.

In the macroscale, the aluminum concentration increases in the direction of solidification from the bottom to top of the samples, see Figure 4. This macrosegregation was previously reported, and qualitatively explained on the base of the phase diagram alloy [17].

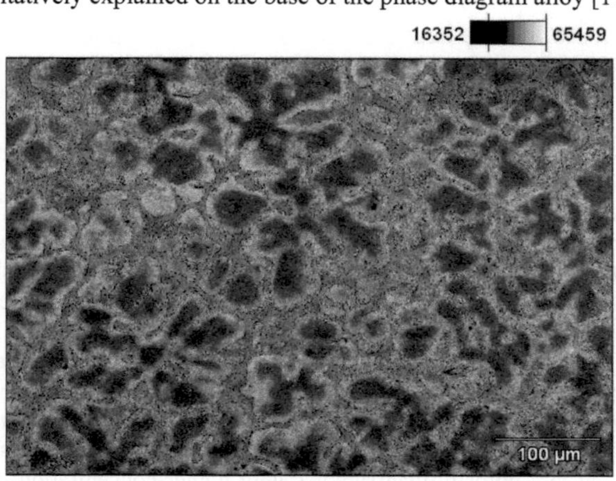

Figure 1. **Microstructure of ZA27 alloys: Primary dendrite surrounded by a laminated eutectic $(\alpha+\eta)$ with typical plate like form of the $(\alpha+\eta)$ sheets in the $(\alpha+\eta)$ eutectoid.**

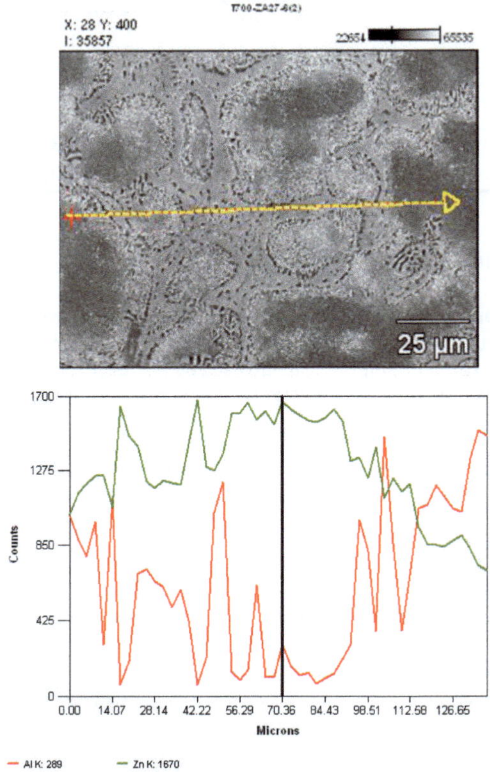

Figure 2. Microstructure of ZA27 alloy and Zn and Al distribution.

The addition of silicon to ZA27 alloy does not change the morphology of the alloy. Silicon appears in the microstructure in two forms: in solution and as precipitates. The precipitates appear mostly in the dendritic cores, as shown in Figure 5 (a). This Figure shows an EDXA of the silicon precipitates indicated in the same figure. The presence of precipitates inside the dendrites and the clustering show that precipitates forms in the melt and the dendrites possibly nucleate on them, in agreement with the ternary phase diagram for the three elements, Zn-Al-Si. As was reported before, same degree of porosity is present in the sample, which increases from the bottom to top of the casting following the increase in number of silicon precipitates [17]. There is a qualitative correlation of porosity with silicon content since samples with lower silicon concentration present less porosity than samples with higher concentration. Composition maps for copper in the alloys show that for a nominal concentration of 1 pct and below, copper is present nearly homogenously distributed in solid solution.

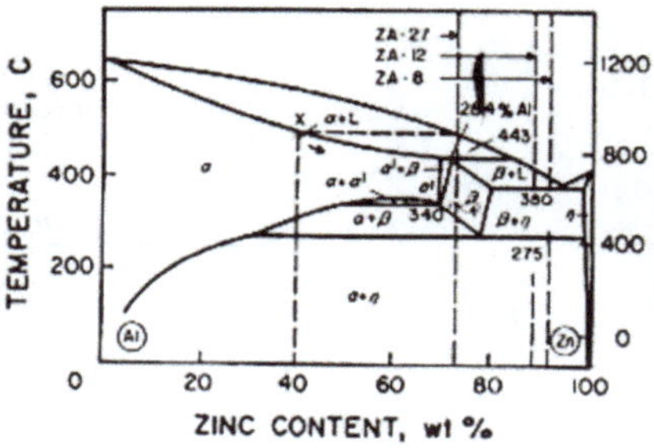

Figure 3. Phase diagram of Zn-Al alloy Alloy [24]

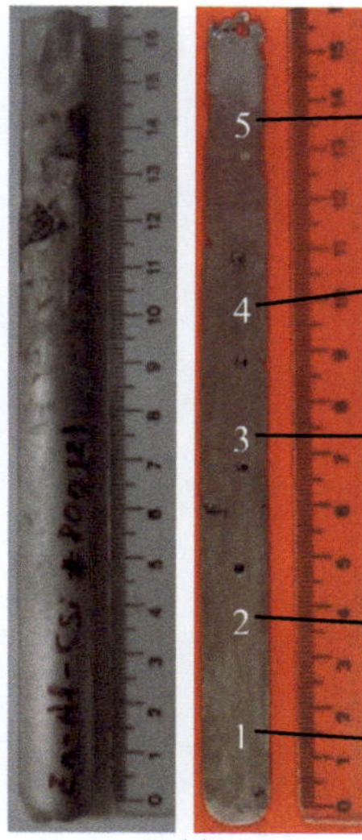

(a)

Section 5

elemento	Mass %
Zn	62.75
Al	32.17
Si	3.85
Cu	1.23

Section 4

elemento	Mass %
Zn	64.53
Al	30.64
Si	3.71
Cu	1.12

Section 3

elemento	Mass %
Zn	66.35
Al	28.92
Si	3.65
Cu	1.08

Section 2

elemento	Mass %
Zn	67.87
Al	27.56
Si	3.56
Cu	1.01

Section 1

Element	Mass %
Zn	69.57
Al	26.36
Si	2.968
Cu	0.985

(b)

Figure 4. (a) Solidified sample 5. (b) EDXA analysis.

The effect of adding particles in the microstructure of ZA27 alloy is to reduce the grain size as the volume fraction of the particle content increases. The particles appear mostly in the interdendritic regions as opposed to the silicon precipitates, as shown in Figure 5 (a). The silicon carbide particles are clearly distinguished from the silicon precipitates in two aspects: on one hand, the precipitates have a light gray color, whereas the particles have a dark gray color; on the other hand, the precipitates have irregular contours, whereas the particles have sharp corners. The average particle size was measured, and an average distribution of 5.27 ± 1.07 μm was found in the reinforced alloys. Further characterization of the reinforced alloys can be found somewhere else [17-22].

The distribution of precipitates increases in the direction of solidification from bottom to top of the sample. The effect of reinforcement by SiC particles addition in ZA27 is made by analyzing the results obtained with one sample. In the first place the measured volume fraction at from locating in sample is not uniform increasing from the bottom to the top of the sample indicating that some flotation has occurred, see Table II. The actual particle size distribution is shown in Figures 6 at the form locations considered. It is observed that the distribution is similar and that there are a larger number of the bigger particles at the top which is consisted with a flotation process. Contrary to what is observed with the silicon precipitates (appear mostly in the dendritic cores, as shown in Figure 6 (b.1 to b.3)), the silicon carbide particles appear mostly in the interdendritic regions (Figures 6 (c.1) to (c.3)). The alloy present porosities, the porosity normally occurs in the low portion of the cast known as underside shrinkage defects (USD) [9-22], see Figures 6 (d.1) to (d.3).

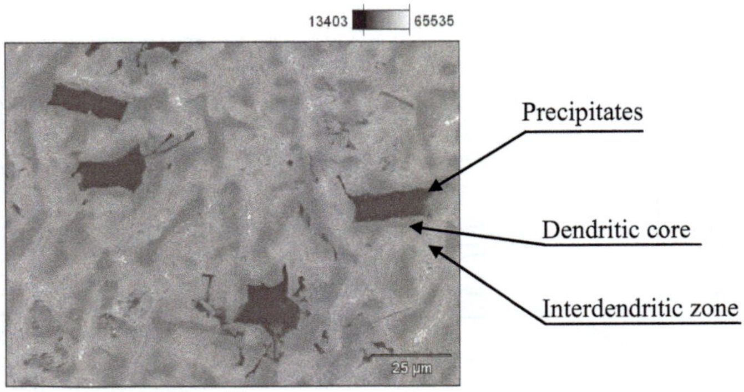

(a)

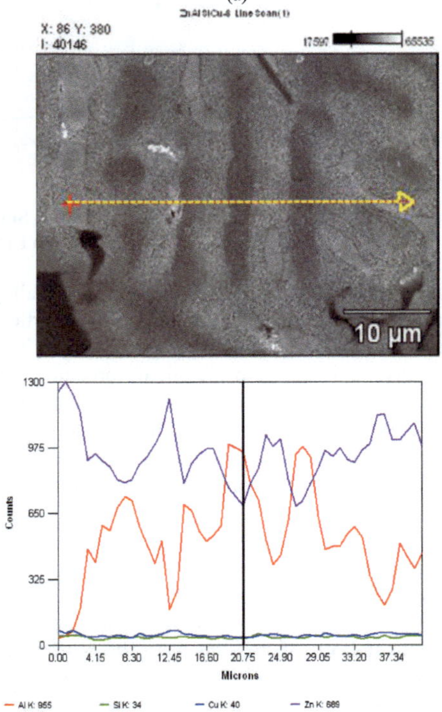

(b)

Figure 4. **(a)** **Microstructure showing silicon precipitates.**
(b) **Mapping the elements. ZA27+7wt pctSi+2.5 wt pct Cu composite.**

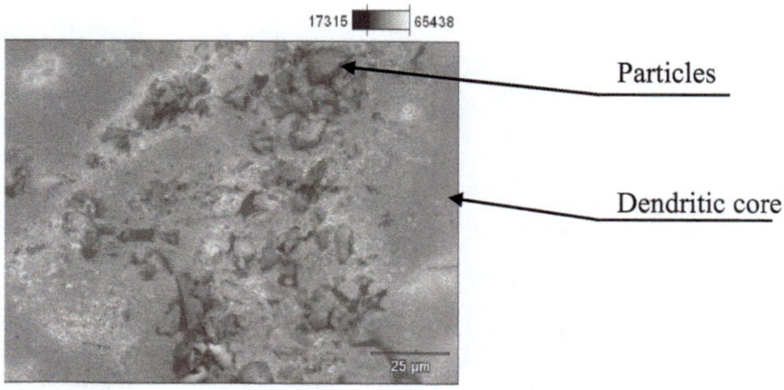

(a)

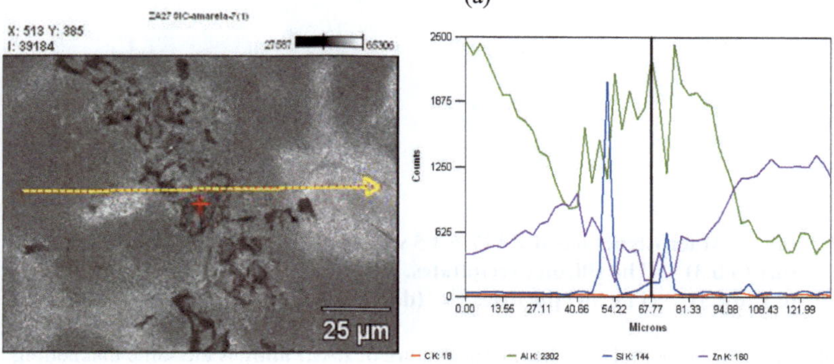

(b)

Figure 5. **(a)** **Microstructure showing SiC particles.**
(b) **Mapping the elements. ZA27 + 3.5 wt pct Si+1wt pct Cu + 16 vol pct SiC.**

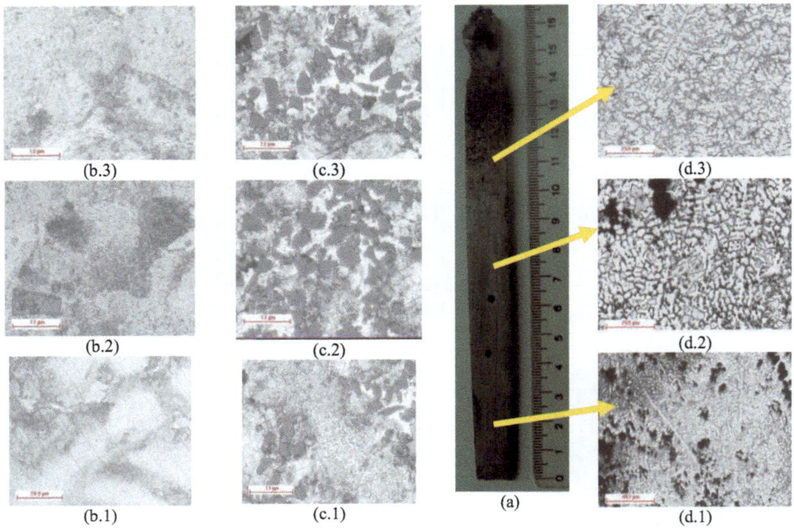

Figure 6. (a) Macrostructure of ZA27 + 3.5 wt pct Si+1wt pct Cu + 16 vol pct SiC. (b.1 to b.3) The silicon precipitates. (c.1 to c.3) The SiC particles. (d.1 to d.3) Porosity effect.

The effect of adding Al_2O_3 particles in the microstructure of ZA27 alloy is the same that adding SiC particles. The particles appear mostly in the interdendritic regions as shown in Figure 7. The Al_2O_3 particles are clearly distinguished from the dendrite structure; the precipitates have a dark gray color. The average particle size was measured, and an average distribution of 4.57 ± 1.05 μm was found in the reinforced alloys.

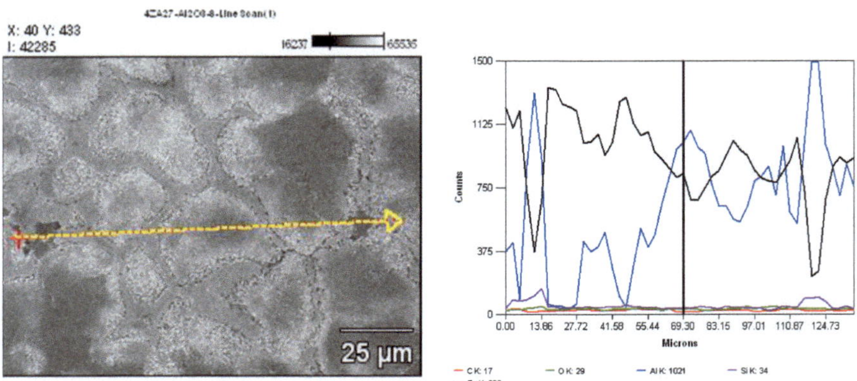

Figure 7. Microstructure and elements distribution of ZA27 Alloy + 16 vol pct Al_2O_3(p).

In position 1 there is no particles higher than 10.5 μm, whereas in the position 2, the size distribution is between 1.75 and 14 μm. In position 3 is possible appreciate particles of 3,51 μm of average diameter (this is about two times with respect to the other positions). Finally in position 4 (near de base of the sample the size of particles reaches 15.8 μm In Table II is presenting the data corresponding to metallographic studies, that is, position, grain size (Abrams), interdendritic volume fraction, particle average diameter (μm), average grain size (μm), in each position.

The grain size decreases with increasing the volume fraction of the particles (Figure 9). The results indicate that the presence of particles comes have an effect on solidification by decreasing the size of the structure. In the present case the variation in size may not be attributed to an increase in solidification velocities since the first portion that solidify in the bottom part which presents the largest size. Moreover, this is indicating that the particles could have a stronger effect on solidification structure.

Table II. **Distribution of SiC particles and microstructure corresponding to ZA27 + 3.5 wt pct Si+1wt pct Cu + 16 vol pct SiC sample.**

Position	Abrams (ASTM N°)	Interdendritic Volume Fraction	Particle Average Diameter (um)	Grain Size Average (um)
1	5,1	28,18	4,00	47.02
2	5,2	36,27	5,37	36,64
3	4,86	29,14	4,87	43,22
4	5,1	27,66	6,35	37,55

Since the dendrite arm spacing is directly related to the rate of solidification, secondary dendrite arm spacings (SDAS) were measured to determine the effect of solidification conditions on the microstructure of composites.

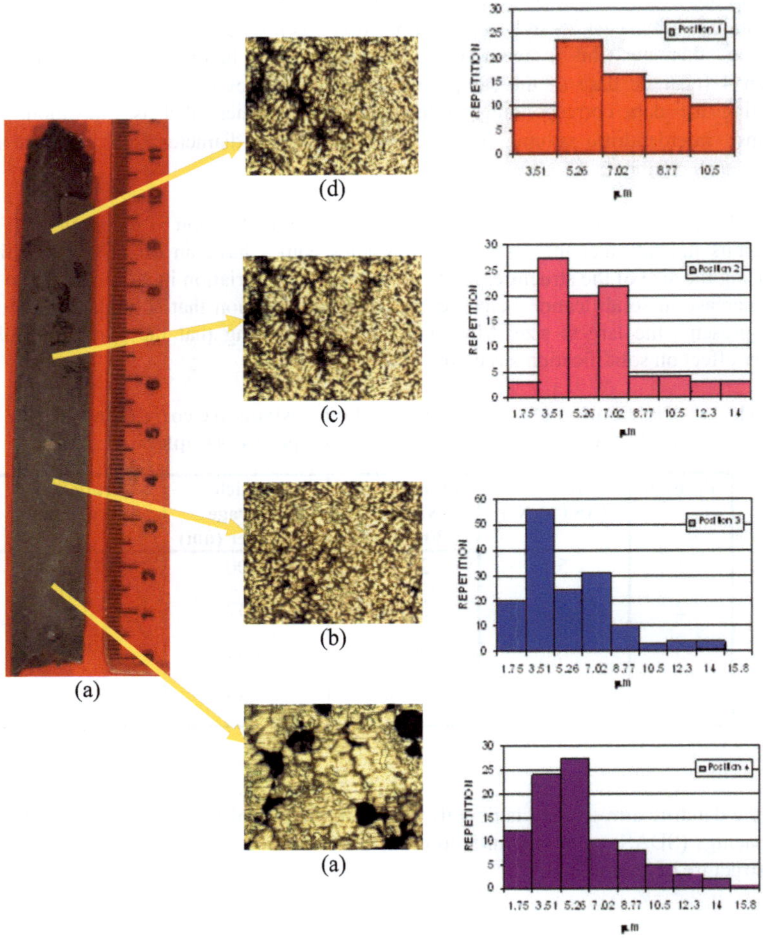

Figure 8. Distribution of particles in function of position and size. ZA27 + 5 vol pct SiC sample.

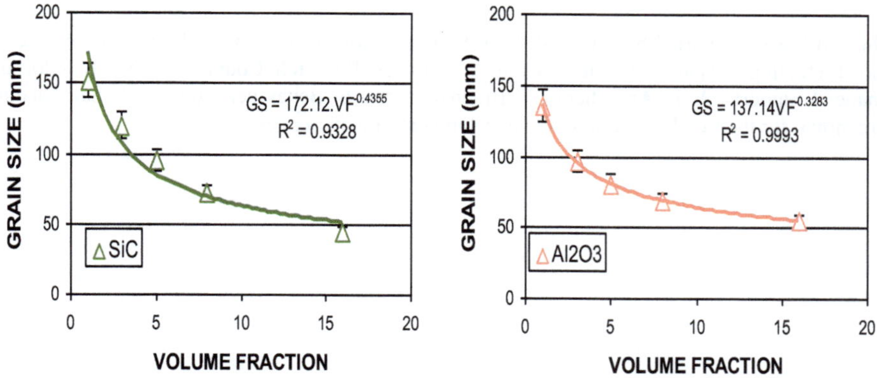

Figure 9. Grain size versus volume fraction. (a) SiC. (b) Al_2O_3.

If the cooling rate increases, $\dot{T}$, causes finer and more homogeneous dendrites as can be seen in Figure 10 (a) and (b) for both types of particles and three different cooling rates.

4. Conclusions

From the results of the previous sections the main conclusions of this investigation on the columnar to equiaxed transition in composites are:

1. The Zinc-Aluminum alloy reinforced with SiC and Al_2O_3 particles were successfully produced by solidification process.
2. The size and evolution of microstructure according to the size of the MMCs particles and the variation of the thermal parameters were analyzed. In general, it was found that the size of the microstructure (secondary dendritic arm spacing) decreases with the increase of particles in the matrix. Also, the segregation which was found in the matrix of the composites was significantly less than in the case of ZA27 alloys.
3. The presence of particles comes have an effect on solidification by decreasing the size of the structure.
4. Considering the variation in cooling rate, if the cooling rate increases causes finer and more homogeneous dendrites.

Acknowledgements

This work was supported by PICT-2011-1378 of the National Agency for Promotion of Science and Technology. Thanks are due to the Argentinean Research Council (CONICET) for the financial support. A. E. Ares thanks E. Brandaleze and E. Benavidez, from UTN San Nicolas, Argentina, for their collaboration in the development of experiences.

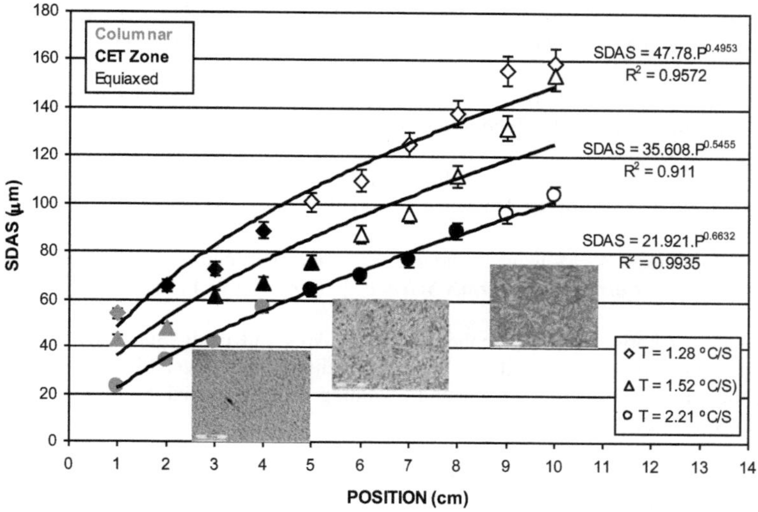

(a)

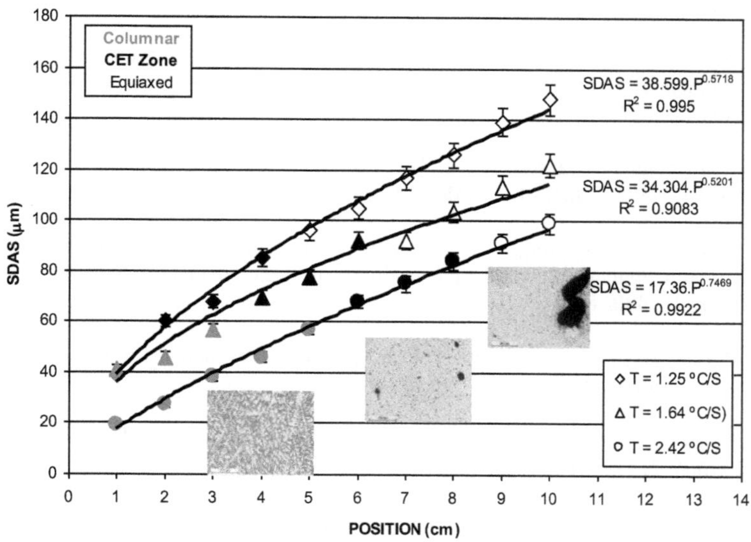

(b)

Figure 10. **SDAS evolution versus position for three different cooling rates. (a) SiC. (b) Al_2O_3.**

References

1. N. Gupta, K. G. Satyanarayana and C. Materials, Symposium Review: Solidification Processing of MMCs, ***Journal of Materials Science***, 58 (2006) 91-93.

2. D. Miracle, Metal Matrix Composites—From Science to Technological Significance, ***Composites Science and Technology***, 65 (2005) 2526-2540.

3. A. Evans, C. Marchi and A. Mortensen, Metal Matrix Composites in Industry: An Introduction and a Survey. Kluwer Academic Publisher, Dordrecht (2003).

4. I.A. Ibrahim, F.A. Mohamed, and E.J. Lavernia, Particulate Reinforced Metal Matrix Composites. ***Journal of Materials Science***, 26 (1991) 1137-1156.

5. E. M. Agaliotis; M. R. Rosenberger; A. E. Ares and C. E. Schvezov, Modelling the Interaction of Particles with a Concave Solidifying Interface, MCWASP International Conference – Modeling of Casting, Welding and Advanced Solidification Processes (2009) 209-217.

6. E. M. Agaliotis; M. R. Rosenberger; A. E. Ares and C. E. Schvezov, Numerical Modeling of the Interaction between Foreign Particle and Solidifying Crystalline Interface, In: CFD Modeling and Simulation in Materials Processing, Editors: L. Nastac, L. Zhang, B. G. Thomas, A. Sabau, N. El-Kaddah, A. C. Powell, H. Combeau. John Wiley & Sons, Inc. (2012) p. 171-178.

7. E. M. Agaliotis, M. R. Rosenberger, A. E. Ares y C. E. Schvezov, Analysis of the Interaction of Particles with a Non-Planar Solidifying Interface, TMS Light Metals, (2013) 823-827.

8. R. Yamanoglu, M. Zeren, R. M. German, Solidification Characteristics of Atomized AlCu4Mg1-SiC Composite Powders, ***J. Min. Metall. Sect.*** B-Metall. 48 (1) B (2012) 73 – 79.

9. M. Sahoo; L. V. Whiting; D. W. G. White, Control of Underside Shrinkage in Zinc – Aluminum Foundry Alloys by the Addition of Trace Elements, ***AFS Transactions*** 93 (1985) 475- 480.

10. P. P. Lee, T. Savaskan, E. Laufer, Wear resistance and microstructure of Zn–Al–Si and Zn–Al–Cu alloys, ***Wear*** 117 (1987) 79 - 89.

11. S. Murphy, Solid-phase reactions in the low-copper part of the Al–Cu–Zn system, ***Z. Metallkde*** 71 (1980) 96-102.

12. J.A: Cornie, R. Guerriero, L. Meregalli and I. Tangerini, in Proceedings of Cast Reinforced Metal Composites Conference, Chicago, 24-30 September 1988, edited by S. G. Fishman and A. K. Dhingra, p. 155.

13. N. Karni, G. B. Barkay; M. Bamberger, Structure and properties of metal-matrix composite, ***Journal of Materials Science Letters*** 13 (1994) 541-544.

14. S.C. Tjong and F. Chen, Wear Behavior of As-Cast ZnAl27/SiC Particulate Metal-Matrix Composites under Lubricated Sliding Condition, ***Metallurgical and Transactions*** 28A (1997) 1951-1955.

15. H. X. Zhu and S. K. Liu, Mechanical Properties of Squeeze Cast Zinc Alloy Matrix Composites Containing α-Alumina Fibers, ***Composites*** 24 (1993) 437-442.

16. S.C. Sharma; B.M. Girish; Rathnakar Kamath, B.M. Satish, S.C. Sharma; B.M. Girish; Rathnakar Kamath, B.M. Satish. ***Wear*** 213 (1997) 33-40, Wear 213 (1997) 33-40.

17. R. Auras and C. Schvezov, Wear behavior, microstructure, and dimensional stability of as-cast zinc-aluminum/SIC (metal matrix composites) alloys, ***Metallurgical and Transactions*** 35A. (2004) 1579-1590.

18. A. E. Ares; R. C. E. Schvezov, Solidification Structures and Properties of Zinc-Aluminum /SiC (MMC) Alloys, TMS Annual Meeting (2006) 183-193.

19. A.E. Ares and C.E. Schvezov, Experimental Study of the Columnar-to-Equiaxed Transition During Directional Solidification of Zinc-Aluminum Alloys and Composites, ***World Journal of Engineering*** (WJOE), 7 (2010) 13-29.

20. A. E. Ares, L. M. Gassa, C. M. Mendez, Corrosion Resistance of Directionally Solidified Casting Zinc-Aluminum Matrix, In: Corrosion Resistance, Chapter 2, Pages: 35-54, Editor: Hong Shih. INTECH Publishers, (2012).

21. A.E. Ares and C.E. Schvezov, Columnar-to-Equiaxed Transition in Metal Matrix Composites Reinforced with Silicon Carbide Particles, Journal of Metallurgy, 2013 (2013), Article ID 628495, 12 pages, http://dx.doi.org/10.1155/2013/628495, Hindawi Publishing Corporation.

22. A.E. Ares and C.E. Schvezov, Metal Matrix Composites Directionally Solidified, in Proceedings of Advanced Composites for Aerospace, Marine, and Land Applications, 2014 TMS Annual Meeting, edited by T. Sano, M. Peretti, T. Srivatsan, p.115-124.

23. G.F. Vander Voort, Metallography, Principles and Practice, Editorial ASM International (1984).

24. Binary Alloy Phase Diagram, Thaddeus B. Massalski ed., ASM, Metlas Park, OH, c1986.

Advanced Composites for Aerospace, Marine, and Land Applications II
Edited by: T. Sano and T.S. Srivatsan
TMS (The Minerals, Metals & Materials Society), 2015

NANOPARTICULATE REINFORCED ALUMINUM ALLOY COMPOSITES PRODUCED BY POWDER METALLURGY ROUTE.

Kaspar Kallip[1, 2], Lauri Kollo[2], Marc Leparoux[1], Christopher Bradbury[1]

[1]EMPA Swiss Federal Laboratories for Materials Science and Technology
Feuerwerkstrasse 39
Thun CH-3602
Switzerland

[2]Tallinn University of Technology
Ehitajate-Tee 5
Tallinn EE-19086
Estonia

Keywords: MMC, CNT, Powder Metallurgy, Aluminum.

Abstract

High energy planetary ball-milling was used to effectively disperse 3, 6 and 9 wt. % multiwall carbon nanotubes (MW-CNTs) into commercially available aluminum alloys (Al6061, AlMg5, S250 and S790). Composite bulks were manufactured by uniaxial hot pressing. For the Al6061- CNT composites, standard heat treatments (T4, T5 and T6) were performed and their influence on the structural evolution (grain coarsening, CNT reaction) and hardness was recorded. Scanning Electron Microscopy (SEM), X-Ray Diffraction (XRD) and Raman spectroscopy were used to characterize the produced composites. The study shows that CNTs can be effectively mixed with high-strength aluminum alloys. Up to 5 fold increase in hardness was achieved compared to unreinforced alloys ranging up to 390 HV20 for the S250 alloy with 6 wt. % of MW- CNTs. The applied standard heat treatments did not lead to any improvements of the mechanical properties. The developed nanocomposite materials could find applications where high hardness of aluminum is needed, or in functionally graded composites.

1. Introduction

The need for high strength, lightweight materials has encouraged pursuing studies on aluminum composite materials technology. This is evident in fields such as renewable energy, aerospace, electronics and alternative transport, where metal matrix composites (MMCs) have found usage [1]. Aluminum alloys, due to their good specific strength, have been extensively investigated for increasing their strength by dispersing nanofibers and nanoparticles into their ductile matrices by the means of mechanical alloying [2]. Since the unprecedented properties of carbon nanotubes (CNTs) were reported by Iijima in 1991 [3], the CNTs have been investigated as reinforcement for MMCs [1].

To harvest the highest properties of Al matrix and CNT reinforcement, homogeneous distribution of reinforcement and adequate interfacial bonding are essential. Initially short ball- milling times (5-10 min) were used in order to avoid destroying CNTs [4. 5]. As an alternative method rubber assisted dispersion of CNTs has been introduced [6]. However longer milling times were needed to increase the strength of MMC, as shown by Choi et al., who have carried out milling optimization for CNT reinforced aluminum matrix composites. Authors have stated that insufficient milling intensity, i.e. speed, and lead to the formation of CNT layers on top of the aluminum grains. These CNTs are however more prone to be damaged during milling. Extensive milling intensity induces a covering of the CNTs with ductile aluminum. In this case, the soft matrix provides a protection from direct shocks of milling balls and the CNTs are less damaged [7]. The damaging of the CNTs during milling has also been discussed by Liu et al. who showed degradation of CNTs during prolonged milling using Raman spectroscopy [8]. In compliance with extended milling times, increased reinforcement content has also been investigated. Esawi et al. have successfully dispersed 5 wt.% CNTs into a pure aluminum matrix using high energy milling and hot extrusion and thus achieving an ultimate tensile strength up to 250 MPa [9]. Bradbury et al. have even gone further being able to disperse up to 9 wt. % CNTs into an aluminum matrix using high energy planetary ball-milling. The authors reported up to 4 fold increase in hardness (up to 150 HV20) using 6 wt. % of CNTs in a pure aluminum matrix [10]. The work of Bradbury et al. was taken as a baseline in the current work to test the strengthening ability of large amounts of CNT reinforcements in the case of selected aluminum alloys.

This work then explores the MW-CNT reinforcement of aluminum alloy matrices by powder metallurgy routes involving high energy planetary ball milling and uniaxial hot compaction.

2. Experimental

2.1 Starting materials

Al-powder (ECKA Atomized Aluminium Powder Co) with a purity level of 99.5% [DIN EN 576] had a mean particle size smaller than 63 μm (Fig. 1A). The 6061 alloy (AlMg1SiCu) with mean particle size below 45 μm, was provided by TLS Technik GmbH & Co. (Fig. 1B). The AlMg5 alloy (Al5019) provided by ECKA Atomized Aluminium Powder

Co had a mean particle size smaller than 63 μm (Fig. 1C). The S790 (7xxx series) alloy (AlZn11Mg2Cu) (Fig. 1D) as well as the S250 (4xxx series) alloy (AlSi20Fe5Ni2) (Fig.1E) were both provided by Peak Werkstoff GmbH.

The multiwall CNTs (MW-CNTs) supplied by Bayer Material Science GmbH under brand name of Baytubes® C150P had a purity over 95% (Fig. 1F). These CNTs have a length of 1-10 μm and an average density of 1.8 g/cm^3. They were supplied as agglomerates with sizes ranging between 0.1 and 1 mm. A detailed analysis of the Baytubes CNTs has been conducted by Tessonnier et al. [11].

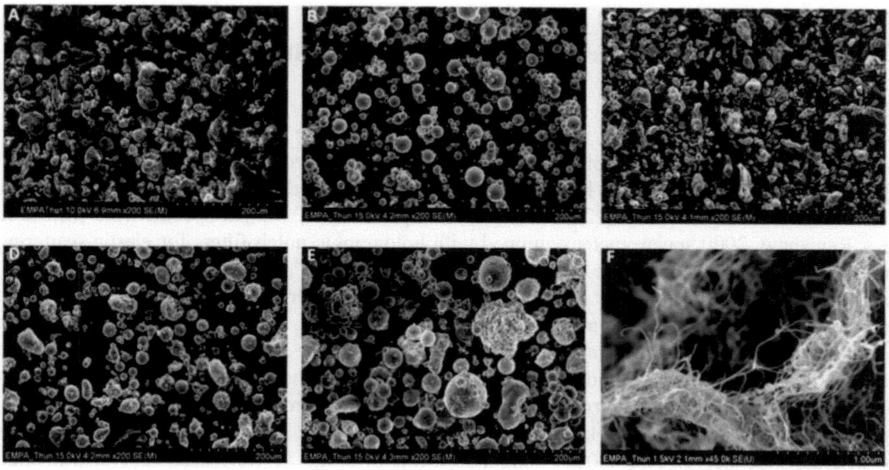

Figure 1. Raw materials. A) Aluminum pure, B) AL6061, C) AlMg5, D) S250, E) S790, F) Baytubes C150P MW-CNTs.

2.2 Materials processing

A laboratory scale planetary ball mill (Retsch PM 400) was used to disperse different amounts of CNTs into aluminum alloys. Milling was conducted at a vial rotational speed of 360 rpm for 6h and 20 wt. % of heptane was added as process control agent. Further details on milling parameters are described in [12]. Uniaxial hot pressing under air was used with pressure of 570 MPa to produce samples with diameter of 30 mm and height of about 6 mm. Three standard heat treatments were performed on pure Al6061 alloy and Al6061 alloy based nanocomposites [13]. T4 treatment; 1h solution phase in α zone at 530°C followed by immediate water quenching and natural aging at room temperature for at least one week. T5 treatment; compacting of MMC and artificial aging for 8h at 177°C. T6 treatment; 1h solution phase at 530°C followed with immediate water quenching and artificial aging for 8h at 177 °C, followed with air cooling to room temperature. (Fig. 2)

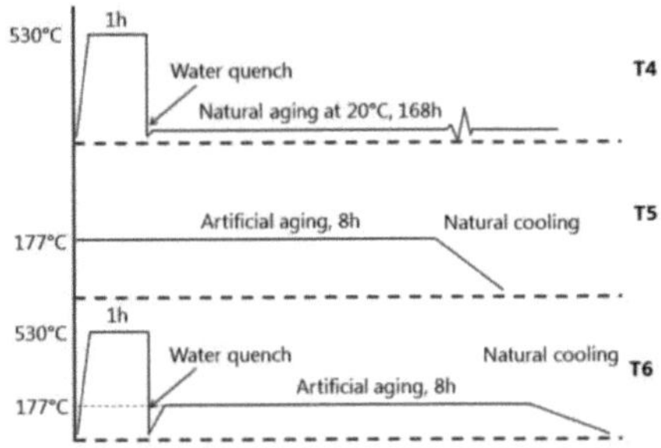

Figure 2. Heat treatments carried out for Al6061 matrix composites.

2.3 Characterizations

The density of the samples was measured by Archimedes' method according to ISO 3369:1975. The Vickers macro hardness (HV20) was measured according to EN ISO 6507-1 with a load of 20 kg for 15 s (220, GNEHM Härteprüfer AG). Raman spectra were measured on a Renishaw 2000 spectrometer equipped with holographic notch filters for elastic scattering and a CCD array detector. The samples were excited with a helium–neon laser (632.8 nm) with the laser focused onto the sample using the 50× lens through microscope. Each sample was measured in twelve different places and the spectrum was corrected for the baseline. The instrument was calibrated with a Si single crystal. The electron microscopy images were taken with a Hitachi S-4800 cold field emission scanning electron microscope (CFE-SEM). The X-ray diffractometer (XRD) pattern was obtained with X'Pert Pro diffractometer (PANAlytical) using Cu Kα ($\lambda = 1.5148$ Å, 35 kV and 40 mA) in the 2θ range 25–140° using a linear detector (X'Celerator). A step size of 0.0167° and a scan rate of 0.021 °/s were used. The instrumental line broadening was determined with an annealed sample of Y2O3 (Alfa Aesar, purity 99.999%). The Rietveld refinement was done using a program named Maud, developed by Prof. Lutterotti at the University of Torento [14].

3. Results

Four commercially available aluminum alloys were chosen for the study. The results can then be compared to the previous study performed in our group on pure aluminum matrices [10]. All the alloys were milled and compacted under similar conditions that proved to be adequate for achieving dense bulk material. Indeed, all the samples possessed a density higher than 92% of the theoretical density. In figure 3A hardness changes can be observed as a function of the reinforcement weight fraction.

The highest hardness increase was observed with Al6061. Fivefold increase in hardness was obtained when adding only 3 wt. % of CNTs, reaching 230 HV20 compared to about 40-50 HV20 for the unreinforced alloy. Increasing further the reinforcement content did not lead to higher mechanical properties of the MMC, and the density even decreased slightly. Therefore for Al6061, a maximum amount of 3 wt. % CNTs could be successfully dispersed. Fourfold hardness increase was noted with AlMg5 alloy based MMCs and the hardness increased up to 260 HV20. However in the case of AlMg5 MMCs, a more pronounced decrease in hardness was noted when increasing the CNT content over 3 wt.%. The highest hardness of the present study was achieved with the S250 alloy reaching up to 390 HV20 for 6 wt. % CNT. However the starting alloy having already 175 HV20, so the growth factor of hardness was the lowest (Fig. 3A). Among the investigated parameters, the MMCs produced with AlMg5 and Al6061 matrices possessed their highest hardness with 3 wt. % CNT reinforcement. Using higher content led to decrease of hardness up to 6% and 14% from maximum values in the case of Al6061 and AlMg5 respectively. With composites based on S250 and S790 matrices the hardness decrease for 9 wt. % CNT was more evident being 22% and 35% lower than the maximum values respectively.

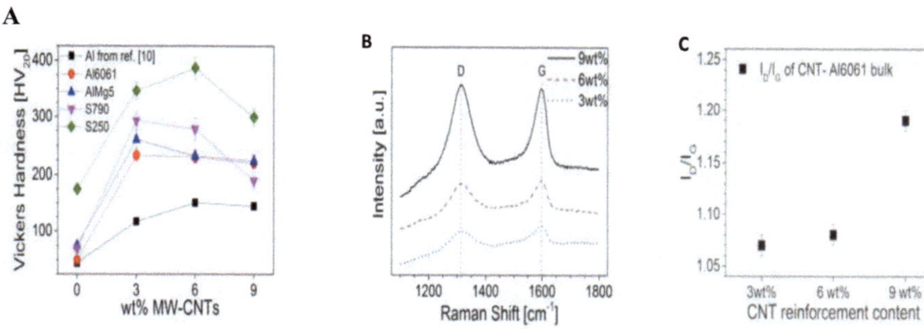

Figure 3. A) Hardness of MW-CNT reinforced Aluminum alloys.
B) Raman Spectra of Al6061 matrix MMCs.
C) Intensity ratio ID/IG of Al6061 matrix MMCs.

All four alloys could be reinforced with CNTs leading to significant hardness increases however, in order to focus the study, only Al6061 matrix composites were further investigated. The hot pressed CNT-Al6061 bulks were investigated by Raman spectroscopy. Typical spectra for the Al6061 composites can be seen in Fig. 3B. The spectra show two distinctive bands D (Disordered) and G (Graphite) at 1310 cm^{-1} and 1597 cm^{-1} respectively. The D-band seems to increase with the CNT content indicating a higher content of damaged CNTs. This is confirmed by the evolution of the integrated intensity ratio between the D and G-band in Fig. 3C. The ID/IG ratio increases with increasing CNT content especially at the highest investigated CNT content, 9 wt.% .

In the following, the influence of the hot pressing temperature was investigated. Thus the increase of the compaction temperature from 350°C to 450°C resulted in higher hardness and density of the MMCs (Fig. 4A). However, using 450°C induced the formation of Aluminum carbide (Al4C3) as shown in Fig.4B.

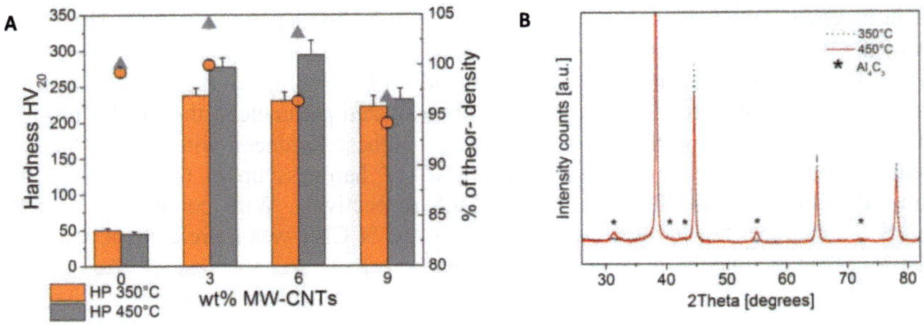

Figure 4. **A)** **Influence of compaction temperature on hardness and density of Al6061 based MMCs.**
B) **XRD indicating Aluminum carbide (Al_4C_3) formation on 6 wt.% CNT reinforced Al6061 MMC after increasing compaction temperature to 450°C.**

The 6xxx series alloys are normally heat treated to improve their mechanical performances. Thus standard heat treatments T4, T6 and T5 were performed on the investigated Al6061 matrix nanocomposites. Figure 5A describes the hardness response after these heat treatments. No noticeable increase was observed. On the contrary, for 3 and 6 wt. % CNT, the heat treatments (T4 and T6) lead to even a decrease of the composite hardness. For 9 wt. % CNT, none of the heat treatments used changed the hardness of composite. Furthermore, the relatively high temperature used during solution phase heat treatment (T4 and T6) seemed to contribute to the formation of Al_4C_3 (Fig. 5B).

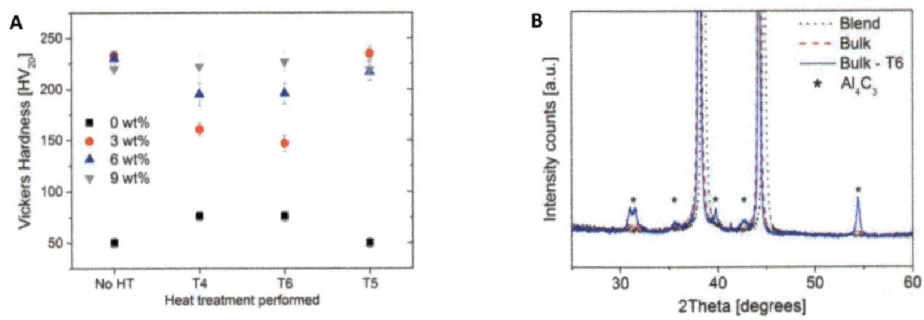

Figure 5. **A)** **Hardness of Al6061 matrix MMCs regarding different heat treatments.**
B) **XRD pattern of 3wt% CNT reinforced Al6061 depending on different heat treatments.**

Rietveld refinement was performed to assess the crystallite size of heat treated CNT-Al6061 MMCs. For solution phase heat treatments only T6 will be presented as there was no significant difference on crystallite size between T4 and T6 treatments. The crystallite size increased while using high temperature solution phase heat treatment for all the reinforcement quantities (Fig. 6).

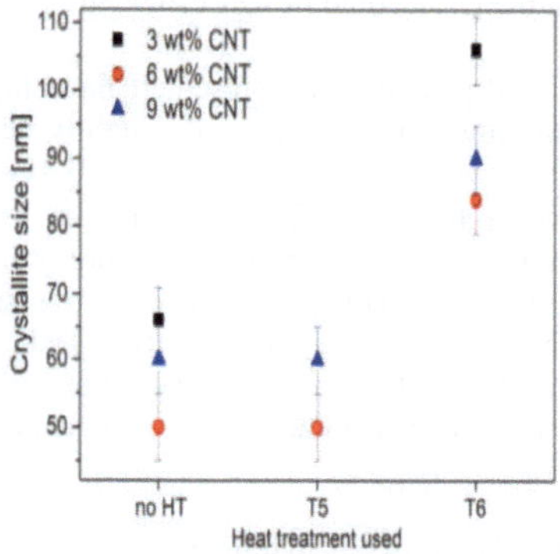

Figure 6. Calculated crystallite size of Al6061 matrix CNT composite with different heat treatments used.

4. Discussion

A powder metallurgy route was used to disperse high quantities (3, 6, 9 wt. %) of MW-CNTs into 4 different aluminum alloys. A previous work on dispersing similar contents of CNTs into pure aluminum matrix carried out by Bradbury et al. [10] showed high hardness values and capability of dispersing up to 9 wt. % CNTs into Aluminum matrix. The macro hardness of the Al-6 wt. % CNT composite increased more than threefold compared to the unreinforced Aluminum bulk compacted under the same conditions. A further increase of the CNT content up to 9 wt.% did not lead to higher hardness, on the contrary the hardness decreased.

Similar trend was observed with the investigated aluminum alloys in this present study. All the alloys milled and compacted under similar conditions produced dense bulk samples and their hardness increased by a factor of 2 to 5 compared to the respective pure unmilled alloy. A preliminary short investigation regarding the milling time showed that no significant improvement of the hardness response is observed after 4-6 hours. Therefore in the present study, the milling time was shortened from 20 to 6 hours. The highest hardness increase was observed for the Al6061 based composite while only 3 wt. % CNTs still increased the hardness up to 230 HV20 that is fivefold over the unmilled alloy (Fig. 3A).

The CNTs were damaged during processing as can be deduced from the relatively high Disorder peaks seen on Raman spectra (Fig. 3B) and also by the increase of the ID/IG ratio

(Fig. 3C). The principal mechanism of damaging the CNT is due to the high energy ball impacts during milling. Higher content of CNTs will lead to more damaged CNTs. On the opposite, Bradbury et al. [10] showed that the ID/IG intensity ratio was almost constant for CNT content between 3 and 9 wt. % in aluminum. They then suggested that the CNTs are being covered by the soft aluminum during milling and this layer protect the CNT from further damages as also described by Choi et al. [7]. In present work, the raw alloys are significantly harder than pure aluminum and the coverage of the CNT could be less efficient.

Three alloys Al6061, AlMg5 and S790 presented their highest hardness for 3 wt. % CNT. At higher CNT concentrations, the hardness only slightly decreased. For the S250 alloy, the highest hardness was obtained with 6 wt. % CNTs (390 HV20), as in the case of pure aluminum matrices [10]. However for higher CNT contents in the S250 alloy, a considerable decrease in hardness was observed while in the case of pure aluminum the hardness did not decrease so much for 9 wt. % CNT. This could be due to insufficient milling energy for dispersing high content of CNT into S250-s hard matrix. This content is also the highest refer in the literature as far as the authors know. Therefore a further study is ongoing to optimize the milling parameters for dispersing higher content of reinforcement materials into hard aluminum alloy matrices.

In order to optimize the hot pressing parameters, higher compacting temperature was investigated only with the Al6061 matrix. Increase in hardness and density of MMCs was noticed while increasing compaction temperature from 350 to 450°C (Fig. 4A). However using 450°C induced the formation of Al_4C_3 (Fig. 4B) that could lead to unwanted brittleness or corrosion issues of the produced MMC. Therefore 350°C was selected for further investigations on the aluminum alloys.

The 6xxx series alloys are normally heat treated to improve their mechanical performances. However, no hardness improvement was observed after heat treatments on our Al6061-CNT composites. On the opposite, for lower reinforcement contents (3 wt. % and 6 wt. %), the hardness even decreased after solution phase heat treatments. For 9 wt. % reinforcement, the heat treatments had almost no influence on the mechanical response. It has been demonstrated that the aluminum crystallite size increased after T4 and T6 treatments. The treatment temperature 530°C is even higher than the compaction temperature. Considering the Hall-Petch effect, it is not surprising that the hardness decreases after the heat treatments. For high CNT concentration, i.e. 9 wt. %, even if the crystallite size increases after T4 and T6, the hardness remains constant. No explanation could be proposed so far and further investigations, especially structural characterizations are necessary. The lower temperature artificial aging (T5) did not induce any change neither of the crystallite size nor of the hardness whatever the reinforcement fractions (Fig. 5A).

5. Conclusion

High content of MW-CNTs were successfully dispersed in 4 different aluminum alloys by high energy planetary ball milling. The resulting composite materials are 3 to 5 times harder than the pure alloy compacted under the same conditions.

The influence of hot pressing temperature was investigated. The higher pressing temperature of 450°C led to higher hardness and density of MMC but also enhanced the formation of Al_4C_3.

The standard heat treatments (T4, T5, and T6) performed typically on Al6061 was found to have no positive influence on the hardness of the investigated nanocomposites. On the contrary in the case of 3 wt. % and 6 wt. % reinforcement the hardness decreased after solution phase heat treatments (T4 and T6) possibly due to crystallite size growth induced by high temperature. For 9 wt. %, the hardness remained constant at a high level after the high temperature treatments even if the crystallite size increased. This requires more investigations but could offer new opportunities for "high" temperature applications.

References

1. "Metal matrix composites" G. Ibe (Ed.), Ullmann's Encyclopedia of Industrial Chemistry, Wiley-VCH, Weinheim (2003)

2. Suryanarayana, C., Nasser Al-Aqeeli. ***Progress in Materials Science*** 58, no. 4 (2013): 383-502.

3 S. Iijima, Nature, 354 (1991), pp. 56–58

4. R. George, K.T. Kashyap, R. Raw, S. Yamdagni, ***Scr. Mater.***, 53 (2005), pp. 1159–1163

5. C.F. Deng, D.Z. Wang, X.X. Zhang, A.B. Li, ***Mater. Sci. Eng., A***, 444 (2007), pp. 138–145

6. H. Kwon, A. Kawasaki, ***J. Nanosci. Nanotechnol.***, 9 (2009), pp. 6542–6548

7. Choi, H. J., J. H. Shin, and D. H. Bae. *Composites Part A: Applied Science and Manufacturing* 43.7 (2012): 1061-1072.

8. Z.Y. Liu, S.J. Xu, B.L. Xiao, P. Xue, W.G. Wang, Z.Y. Ma, **Composites Part** A, 43.12, (2012), P 2161-2168,

9. Esawi, A. M. K., Morsi, K., Sayed, A., Taher, M., & Lanka, S. (2010). ***Composites Science and Technology***, 70(16), 2237-2241.

10. Bradbury, C. R., Gomon, J. K., Kollo, L., Kwon, H., & Leparoux, M. (2014). ***Journal of Alloys and Compounds***, 585, 362-367.

11. Tessonnier, J. P., Rosenthal, D., Hansen, T. W., Hess, C., Schuster, M. E., Blume, R., Schlögl, R. (2009). ***Carbon,*** 47(7), 1779-1798.

12. Kollo, L., Leparoux, M., Bradbury, C. R., Jäggi, C., Carreño-Morelli, E., & Rodríguez-Arbaizar, M. (2010. ***Journal of Alloys and Compounds***, 489(2), 394-400.

13. Handbook, ASM Speciality. "Aluminum and aluminum alloys." ASM International (1993)

14. Lutterotti, L. (2010). Nuclear Instruments and Methods in ***Physics Research Section B***: Beam Interactions with Materials and Atoms, 268(3), 334-340.

GROWTH KINETICS OF MAGNESIO-ALUMINATE SPINEL IN Al/Mg LAMELLAR COMPOSITE INTERFACE

Yasser Fouad[1]
Bakr Mohamed Rabeeh[2]

[1]- Associate Professor, Engineering and Materials Science,
German University in Cairo, GUC.
[2]- Professor, Engineering and Materials Science,
German University in Cairo, GUC.

Keywords: Diffusion couples, Spinel, HIPing, Hybrid, Composite; interface, interphase, micro-laminated, Nano composite.

Abstract

The synthesis of Mg-Al_2O_3 double layered interface is introduced via the application of hot isostatic pressing, HIPing, in Al-Mg foils. Polycrystalline spinel layers are grown experimentally at the interfacial contacts between Al-Mg foils. The growth behavior of the spinel layers along with the kinetic parameters characterizing interface motion and long-range diffusion is established. Low melting depressant (LMD), Zn, and alloying element segregation tends to form micro laminated and/or Nano structure interphase in a lamellar composite solid state processing. Nano composite ceramic interphase materials offer interesting mechanical properties not achievable in other materials, such as superplastic flow and metal-like machinability. Microstructural characterization, mechanical characterization is also established via optical microscopy scanning electron microscopy, energy dispersive X-ray spectroscopy and tensile testing. Chemical and mechanical bonding via inter diffusion processing with alloy segregation are dominant for interphase kinetics. Mechanical characterization with interfacial shear strength is also introduced. HIPing processing is successfully applied on 6082 Al-alloy and AZ31 magnesium alloy for either particulate or micro-laminated interfacial composite processing. The interphase kinetic established through localized micro plasticity, metal flow, alloy segregation and delocalized Al oxide and Mg oxide. The kinetic of interface/interphase induce new nontraditional crack mitigation a long with new bridging and toughening mechanisms.

Introduction

Mg alloy along with Aluminum alloy is being introduced for a lamellar composite structure processing that recommended for many applications. Mg alloys are being increasingly employed in the aerospace, automobile and 3C (Computer, Communication, Consumer electrics products) industries as important structural materials because of their low density, high strength-to-weight ratio, good damping characteristics and low casting costs, etc.[1]. One of the major drawbacks of magnesium alloys in many applications is their low corrosion resistance [2,3] reported that introducing an outer ply of aluminum alloy could improve the corrosion resistance of the magnesium substrate effectively, while maintaining the low density and high specific strength of the assembly. Al/Mg/Al foils are symmetrically hot isostatic pressed, HIPed at different parametric study. Temperature, pressure as well as holding time are being controlled and studied for lamellar metal matrix composite processing. The Al/Mg-based laminated composites can be comprised to meet a wide variety of demands. Temperature ranges are selected from 450°C to 560°C for different holding times 10, 20 and 30 minutes in a working chamber of 30 Ton capacity hydraulic press. However, the bonding processes mentioned above require special facilities, such as vacuum system, complex equipment or high-cost instrument, etc. The present work introduced with new emerged facility that induce metal flow, air bleeding as well as diffusion processing [4-10]. The kinetic of interphase is also introduced with scanning electron microscopy along with energy dispersive X-ray spectroscopy and mechanical characterization. Hot Isostatic Pressing is now being applied to fabricate Al (6082)/Mg–3Al–1Zn (AZ31) laminated composites as an emerging method for bonding of dissimilar materials in a solid state processing. In particular, Bakr et al. demonstrated the feasibility to produce a lamellar MMC with controlled interphase/interface by hipping method [11]. The kinetic of interface/interphase at varying experimental settings (load, temperature and experiment duration) is established. Interfacial twining and Microstructural and textural evolution of magnesio-aluminate spinel reaction are formed. Deformation fracturing and twinning of constituent interface, low melting phase segregation, along with oxidation induce the formation of localized magnesio-aluminate structure. This internal deformation of constituent interface disturbed the reaction interphase and introduced localized zones of concentrated stress and directional solidification in a semisolid state [12-13]. Toughening mechanisms are introduced along interphase by delocalized Nano particulate of ceramic phase (Alumina and magnesia). The kinetic of ceramic phase that hinder dislocations and slipping induce crack mitigation and toughening mechanisms. The fabrications of Al 6082-AZ31-Al 6082intermetallic laminated composites combining using Hipping techniques and their influence on the microstructures, phase constitutions, the joining interfaces of composites and the bonding strength and tensile strength of the fabricated laminated composite are introduced.

Experimental Procedures

The extrinsically modified Al/Mg alloys are introduced via HIPing at different parametric study. Chemical compositions of the intrinsically alloyed monolithic materials; Al 6082 and Mg Az31 are presented in table 1.

Table 1. Chemical Composition of Al 6082 and AZ31 Magnesium Alloy.

Alloys	Cu	**Zn**	Mn	Si	Mg	Cr	Fe	Al
Al 6082	0.04	0.02	0.05	1.92	0.79	0.15	0.24	Bal.
						Ca	**Ni**	**Al**
AZ31	< 0.05	1.05	0.20	< 0.15	Bal.	< 0.04	< 0.005	3.10

Al and Mg alloy are prepared as thin foil/sheet of 3 mm thickness and different dimensions of 100X50 mm (lengths x width). A hydraulic press with hot working chamber is applied with special fixture that promotes metal flow and inters diffusion kinetics. Al/Mg foil/foil technique symmetrically HIPed at different parametric study of temperature 450°C to 640°C, and press of 30 Ton, with holding time varied from 10 to 60 minutes. Different HIPed samples are tested for microstructural observation to investigate bonding characteristics, interphase kinetics as well as delamination behavior. Scanning electron microscopy and energy dispersive X-ray spectroscopy is applied for microstructural characterization and chemical analysis. The best results of parametric study are applied for tensile test samples. Mechanical characterization is applied for both composite and its constituents. Al alloy, Mg alloy as well as Al/Mg metal matrix composite (MMC) samples are separately tested using servo hydraulic tensile testing machine. The kinetic of interface/interphase is established monitoring its micro constituents.

Results and Discussion

Interfacial bonding of laminated composite materials is an interesting point of study. Al/Mg/Al lamellar structure is introduced by Hot Isostatic Pressing, HIPing, in three symmetrically applied plies for structural applications. A new emerging solid state processing is introduced to study the kinetic of interface / interphase under HIPing conditions. Temperature, pressure and holding time are controlled with microstructural analysis and EDX to introduce interfacial behavior and mechanical characterization. The kinetic of interphase in a lamellar composite structure introduces in steps of formulation with microstructural observations along with alloying elements effect that induce either micro-laminated structure or delocalized oxides that induce toughening mechanisms. However, delamination is a major limitation in a lamellar composite structure; this work introduces a mitigation behavior of delamination along with new introduced toughening mechanisms along interphase/interface. Extremities are captured along with the commencement of delocalized liquid phase in a semisolid reaction.

Compared with other bonding methods, the bonding strength of the experimental laminates prepared by hipping bonding was medium due to delamination. For diffusion bonding, the maximum bonding strength of pure Al/pure Mg laminate with or without interlayer was 86.1 MPa[4] and 22.7 MPa[6], respectively; and that of 2024 Al alloy/AZ31B Mg alloy and 6061 Al alloy/AZ31BMg alloy was 56 MPa[8] and 41.3 MPa[4], respectively. The maximum bonding strength of 7075 Al alloy/AZ31B Mg alloy fabricated by explosive welding was 70 MPa [10]. The maximum bonding strength of Al 6082/AZ31/ Al 6082 laminated composite fabricated by hipping in this work was 95.4MPa. The bonding strength depends on the thermo mechanical processing conditions of the laminated composite. Mechanical bonding occurs instantaneously or over a very short time period and depends on the forces of attraction between the atoms. It is believed that not only mechanical bond played a major role in the bonding strength in this research but also kinetic of interphase/interface. As interfacial bonding started at a HIPing temperature, elemental concentration gradients played a role in driving certain elements from the internal layers to the interfaces. Metal flow, along with micro-laminated interface is induced by diffusion mechanism. Delocalized liquid phase in a semisolid reaction introduced as reaching the eutectic concentration that induce dendritic whiskers due to directional solidification. In addition to new delocalized interphase flow from interface with aluminum oxide, alumina, and magnesium oxide, magnesia particulate/whiskers that induces toughening mechanisms with crack mitigation. Magnesio-aluminate spinel in al/mg lamellar composite interface is introduced with nontraditional processing.

Microstructural Analysis

During HIPing, Al/Mg/Al 3-ply introduced at different parametric study of temperature, pressure and holding time. Aluminum ply reveal more ductility compared with Mg ply that produced transverse micro cracking along interface in Mg ply and presented in Figure 1. Figure 1 presents Al/Mg metal matrix composite with clear interfacial bonding and transverse micro cracking HIPed at 530°C for 30 minutes and high press 30Ton. This represents the upper bound of press that induces transverse cracking along interface. Figure 2 presents Al/Mg MMC with lateral micro cracking, that HIPed at 620°C for 60 minutes and high press 30Ton. The increase in HIPing temperature and pressure induce localized interfacial melting and complete consumption of Mg with lateral cracking as upper bound of temperature and pressure. Figure 1 and Figure 2 represent the two extremities of parametric study (pressure and temperature). Figure 3 presents Al/Mg metal matrix composite with clear bonding interface at 500°C for 30 minutes and high press 20Ton. The commencement of interfacial bonding with no delamination or micro voids established at 20 minute with 20Ton press along Al/Mg interface. The upper bound of interfacial growth that produces a clear interphase with delocalized zone of interests established in Figure 4. This specimen is HIPed at 600°C for 20 minutes and high press 20Ton and presents growth kinetic, delamination with lateral cracking along interface. The kinetic of interface/interphase is detected and presented in Figure 5 to 9. Figure 5 presents Al/Mg metal matrix composite with clear interfacial bonding but with a minor transverse cracking of a specimen HIPed at 540°C for 20 minutes and high press 15Ton. Interfacial control introduced with micro-laminated interface and micro voids in aluminum (Figure 6). Figure 6 presents Al/Mg metal matrix composite with micro laminated interphase but no evidence of interphase at 540°C for 20 minutes and low press 10Ton. Slight increase in temperature and pressure at the same holding time induces interfacial growth. Figure 7 introduces Al/Mg metal matrix composite with micro laminated interphase, bulk interfacial growth and bulk metal flow that induces interphase at 620°C for 20 minutes and high press 15Ton. However, interphase is introduced, minor voids are captured along laminated interface may due to the depletion of diffused elements. Interface/interphase growth kinetics established with more delocalized zone of interests and bulk delocalized eutectic melts (Figure 8 and Figure 9). Micro laminated interface and delocalized Nano particulate/whiskers interphase established at 620°C for 30 minutes and high press 10Ton (Figure 8). High magnification of the interphase established with delocalized whiskers enriched with alumina as well as magnesia in bulk aluminum enriched interphase (Figure 9). Solid state processing with HIPing produces a semi-solid eutectic reaction with delocalized melts and directional solidification whiskers. The hipping temperature ranges from 500°C to 620°C to control interface with no delamination to interphase with localized melting phases. In the case of deformation twinning, the decreased load along with increase holding time indicates later commencement of interfacial growth. Next to twins, the reaction interface is characterized by tight physical contact on the Mg side and less tight contact on the other side of Al. However, cracks formed in the initial stages of an experiment, deformation and twining, spinel formed along these interface segments at high temperature, high holding time and low press.

Energy dispersive X-ray spectroscopy EDX

Energy dispersive X-ray spectroscopy EDX is applied on a specimen (Figure 9) to study micro constituents of composite and the semi-solid eutectic reaction with delocalized melts and interfacial front. Fig. 10 presents elemental distributions across Al 6082/AZ31/Al 6082 interface of the laminated composite: (1) bulk Mg ply (Figure 10a), interphase (Figure 10 b, and c), and bulk Al ply (Figure 10d). The diffusion characteristics of aluminum through interface and magnesium induce delocalized interphase with magnesium oxide and aluminum oxide. Alloy segregation induces delocalized zones with different composition, bulk zones with 80% Mg, 16% Al and 4 % O_2 (Figure 10 b), and branched frontal interphase with Mg 56%, 40% Al and 4% O_2 (Figure 10 c). Oxygen content induces the formation of alumina, and magnesia with different weight percent. Magnesium aluminate spinel is delocalized along interphase. Interphase growth indicates aluminum diffuse along Mg interface with less Al % at frontal interface.

Mechanical testing

Mechanical characterization established for both micro-constituents and bulk composite at different conditions. Stress-strain diagram applied for both Al-6082 and Mg-AZ31ply separately and presented in Fig. 11a and b for both Al and Mg respectively. Fig.12 presents typical stress–strain diagram of the Al 6082/AZ31/Al 6082 laminated composite that HIPed at 620°C for 30 minutes and high press 10Ton. The modification established at interphase induces stress-strain diagram that presented in Figure 13. Nonlinear stress-strain diagram (Fig. 13) reveal hardening exponent and minor delamination established at sample HIPed at 620°C for 30 minutes and low press 10Ton. Crack mitigation, toughening mechanisms and more ductility established for the new sample that presented in Figure 9. The maximum bonding strength of Al 6082/AZ31/ Al 6082 laminated composite fabricated by hipping was 95.4MPa. The bonding strength depends on the thermo mechanical processing conditions of the laminated composite. The application of Al/Mg/Al lamellar composite reveals more toughening with modified interface/interphase (Figure 13). Crack mitigation introduced with an increase of both strength and toughness in stress-strain diagram. The enhancement of interfacial shear stress established with nonlinear stress-strain diagram induced more toughening behavior. Magnesium aluminate spinel is that delocalized along interphase induce toughening mechanism.

Conclusions

Al/Mg/Al lamellar structure is established by Hot Isostatic Pressing, hipping, in three plies for structural applications. Parametric study with a new emerging solid state processing is introduced to study the kinetic of interphase/interface under HIPing conditions. The hipping temperature ranges from 500°C to 620°C to control interface with no delamination to interphase with localized melting phases. The extremities are controlled for the optimal conditions. The localized liquid phases are delocalized along interface induce directional solidification and branching with oriented shape. High press (30 ton) induces the formation of cracks in Mg ply. In the case of deformation twinning, the decreased load along with increase holding time indicates later commencement of interfacial growth. Next to twins, the reaction interface is characterized by tight physical contact on the Mg side and less tight contact on the other side of Al. The interphase variations are less pronounced in high load (30 Ton). In contrast, low load (20 ton) experiments with low temperature and high holding time show large lateral cracking and tearing in the Aluminum interface, however no spinel formed along these interface segments.

The spinel commencement increase with increasing temperature and time and they show delocalized zone manifested by deviation from low melting phase, oxidation, and directional solidification. Delocalized nucleation and directional growth were identified as the main mechanism of texture formation.

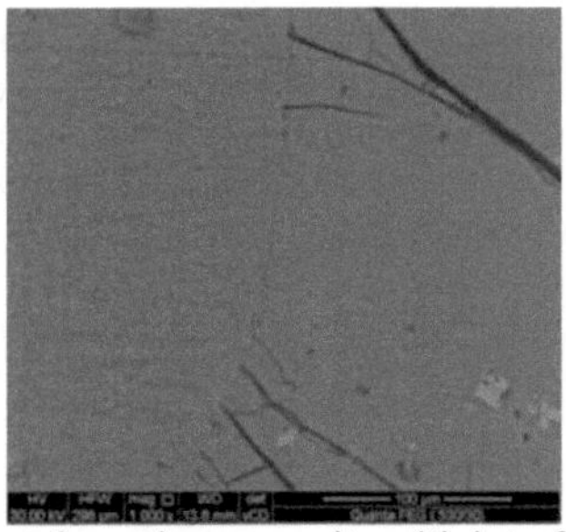

Figure 1: Al/Mg metal matrix composite with interface micro cracking at 530oC for 30 minutes and high press 30Ton.

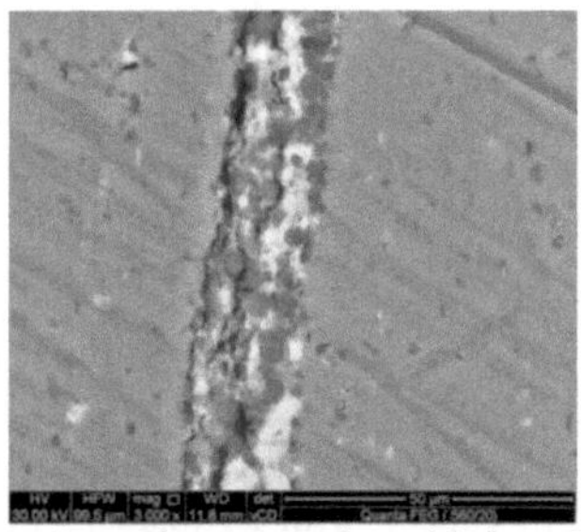

Figure 2: Al/Mg metal matrix composite with interface lateral cracking at 620°C for 60 minutes and high press 30Ton

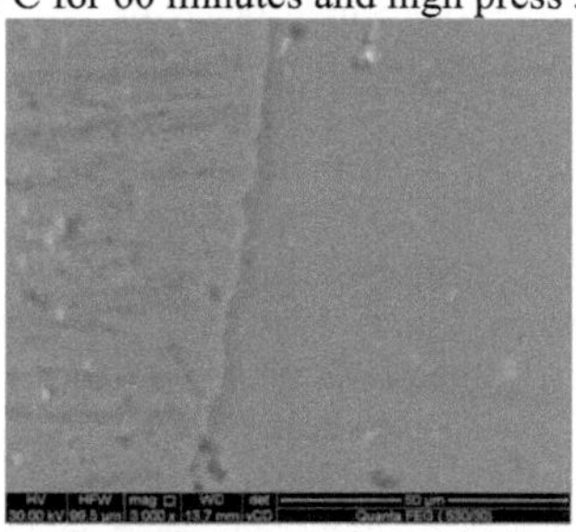

Figure 3: Al/Mg metal matrix composite with clear bonding interface at 500°C for 30 minutes and high press 20Ton.

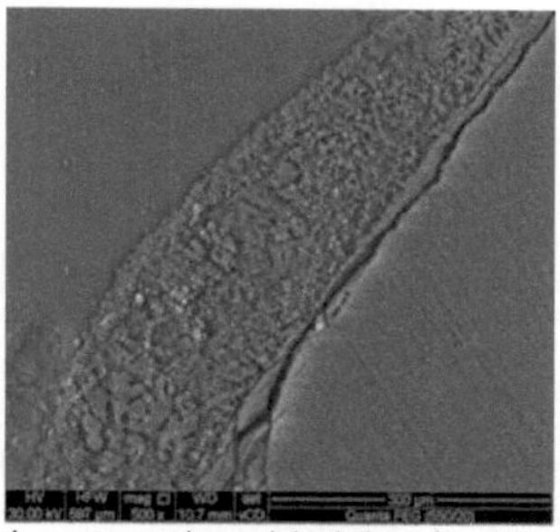

Figure 4: Al/Mg metal matrix composite with delocalize interphase and lateral cracking at 600oC for 20 minutes and high press 20Ton

Figure 5: Al/Mg metal matrix composite with clear bonding interface with minor transverse cracking $540^{\circ}C$ for 20 minutes and high press 15Ton

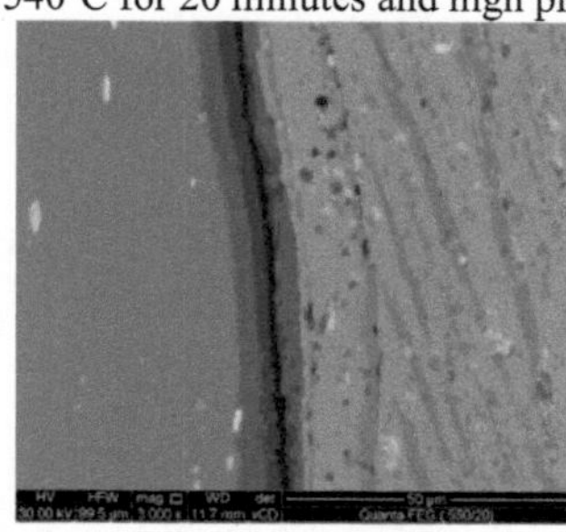

Figure 6: Al/Mg metal matrix composite with micro laminated interphase and bulk flow interface at $540^{\circ}C$ for 20 minutes and high press 10Ton.

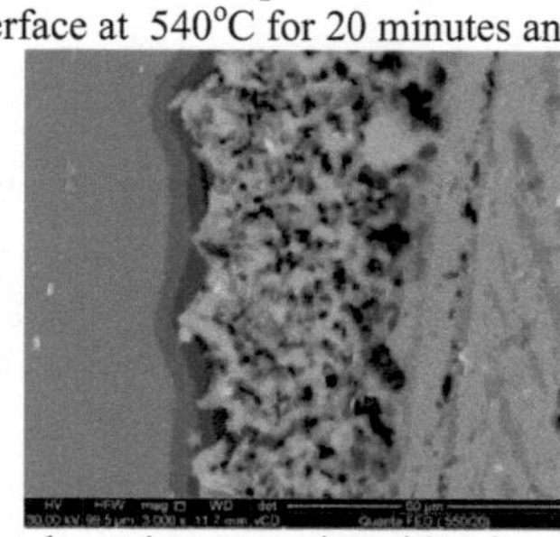

Figure 7: Al/Mg metal matrix composite with micro laminated interphase and bulk material flow interphase at $620^{\circ}C$ for 20 minutes and high press 15Ton

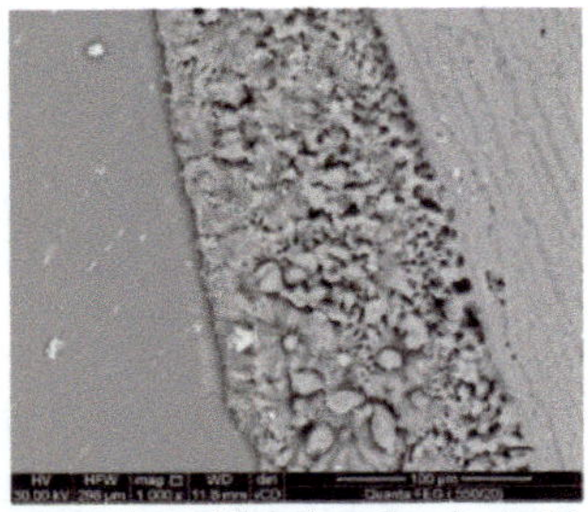

Figure 8: Al/Mg metal matrix composite with micro laminated interphase and delocalized Nano particulate/whiskers interphase at 620°C for 30 minutes and high press 10Ton.

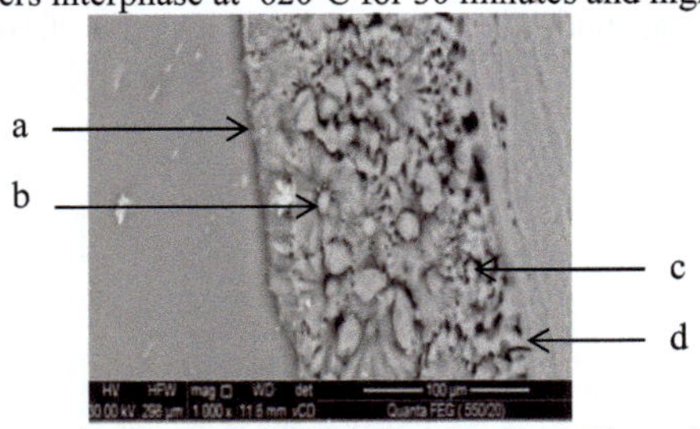

Figure 9: Al/Mg metal matrix composite with micro laminated interphase and delocalized alumina and magnesia interphase at 620°C for 30 minutes and high press 10Ton.

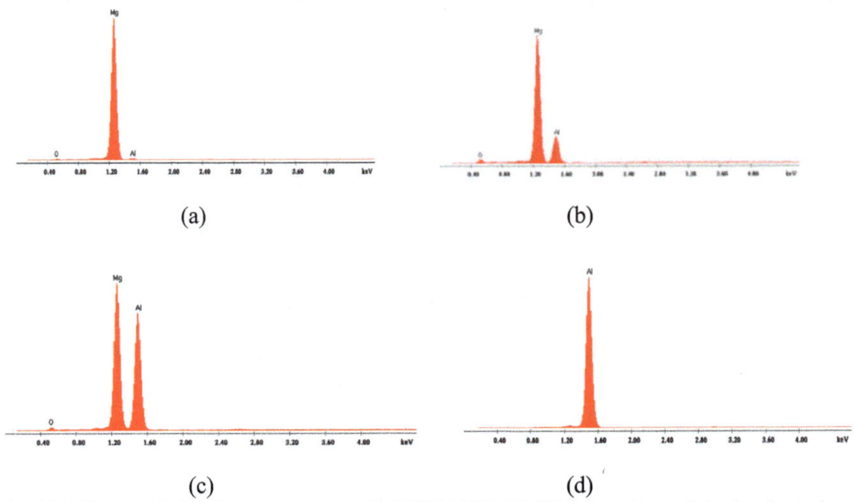

Fig. 10: Elemental distributions across Al 6082/AZ31/Al 6082 interface of the laminated composite: (a) bulk Mg ply, (b), and (c) interphase , and (d) bulk Al ply.

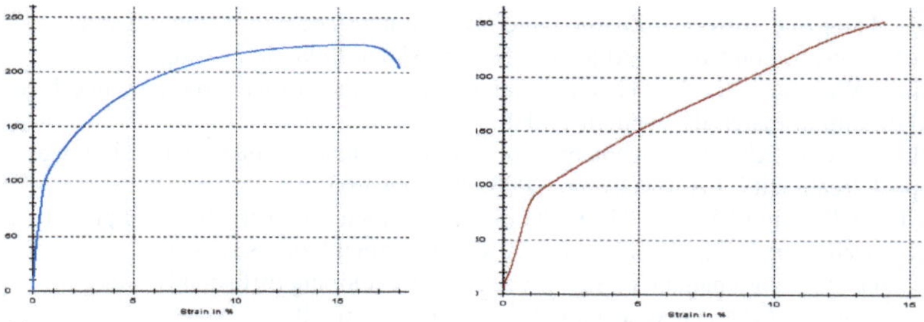

(a) (b)

Fig. 11: Typical stress–strain curve of (a) Al 6082 (b) AZ31 before hipping

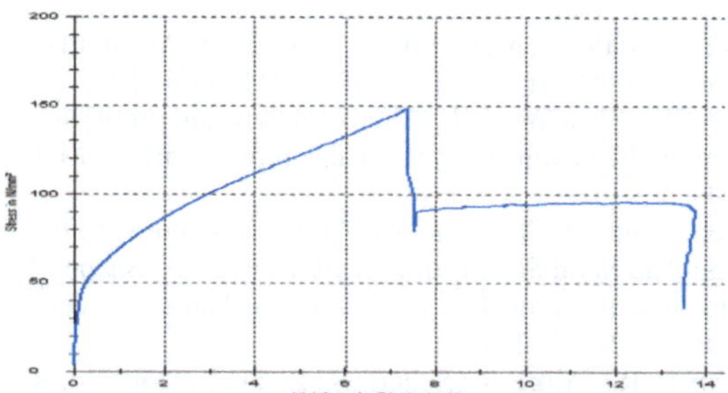

Fig.12: Typical stress–strain curve of the Al 6082/AZ31/Al 6082 laminated composite after the shear strength test:

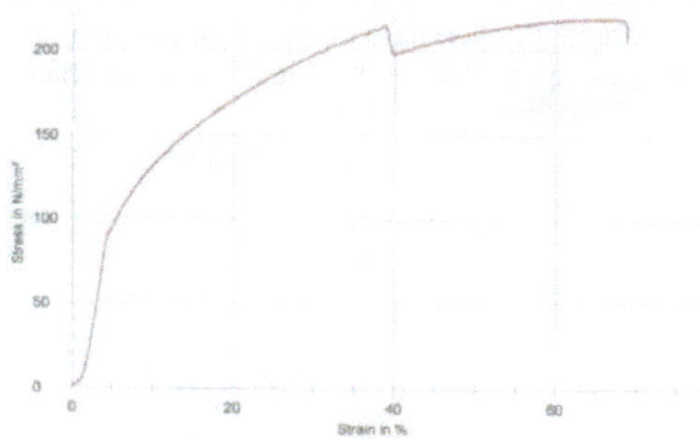

Fig.13: Typical stress–strain curve of the Al /Mg/Al laminated composite after the modified interphase.

References

1. Hidetoshi Somekawa, Hiroyuki Watanabe, Toshiji Mukai, Kenji Higashi, "Low temperature diffusion bonding in a superplastic" Scripta Materialia 48 (2003) 1249–1254.
2. Zhang XP, Zhao ZP, Wu FM, et al. "Corrosion and wear resistance of AZ91D magnesium alloy with and without micro arc oxidation coating in Hank's solution", J Mater Sci. 2007;42:8523–8.

3. Zhao LM, Zhang ZD. "Effect of Zn alloy interlayer on interface microstructure and strength of diffusion-bonded Mg–Al joints", Scripta Mater 2008; 58:283–6.
4. Li YJ, Liu P, Wang J, et al. "XRD and SEM analysis near the diffusion bonding interface of Mg/Al dissimilar materials Vacuum", 2008;82:15–9.
5. Liu FC, Liang W, Li XR, "Improvement of corrosion resistance of pure magnesium via vacuum pack treatment", J Alloys Comp. 2008; 461:399–403.
6. Liu LM, Tan JH, Zhao LM, et al. "The relationship between microstructure and properties of Mg/Al brazed joints using Zn filler metal", Mater Charact2008;59:479–83.
7. Mahendran G, Balasubramanian V, Senthilvelan T. "Developing diffusion bonding windows for joining AZ31B magnesium–AA2024 aluminum alloys", Mater Des 2009;30:1240–4.
8. Liu XB, Chen RS, Han EH. "Preliminary investigations on the Mg–Al–Zn/Al laminated composite fabricated by equal channel angular extrusion", J Mater Process Technol 2009; 209:4675–81.
9. Del Valle, J.A., Pérez-Prado, M.T., Ruano, O.A., 2005. Accumulative roll bonding of a Mg-based AZ61 alloy, Mater. Sci. Eng. A 410–411, 353–357.
10. Yan YB, Zhang ZW, Shen W, et al. "Microstructure and properties of magnesium AZ31B–aluminum 7075 explosively welded composite plate", Materials Sci. Eng. A 2010; 527:2241–5.
11. Bakr M. Rabeeh, Mona M. El Batanouny, Ali E. El Ashram "Microstructural Characterization And Solid State Processing Of Cu-Zn-Al Shape Memory Alloy In A Metal Matrix Composite" Canadian Journal on Mechanical Sciences and Engineering Vol. 2, No. 2 February 2011, P.P. 11-17.
12. Götze, L. C.; Abart, R.; Milke, R.; Schorr, S.; Zizak, I.; Dohmen, R.; Wirth, R.," Growth of magnesio-aluminate spinel in thin-film geometry: in situ monitoring using synchrotron X-ray diffraction and thermodynamic model", Physics and Chemistry of Minerals, Springer, Verlag Berlin Heidelberg , 5-2014.
13. Jerabek, Petr; Abart, Rainer; Rybacki, Erik; Habler, Gerlinde," The effects of stress concentrations on reaction progress: an example from experimental growth of magnesio-aluminate spinel at corundum - periclase interfaces under uniaxial load", EGU General Assembly 2014, held 27 April - 2 May, 2014 in Vienna, Austria, id.11155.

COMPOSITE MICROSTRUCTURE and MECHANICAL PROPERTIES

Advanced Composites for Aerospace, Marine, and Land Applications II
Edited by: T. Sano and T.S. Srivatsan
TMS (The Minerals, Metals & Materials Society), 2015

PREDICTION OF CRACK INITIATION SITE IN FASTENER HOLE OF COMPOSITE LAMINATE

Hossam El-Din M. Sallam[1] and **Amr A. Abd-Elhady** [2]

[1]*Civil Engineering Department*
Jazan University
Jazan 706, **Kingdom of Saudi Arabia**
on sabbatical leave from
Materials Engineering Department,
Zagazig University
Zagazig, 44519, **Egypt**

[2]*Mechanical Engineering Department*
Jazan University
Jazan 706**, Kingdom of Saudi Arabia**
on sabbatical leave from
Mechanical Design Department,
Faculty of Engineering
Helwan University
Cairo 11718, **Egypt**

Keywords: Composite material, GFRP, Bearing stress.

Abstract

Bolted composite joints are the most common forms of connections in engineering structures. A numerical study has been carried out to investigate the tensile stress, shear stress and the bearing stress behavior of a plate containing bolted composite joints. It is necessary to determine stress around the joints in order to determine the site of crack initiation, which can severely reduce the overall strength of the structure. Finite element model for single-lap composite joint consists of eight laminates and bolt. The effect of alignment of fibers with respect to the loading direction on the site of crack initiation of composite bolted joints will be investigated. It can be concluded that, the site of crack initiation depends of fiber orientation. Furthermore, the site of crack initiation is almost interior the plate thickness.

1. Introduction

Fiber reinforced composite materials have been widely used in engineering structures due to advantageous properties such as high specific strength, low weight, and corrosion resistance. Bolted joints are widely used in engineering structures, such as aerospace, land vehicles, petroleum pipe lines, civil engineering applications, and pressure vessels. Bolting components allows for nondestructive disassembly of components for maintenance, repair, and proves to be sufficient in strength. The composite joints have become a very important aspect because the structural properties and efficiency of the composite structure are determined by its joints [1], and 70% Of composite structure damage occurs in joint [2]. The problem associated with mechanical joints is stress concentration. Structural discontinuities such as holes reduce the efficiency due to high stress concentrations. The strength of a structure depends on the strength of the joint.

Bolted joint properties are very sensitive to laminate layup, and therefore it is not possible to define non-dimensional quantities such as bearing strength that scale to different laminates or bolt dimensions [3]. As the applied load increases and the joint deforms, the interface and its force distribution between the fastener and the laminate changes [4].To generate joint stiffness and failure loads for new bolt and laminate combinations, a completely new set of tests needs to be conducted; this can be prohibitively expensive for all but the simplest design challenges.

Cracks and other types of damage modes often exist at the joints because of stress concentration, contact interaction between the hole and the bolt, or the manufacturing process, which severely reduce the strength of the structure and lead to final failure [5, 6].It is necessary to determine stress around the joints in order to determine the site of crack initiation.

Owing to the complexity of the joint failure behavior, design engineers often rely on tests. Compared with numerical methods, however, tests generally require more time at a higher cost. Therefore many attempts have been made to understand and predict the failure of composite joints [7] using numerical and analytical methods to reduce the experimental effort required. So that, It is highly preferable to have validated Finite Element (FE) models that can be used to assess basic properties of a new joint configuration such as stiffness, failure load and failure mode [8].

2. Objective

A major goal of present research is to determine the effect of laminates sequence on the tensile stress, shear stress and bearing strength of the joint. Furthermore, the prediction of emanating crack in an laminate composite plate, with a circular hole, which was subjected to a traction force by a pin, was be investigated numerically to predict the failure early.

3. Numerical modeling

A numerical model producing a bolted joint is utilized in the present work using FE ABAQUS/Standard code [9] considering non-linear stress calculation due to large displacements and composite material behavior. A clear example of this kind of nonlinear behavior is a model that contains contact between different parts as is the case for bolted connections. A general static analysis with displacement control was employed in the present study. In a nonlinear analysis, the solution is found by applying the loads in increments until the final solution. The composite laminates that has different stacking sequences were examined numerically with a single-pin joint. Then, the bearing stress of the single-pin joint at circumferential hole with different rotation angle, θ, through the plat depth at each laminate has been evaluated. From this evaluation, the site of crack initiation can be predicated.

The composite plate containing a fastener hole manufactured by unidirectional glass fiber-epoxy prepregs and consists of 8 laminas were simulated. The angle-ply $[0/90]_{4s}$, $[0/45/-45/90]_{2s}$, $[45/90/0/-45]_{2s}$ and $[90/45/0/-45]_{2s}$ glass fiber reinforced epoxy (GFRE) composite laminate with 4 mm thickness were used to determine the stress around fastener hole. The laminates with stacking sequence of the composite material specimen consisted of four types, $[0°/-45°/90°/45°]_{2s}$, $[45°/0°/-45°/90°]_{2s}$, [-45°/90°/45°/0°]2s and $[90°/-45°/0°/45°]_{2s}$ are selected to prediction the modes of crack initiation. The value of fiber volume fraction, V_f, was 39 %.

The 0° direction in the stacking notation denotes the *x*-axis. Also, the 90° direction in the stacking notation denotes the *y*-axis. A unidirectional fiber reinforced lamina with global and material coordinate systems is also shown in Figure 1.

3.1 Geometry

A laminated composite plate used in this study shown in Figure 1, where *D* is diameter of fastener-hole, *W* is the width of plate, and *E* is the distance of bolt-hole center to the edge. *t* is the thickness of plate. The pin located at the center of the hole and a uniform load P is applied to the pin as shown in figure. the values of *E and W* in the specimen are fixed value equal 18 mm, 24mm respectively. The length of plate, *L,* is 135 mm, the hole diameter, *D,* is equal 6 mm, plate thickness, *t, is* 4 mm.

3.2 Material Properties

The Steel pin was modeled with isotropic material properties, with young's modulus is equal 200 GPa, and Poisson's ratio is equal 0.3. The mechanical properties of unidirectional GFRE composite laminate are determined from standard experimentally test. as example, to obtain the modulus of elasticity (E_1) and Poisson's ratio (ν_{12}), a rectangular test specimen was loaded in tension and two strain gauges were stuck on the specimen, one in direction of loading and the other in transverse direction. The mechanical properties of the GFRE composite plate which were used in the present work have been given in Table 1.

Table 1. The Experimental Mechanical Properties of Glass Fiber – Epoxy Composite

E_1 (GPa)	E_2 (GPa)	E_3 (GPa)	ν_{12}	ν_{13}	ν_{23}	G_{12} (GPa)	G_{13} (GPa)	G_{23} (GPa)
32	7	7	0.27	0.25	0.25	2.560	1.800	1.800

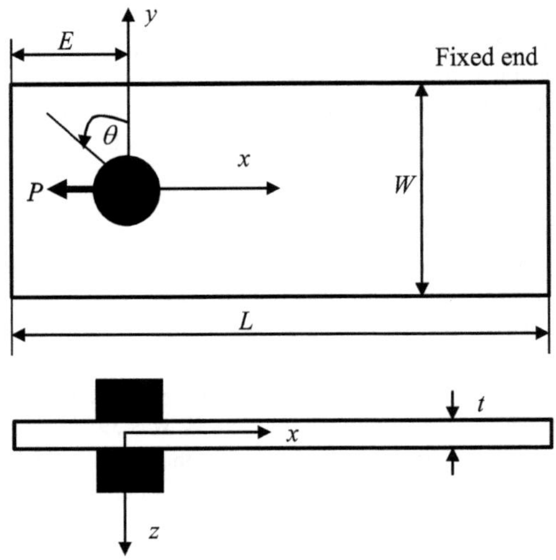

Figure 1. Geometry of a laminated composite plate subjected to pin loading

3.3 Material Modeling

The composite plates were modeled as a composite layup in property module in ABAQUS/Standard code by 8 ply. The orientation of each ply was define as 0°, or ±45° or 90° dependent on laminates sequence. The mechanical properties of composite plate define as engineering constant as table 1.

3.4 Contact and Boundary Conditions

Contact was modeled using the master–slave algorithm in ABAQUS[9]. The order in which the master and slave are defined is important. The master surface is defined as the surface with the stiffer material or the surface with the coarser mesh. In this analysis the contact between the hole and the pin surface had to be solved. The pin (i.e. the stiffer body) was chosen as the master contact surface. For contact between the hole and the rigid surface, the classical surface-to- surface contact search was used. Also a finite sliding tracking approach was used which means that any arbitrary relative separation, i.e. any finite sliding or rotation of the contact surface is allowed [9]. Penalty friction formulation has been used to simulate the contact properties of different components, the friction coefficient is 0.4, and the contact of each contact surface is 'hard contact'. The clearance between the bolt and the laminate was neglect. the contact between the laminate is assuming complete contact (no adhesive).

The model is loaded by applying a fixed boundary condition to one end, and a prescribed tensile displacement in the x-direction to the pin, as shown in Fig. 1. To avoid rigid body modes, components that are not fully constrained, i.e., the bolt and the plate with fixed end are constrained in x, y and z-direction in initial step.

3.5 Mesh

Each joint component (i.e., the plate and pin) is meshed by using C3D8R: an 8-node linear brick, reduced integration element. the mesh refinement process is carried out to assure that results are not dependent upon the element size. So that, the size of element decreases gradually with distance from the join hole decreasing (as shown in Fig. 2). This means that the finite element meshes in the neighborhood of the join hole are much denser. the steel and composite plate are modeled with 45000 elements and 4000 elements are used to model the pin.

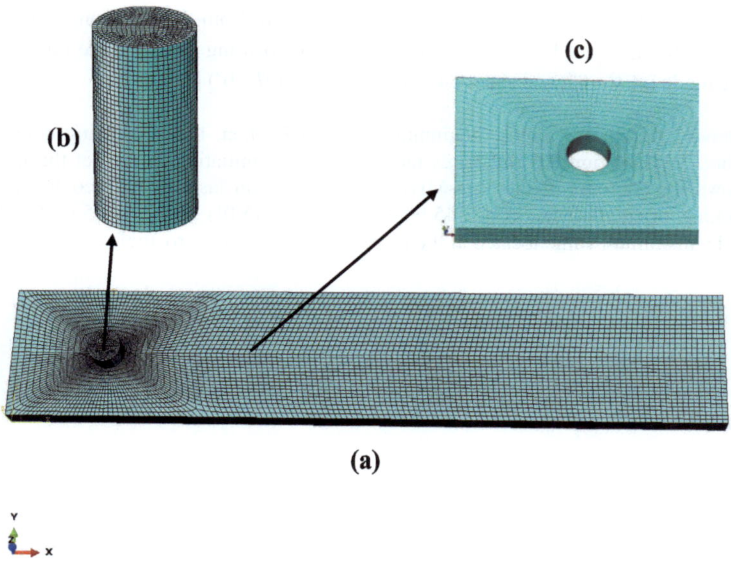

Figure 2. FE model mesh. (a) complete model. (b) pin, (c) fastener hole

4. Result and Discussion

In this study, the bearing stress, σ_{xx}, at circumferential surface of fastener hole of composite plate under pin joint load with different stacking sequence are investigated numerically. By observing the effect of pin joint load on the bearing stress, composite laminated plates are stacked as eight different orientations, $[0/90]_{4s}$, $[0/45/-45/90]_{2s}$, $[45/90/0/-45]_{2s}$ and $[90/45/0/-45]_{2s}$. The results shown in figure 3 indicate the bearing stress at each layer in composite plate and it illustrates that, the pin-bearing stress, σ_{xx}, changes by change the stacking layer orientation. The laminate with ply angle equal 0° have maximum bearing stress wherever its sequence located. furthermore, the laminate with ply angle 0°, and 90° decrease gradually with increasing the hold boundary angle, θ, until θ equal 90 then increase. For ply angle 45°, firstly the value of σ_{xx} increases by increasing the hold boundary angle, θ, to reach a peak value at $\theta = 45^{\circ}$ then decrease to reach a minimum value at θ, = 90 then it repeat again to increases to reach a peak value at $\theta = 135^{\circ}$ ten decreases again. The ply angle -45° has the opposite trend of ply angle

45°. The effect of laminate sequence on σ_{xx} of each ply angle can be neglect as shown in figure 4. From figure 3 and 4 it can be concluded that, the distribution of pin-bearing pressure can be assumed as cosine shape and that confirmed with [10].

It can concluded that, the bearing stress, σ_{xx}, is compression at boundary angle, $\theta = 90°$, and is tension at $\theta = 0°$, for all different orientations, [0/90]4s, $[0/45/-45/90]_{2s}$, $[45/90/0/-45]_{2s}$ and $[90/45/0/-45]_{2s}$ and different layer (as seen in Fig. 5). Furthermore for laminate with angle, 0 and 90, the maximum value of bearing stress in tension and compression located at $\theta = 0°$and 90° respectively and approximately the same values. Figure 5 concerns on boundary angle, $\theta = 0°$ and 90° only because the damage of composite joint beginning from this location. Three modes of failure in composite joints are beginning from this location, bearing failure mode (at $\theta = 90°$), net tension failure mode (at $\theta = 0°$) and shear out failure mode (at $\theta = 0°$) [11- 15].

The modes failure are almost beginning as a crack, then, this emanating crack grows to produce failure. So that, figure 6 illustrates the site of crack initiation to predict the failure early. Figure 6 shows the contours of the Von-Misses distribution on fastener hole composite plate for different stacking orientation angle, [0/-45/90/45]s, [-45/90/45/0]s and [90/-45/0/45]s. In Fig. 6, it is observed a maximum value located at the layer orientation with zero angles.

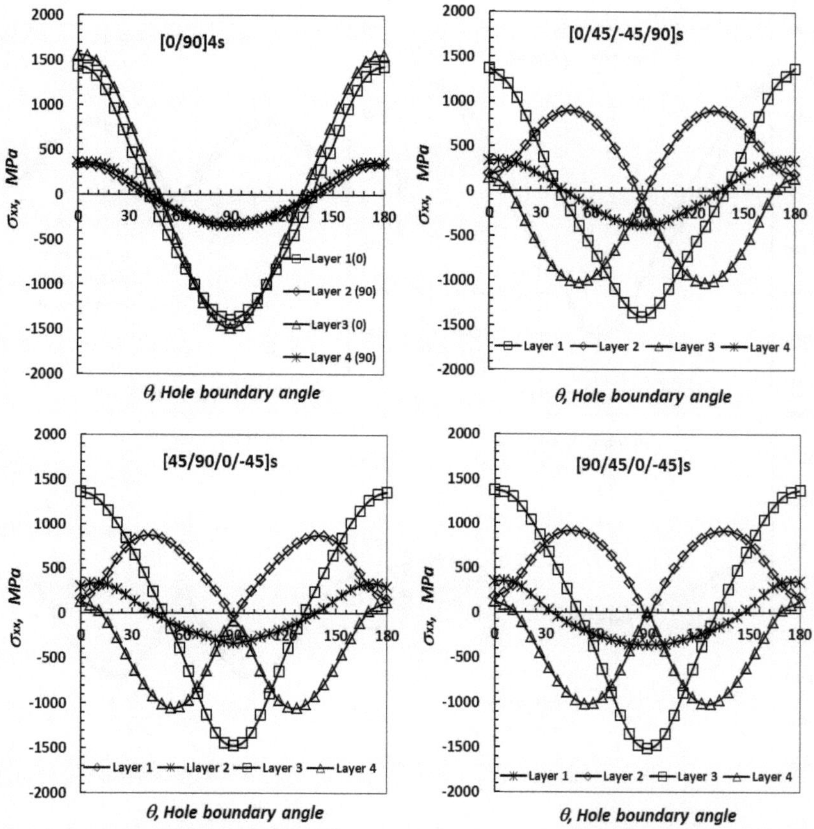

Figure 3. **Stress (σ_{xx}) at each laminate sequence of composite plate contained fastener hole for different composite laminate sequence**

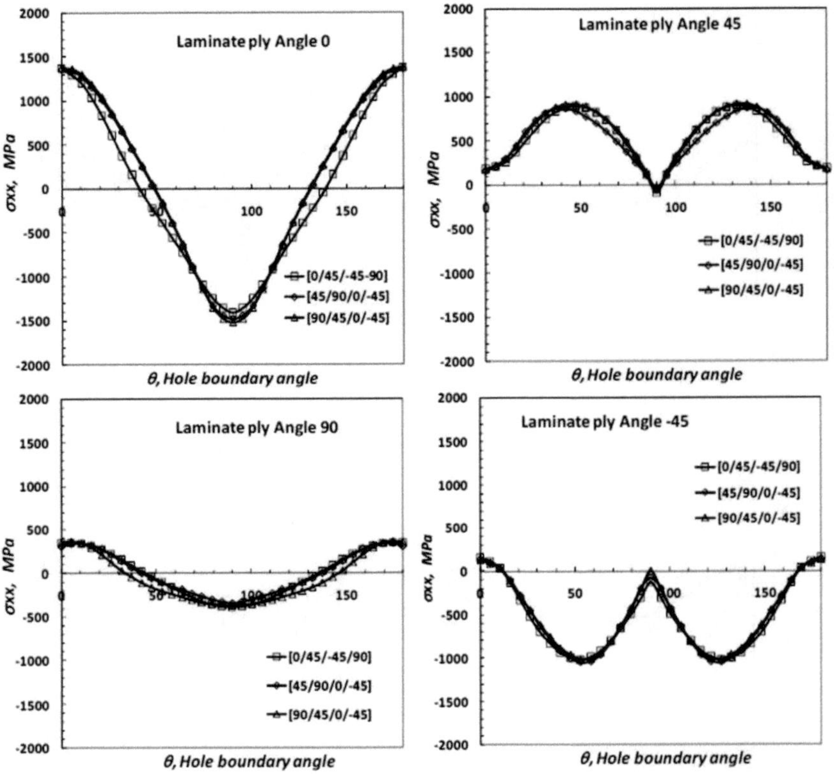

Figure 4. **Stress (σ_{xx}) at each laminate ply angle sequence for different composite laminate sequence**

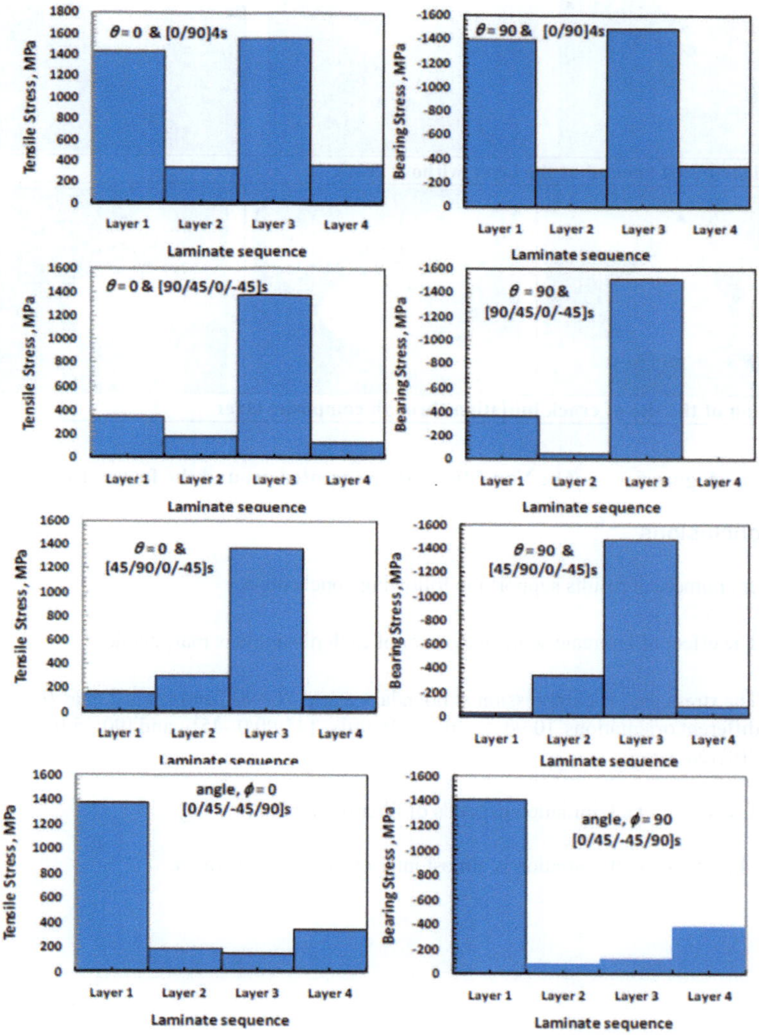

Figure 5. **The values of stress (σ_{xx}) at each laminate sequence of composite plate contained fastener hole for different composite laminate sequence, at $\theta = 0^o$ and 90°**

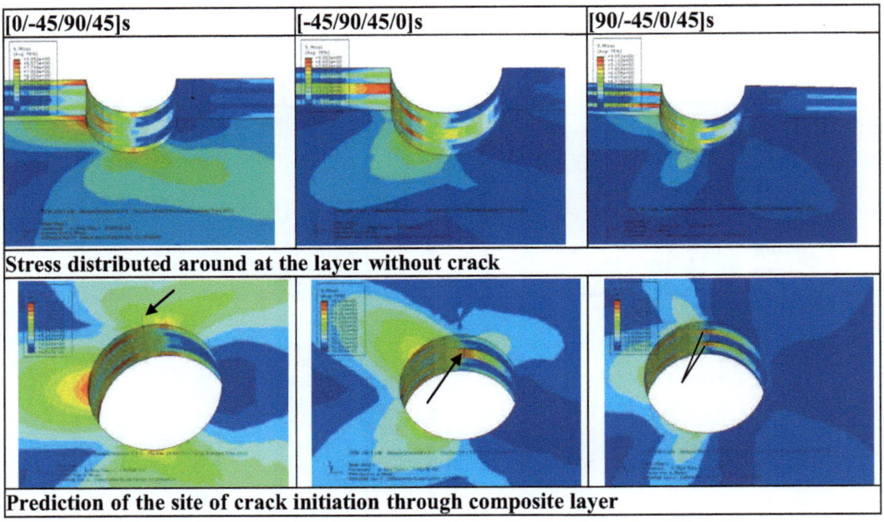

Figure 6. The Von-Misses stress created around the fastener hole

5. Conclusions

The present numerical results support the following conclusions:

1. The effect of laminate sequence on σ_{xx} of each ply angle is marginable.

2. The stress, σ_{xx}, is compression at boundary angle, $\theta = 90^\circ$, and is tension at $\theta = 0^\circ$, for all different orientations, [0/90]4s, $[0/45/-45/90]_{2s}$, $[45/90/0/-45]_{2s}$ and $[90/45/0/-45]_{2s}$ and different layer.

3. The site of crack initiation depends of fiber orientation.

4. The site of crack initiation is almost interior the plate thickness.

6. References

[1] J.-H Choi,. and Y.-J.Chun, "Failure Load Prediction of Mechanically Fastened Composite Joints," ***Journal of Composite Materials***, 2003, no. 37: 2163–2177.

[2] Zhenqing Wang, Song Zhou, Jifeng Zhang, Xiaodi Wub, and Limin Zhou, "Progressive Failure Analysis of Bolted Single-Lap Composite Joint Based on Extended Finite Element Method," ***Materials and Design***, 2012, no. 37:582–588.

[3] G. M. K. Pearce, A. F. Johnson, A. K. Hellier and R. S. Thomson, "A Stacked-Shell Finite Element Approach for Modeling a Dynamically Loaded Composite Bolted Joint Under in- Plane Bearing Loads," ***Appl Compos Mater*** 21March 2013.

[4] H.-S. Ahn, J.-H. Kweon and J.-H. Choi, "Failure of Unidirectional-Woven Composite Laminated Pin-Loaded Joints," ***Journal of Reinforced Plastics and Composites***, 2005, no. 24: 735.

[5] B. Yang, E. Pan and F.G, Yuan, "Three-Dimensional Stress Analyses in Composite Laminates with an Elastically Pinned Hole," ***Int J. Solids Struct.***, 2003 no. 40: 2017–35.

[6] G.S. Padhi, M.A. McCarthy, C.T. McCarthy and boljat, "A Tool for Designing Composite Bolted Joints Using Three-Dimensional Finite Element Analysis, "***Composites: Part A***, 2002, no. 33:1573–84.

[7] Garth Pearce, Alastair Johnson, Rodney Thomson, and Donald Kelly, "Experimental Investigation of Dynamically Loaded Bolted Joints in Carbon Fiber Composite Structures." ***Appl Compos M***ater, 2010, no. 17:271–291

[8] G. M. K. Pearce, A. F. Johnson, A. K. Hellier and R. S. Thomson, "A Stacked-Shell Finite Element Approach for Modeling a Dynamically Loaded Composite Bolted Joint Under in- Plane Bearing Loads," ***Appl Compos Mater*** 21March 2013.

[9] ABAQUS analysis user's manual. Version 6.7. Dassault Systèmes; 2007.

[10] D.J. Cartwright, P. Parker, "Opening mode stress intensity factors for cracks in pin-loads joints" ***International Journal of Fracture***, 1982, Vol. 18, no. 1, January 1982

[11] Mustafa Ozen and Onur Sayman, "Failure Loads of Mechanical Fastened Pinned and Bolted Composite Joints with Two Serial Holes," ***Composites Part B: Engineering***, 2011, V. 42, no. 2:264-274.

[12] R. Li, D. Kelly and A. Crosky, "Strength Improvement by Fiber Steering Around a Pin Loaded Hole," ***Composite Structures*** j., 2002, no. 57: 377–383.

[13] B. Okutan,"The Effects of Geometric Parameters on the Failure Strength for Pin loaded Multidirectional Fiber-glass Reinforced Epoxy Laminate," ***Composites Part B,*** 2002, no. 33: 567–578.

[14] B.M. Çten, and R. Karakuzu, "Progressive Failure Analysis of Pin-loaded Carbon epoxy Woven Composite Plates," ***Composites Science and Technology***, 2002, no. 62: 1259–1271.

[15] B.M Çten, and O. Sayman, "Failure Analysis of Pin-loaded Aluminum-glass-epoxy Sandwich Composite Plates," ***Composites Science and Technology***, 2003, no. 63: 727–737.

Advanced Composites for Aerospace, Marine, and Land Applications II
Edited by: T. Sano and T.S. Srivatsan
TMS (The Minerals, Metals & Materials Society), 2015

FINITE ELEMENT ANALYSIS OF QUANTITATIVE PERCUSSION DIAGNOSTICS FOR EVALUATING THE STRENGTH OF BONDS BETWEEN COMPOSITE LAMINATES

Scott Poveromo[1], Doug Malcolm[2], James Earthman[1]
[1]*Dept. of Chemical Engineering and Materials Science*
University of California, Irvine

[2]MSC Software Corporation
4675 MacArthur Court
Newport Beach, CA 92660

Keywords: composites, kiss bonds, nondestructive testing, finite element analysis

Abstract

Conventional nondestructive (NDT) techniques used to detect defects in composites are not able to determine intact bond integrity within a composite structure and are costly to use on large and complex shaped surfaces. To overcome current NDT limitations, a new technology was adopted based on quantitative percussion diagnostics (QPD) to better quantify bond quality in fiber reinforced composite materials. Results indicate that this technology is capable of detecting weak ('kiss') bonds between flat composite laminates. Specifically, the local value of the probe force determined from quantitative percussion testing was predicted to be significantly lower for a laminate that contained a 'kiss' bond compared to that for a well-bonded sample, which is in agreement with experimental findings. Experimental results were compared to a finite element analysis (FEA) using MSC PATRAN/NASTRAN to understand the visco-elastic behavior of the laminates during percussion testing. The dynamic FEA models were used to directly predict changes in the probe force, as well as effective stress distributions across the bonded panels as a function of time.

Introduction

Bonding composite structures together using adhesives provides many advantages over other joining methods. These advantages include distributing the load over a large bond area, reduced weight, and ability to join dissimilar materials together, higher stiffness and toughness over the bond area and in many cases lower manufacturing cost [1, 2]. However, one of the limitations when using adhesives is the inability to determine non-destructively if a bond joint meets structural requirements. Currently, standard nondestructive testing (NDT) techniques such as ultrasound, thermography, and X-ray are used to detect defects within bonded composite structures which include voids, cracks and delaminations [3]. However, these techniques are not able to detect a kiss bond within a composite structure because it is not an observable defect. A recent method developed by Bossi et al. uses a high peak powered pulsed laser to provide a localized dynamic proof test for bonded structures [4]. This laser bond method has shown to be able to detect differences in adhesive shear strength; it is destructive if the adhesive bond is weak and is non-destructive if the bond is strong. However, this method cannot be used in hard to access areas, is expensive to set-up, and potentially can cause damage within the laminate during repeated testing. The inability to certify bonded structures has lead to a conservative design approach and applying fasteners through the bond to ensure joint integrity. The addition of fasteners increases cost and weight and greatly diminishes the aforementioned advantages. Furthermore, fasteners cannot be the solution to attach materials that are being bonded within composite laminates that act as toughening aids or sensors for creating multifunctional materials.

In order to address this limitation, a new NDT method is needed that can detect adhesive kiss bonds, where the adhesive shear strength is low, typically less than 10% of its ultimate shear strength, due to contamination on the bonding surfaces or improper handling, mixing, or curing of the adhesive. It is believed that most so called kiss bonds result from poor surface preparation of as-molded surfaces due to excess fluorocarbons, silicones, plasticizers, etc. introduced from the manufacturing process [2, 5-6]. If these contaminants are not removed, they will decrease the surface energy which will decrease the contact angle between the adhesive and the bonding surface causing a decrease in shear strength. As a result, the bond that is formed will not be able to carry load as both substrates will in essence be only 'kissing' one another.

A cost effective legacy NDT method that has been used extensively in the past to inspect large composite components is the so-called coin tap test [7]. The coin tap test has been used in the composites industry to differentiate between a well bonded, defect-free composite part to one that has defects [7]. This method uses a coin or the back of a hammer to impact the composite part manually (a light tap). A sound of a 'good' section is very crisp and clear and radiates, while a sound from a 'bad' area has a dull thud. Distinguishing between sounds can be difficult even for a trained technician and the use of this technique over long periods of time is impractical. Also, no property information is gathered using this method that will help determine the extent of damage of the composite being inspected. Besides using sound, a tap test can be used to identify defective areas in a composite by examining different parameters controlling the impact response. This paper characterizes a percussion technique that was used to detect the difference between a well bonded composite structure (high shear strength) and a poorly bonded composite structure (low shear strength).

Technology Behind Quantitative Percussion Diagnostics

Quantitative percussion instrumentation (Periometer®, Perimetrics, LLC) was evaluated in previous work for determining bond integrity and impact damage in composite laminate samples [8-12]. An internal percussion rod containing a force sensor is accelerated to a pre-determined velocity just prior to impact with the specimen (see Figures 1 and 2). Thus, only the kinetic energy of the rod is administered to the specimen. The **low amplitude (~ 10N)** impact of the percussion rod generates a nondestructive stress wave that propagates through the specimen. Inelastic deformation or damping of strain energy during impact is characterized by the loss coefficient, η, a commonly used damping capacity parameter. Custom software on a computer interfaced to the percussion rod determines the mechanical energy returned to the rod based on the resulting force measured versus time. The energy dissipated by the specimen, *D,* is determined from

$$D = U - E_R - D_p \tag{1}$$

where U is the total strain energy, E_R is the mechanical energy returned to the sensor in the percussion rod and D_p is energy dissipated by sources external to the specimen [12]. The loss coefficient is given by

$$\eta = \frac{D}{2\pi U} \tag{2}$$

where the total strain energy, U, for the present percussion testing is approximately equal to the total kinetic energy prior to impact from the rod. In addition to displaying the loss coefficient, the output provided by the custom software is a plot of the probe force measured during each impact event as well as the energy dissipated from the specimen plotted versus time. The average maximum value of probe force is also displayed.

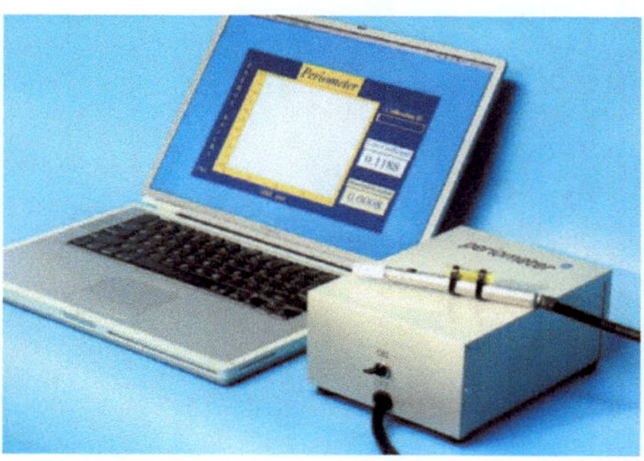

Figure 1. Periometer® instrumentation manufactured by Perimetrics, LLC.

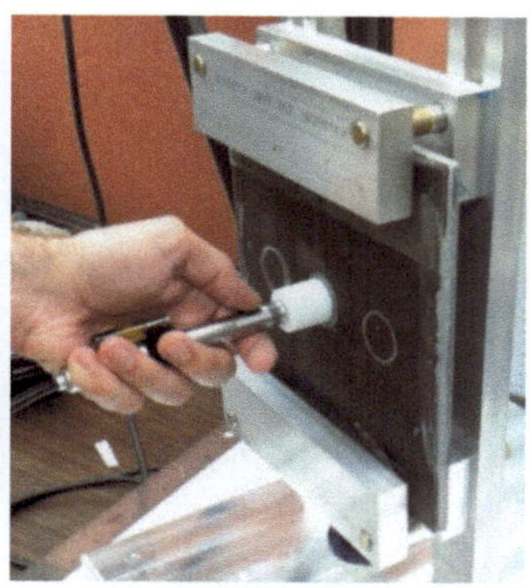

Figure 2. Performing QPD measurements on bonded carbon fiber reinforced composite laminates.

Quantitative percussion diagnostics (QPD) offers the following advantages over the current approaches:

- The Periometer measures the response of a material based on a known input of kinetic energy. Any type of defect that further depletes the kinetic energy applied to the material can be measured and evaluated.

- This percussion technique is rapid (4-5 seconds per measurement), portable, low cost, and easy to operate which lends itself to be used over large components in service.
- QPD can be used to characterize all types of composite materials and components.
- Data are taken from a single location and can be acquired in difficult to access areas that are common in composite structures.

Experimental Procedures

In order to assess the validity of using percussion to detect 'kiss' bonds in composite structures two composite laminates were fabricated; one bonded following a poor surface preparation technique, while the other was bonded per common standard practices. First, two pre-cured carbon fiber/epoxy matrix laminates 305 mm x 305 mm x 1.59 mm (12 in. x 12 in. x 0.0625 in.) were bonded together with a 121 °C (250 °F) cure supported epoxy film adhesive. One specimen had release agent applied in a 152 mm x 152 mm (6 in. x 6 in.) area in the center of the laminate, which simulated a 'kiss' bond while the other bonded laminate had no release agent applied. An illustration of the panel and test fixture is shown in Figure 3. A detailed review of the experimental procedures and testing can be found in reference 8.

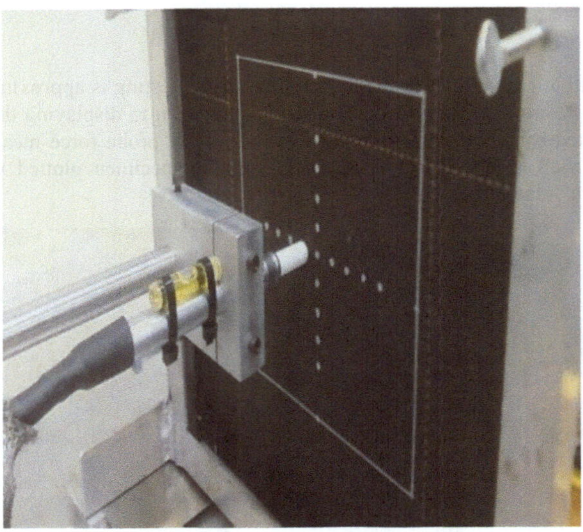

Figure 3. Photographs of fixture holding the 305 mm x 305 mm (12 in. x 12 in.) bonded laminates tested using the present percussion probe.

Finite Element Analysis Details

Finite element analysis using MSC PATRAN/NASTRAN software was completed in order to characterize low shear strength adhesive bonds within secondary bonded composite laminates. The bonded panel was modeled as three 101.6 mm x 101.6 mm (4 in. x 4 in.) rectangular solids where the top and bottom solids (facesheets) had high stiffness (68.9 GPa) and the middle solid (adhesive layer) had low stiffness (2.41 GPa). The model created represents two pre-cured standard modulus carbon fiber/epoxy matrix laminates, each 3.175 mm (0.125 in.) thick, that were bonded together with a 121 °C (250 °F) cure supported epoxy film adhesive that was 0.229 mm (0.009 in.) thick. The bonded panel was meshed using 3-dimensional hexagonal 8-node elements (6 faces and 8 nodes per element). Using an aspect ratio of 10 to 1, the fine mesh created in the model was 178 elements by 178 elements with four elements across the thickness of the adhesive layer and four elements across the thickness of the facesheets of the panel. Properties of the facesheets were modeled as 3d orthotropic and the film adhesive and steel percussion rod were modeled as isotropic. A summary of the properties can be found in Table 1. The low shear strength 'kiss' bond area was simulated by reducing the modulus of the film adhesive layer to 5%, 1%, and 0.5% of the ultimate value (2.41 GPa). Displacement constraints were applied in a three-point configuration from nodes on the back edge of the panel. To reduce the number of elements to ~380,000, a symmetry plane was created at one of the panel edges through the percussion rod as shown in Figure 4.

An implicit nonlinear analysis (SOL 400 in NASTRAN) was used to analyze the impact event at duration of 0.32 ms for 10 μs time increments. The rod was moved to within 0.00254 mm (0.0001 in.) of the panel to reduce model time and an initial velocity of 60 mm/s (2.36 in/sec) was applied. Also, the percussion rod was modeled as two contact bodies; the percussion rod and percussion rod tip, which were glued in contact to one another. The plane of nodes in contact with both bodies (contact surface) represents the piezoelectric sensor used to measure the force of the impact as shown in Figure 5.

Table 1. Summary of material properties used in the finite element model.

Properties	AS4 carbon plain weave fabric/epoxy matrix (assuming ~0.62 F_v)	FM-300-2 epoxy film adhesive	Steel Percussion Rod
E_1 Gpa (psi)	68.9 ($10e^6$)	2.41 ($0.35e^6$)	206.8 ($30e^6$)
E_2 Gpa (psi)	68.9 ($10e^6$)	-	-
E_3 Gpa (psi)	9.65 ($1.4e^6$) Assume = to E2 for AS4 Unidirectional Tape)	-	-
V_{12}	0.06	0.34	0.30
V_{23}	0.06	-	-
V_{31}	$0.007 = v_{13} X (E_3/E_1)$	-	-
G_1 Gpa (psi)	6.89 ($1e^6$)	-	-
G_2 Gpa (psi)	6.89 ($1e^6$)	-	-
G_3 Gpa (psi)	6.89 ($1e^6$)	-	-
P g/cm³ (lbs/in³)	1.61 (0.058)	1.27 (0.046)	7.92 (0.286)

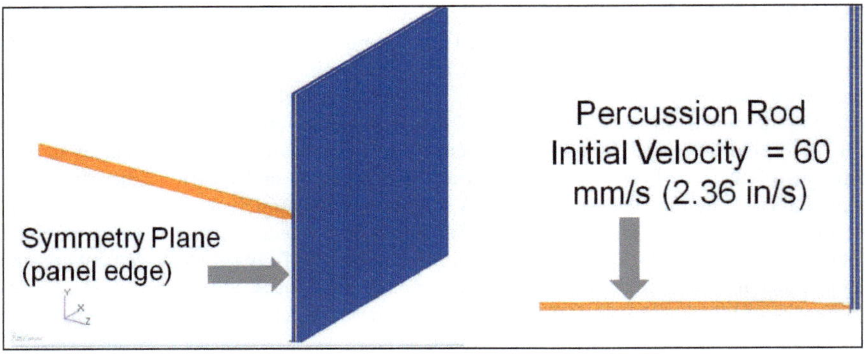

Figure 4. **Illustration created in PATRAN of the half model showing the percussion rod and symmetry plane along the 101.6 mm x 101.6 mm (4 in. x 4 in.) composite panel edge.**

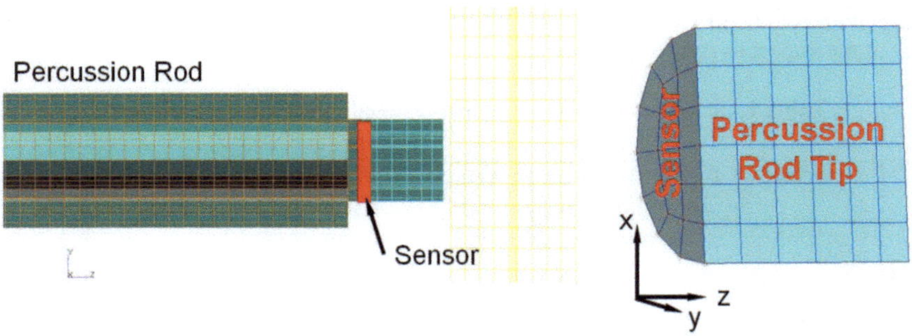

Figure 5. **A detailed illustration of the percussion rod model and piezoelectric sensor used to measure force during the impact.**

Results and Discussion

Impact Force Measurements

Impact measurements were taken at seven different locations horizontally across the middle of the two bonded composite panels as shown in Figure 3. A summary of the results for this testing is illustrated in Figure 6. The maximum impact (probe) force was found to be lower at every location tested for the released ('kiss' bond) panel when compared to the not-released (well bonded) panel. The reduced percussion force for the released panel was observed to vary from 9%-25%. A single factor analysis of variance (ANOVA) was completed for the maximum probe force between data collected for released versus the well bonded specimens. The p values obtained for probe force data was 6.84×10^{-5} indicating significant difference between the results for the released and not-released panels. The significantly lower decrease in probe force can be explained by greater internal friction within the poorly bonded specimens. This explanation is consistent with damping capacity data generated for metal adherends [13-14].

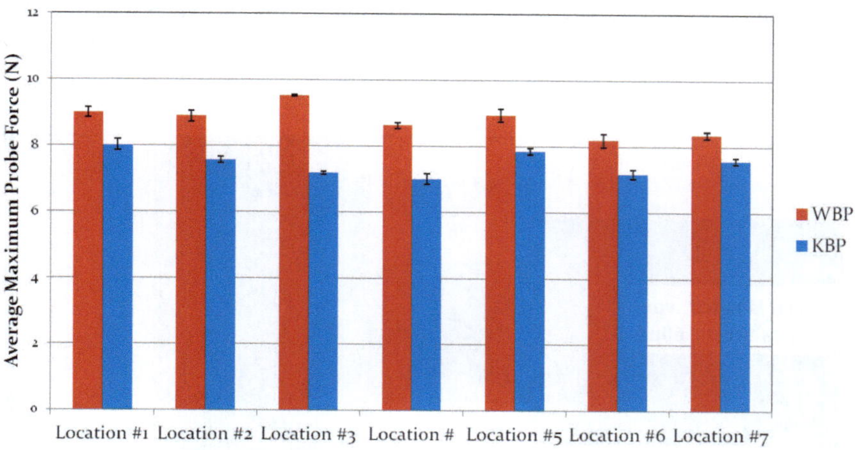

Figure 6. Probe force vs time measured at 7 different locations for the well bonded not released panel (WBP) and the released 'kiss' bond panel (KBP).

Finite Element Analysis Results

An implicit nonlinear analysis was run for four different conditions using the percussion rod/composite model described in the previous section.

- Well Bonded Panel (Adhesive modulus = 2.4 GPa)
- Kiss Bonded Panel 5% (Adhesive modulus = 0.12 GPa)
- Kiss Bonded Panel 1% (Adhesive modulus = 0.024 GPa)
- Kiss Bonded Panel 0.5% (Adhesive modulus = 0.012 GPa)

First the von Mises and Z-component stresses were calculated (psi) vs time and plotted as fringe plots over the percussion rod and composite panel. It was observed that the peak stresses occurred at time increment 60μs during the impact. Snapshots of both stress types at 60μs are shown in Figures 7-10.

The von Mises stress for the well bonded panel is well distributed from the front to back laminate indicating good load transfer across the adhesive. However for the kiss bonded panel (1% modulus) the distribution of stress is mostly localized on the front laminate indicating lower load transfer across the adhesive. Also, the maximum stress is similar for the front and back laminates in the well bonded model. This is not the case for the kiss bonded panel where there is a large difference between the stresses measured for the front and back laminate. These results indicate that a measurement of transmitted force through the panel should result in significantly different values for the well bonded and kiss bonded panels.

The z-component stress at 60μs for the well bonded panel showed good stress distribution from the front to back laminate through the adhesive as well as in the Y direction above and below the percussion rod in the front laminate. This was not the case for the kiss bonded panel (1% modulus) as the stresses were mostly observed in the front laminate adjacent to the adhesive.

Finally, higher stresses were observed at the corners of the percussion rod tip at 60μs for all four outputs. Since the percussion rod was aligned properly with respect to the composite panel in the model, it is believed bowing of the bonded panel occurred during impact caused higher stresses at the corner of the rod tip and lower stresses toward the middle of the rod tip.

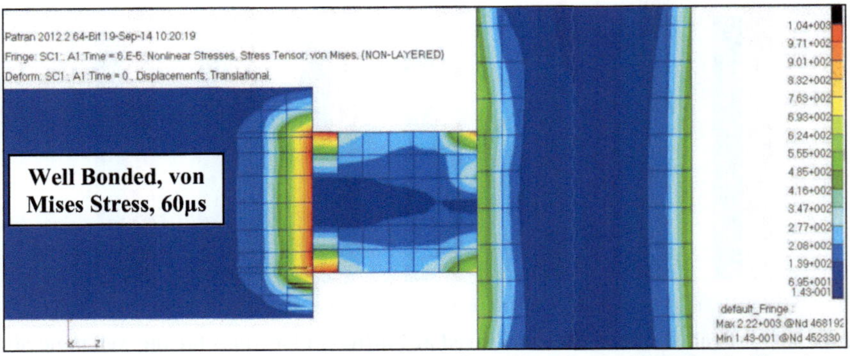

Figure 7. Fringe plot of von Mises Stresses (psi) at 60μs for the Well Bonded Laminate.

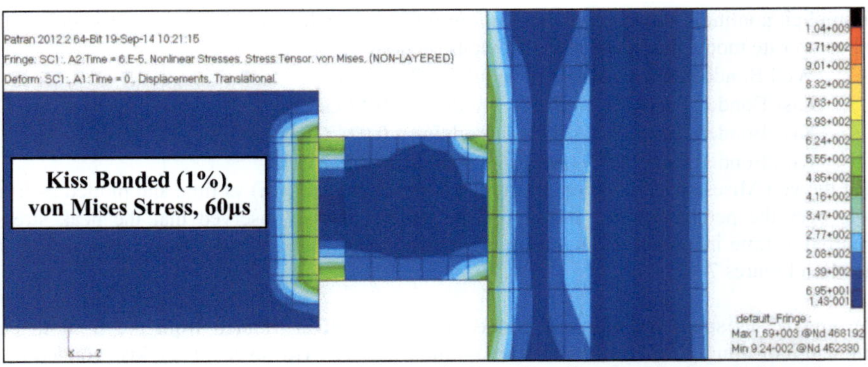

Figure 8. Fringe plot of von Mises Stresses (psi) at 60μs for the Kiss Bonded Laminate (1% modulus).

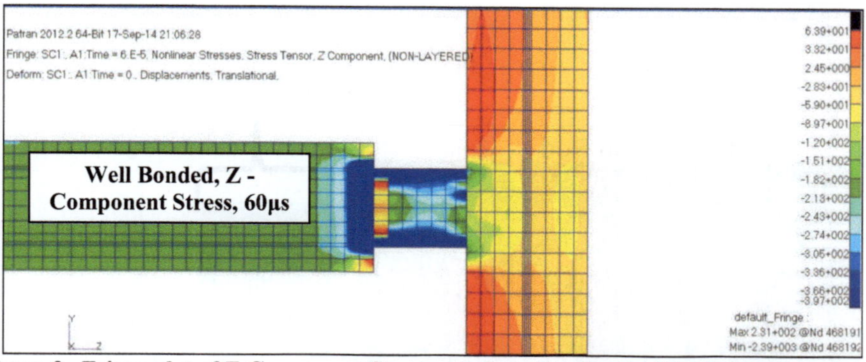

Figure 9. Fringe plot of Z-Component Stresses (psi) at 60μs for the Well Bonded Laminate.

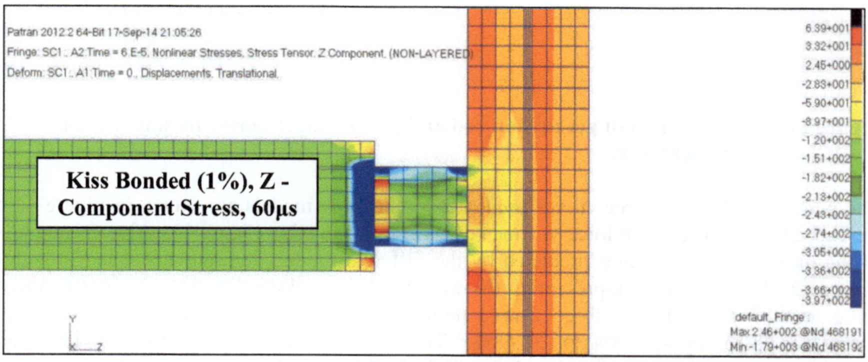

Figure 10. Fringe plot of Z-Component Stresses (psi) at 60μs for the Kiss Bonded Laminate (1% modulus).

The next output analyzed was the force measured at the sensor area in between the percussion rod body and tip. The force at the sensor area was calculated by first measuring the Z-component stress at all 24 nodes representing the piezoelectric sensor as illustrated in Figure 11. The total stress for each time increment was then divided by 24 to get an average stress and then multiplied by the area of the percussion rod tip which is ~2 mm in diameter. The resulting force/time data was then plotted and compared to the experimental data.

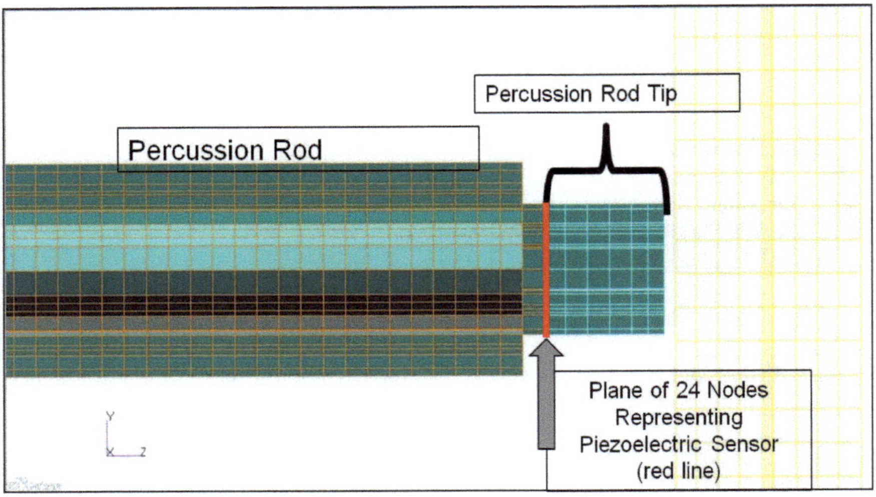

Figure 11. Illustration of percussion rod and piezoelectric sensor used to measure impact force.

The percussion force vs time curves predicted by the FEA models are shown in Figure 12 and the experimental percussion force vs time curves are illustrated in Figure 13. Also, a comparison of the maximum force values predicted by the different models is summarized in Table 2. In Figure 12, the force curve representing the well bonded panel displayed the highest percussion force at 9.77 N while the force curves representing the kiss bond models showed lower percussion forces by -26.58% to -37.48%. The maximum force values for the well bonded model (9.77 N) and the 5% kiss bonded model (7.17 N) match very close to the experimental data. The impact time response for the kiss bond models all are longer in time than the well bonded model. As the modulus increased for the kiss bond models, the impact time duration decreased. This was not the case in the experimental data which did not show a significant difference in the shape of the curves. Also, the experimental impact time response was around 0.3ms where as the models predicted time responses ranging from 0.15ms to 0.25ms. This discrepancy may be due to the fact that damping was not included in the model material properties. Future model work will include damping factors which we expect to have an effect on the impact time responses.

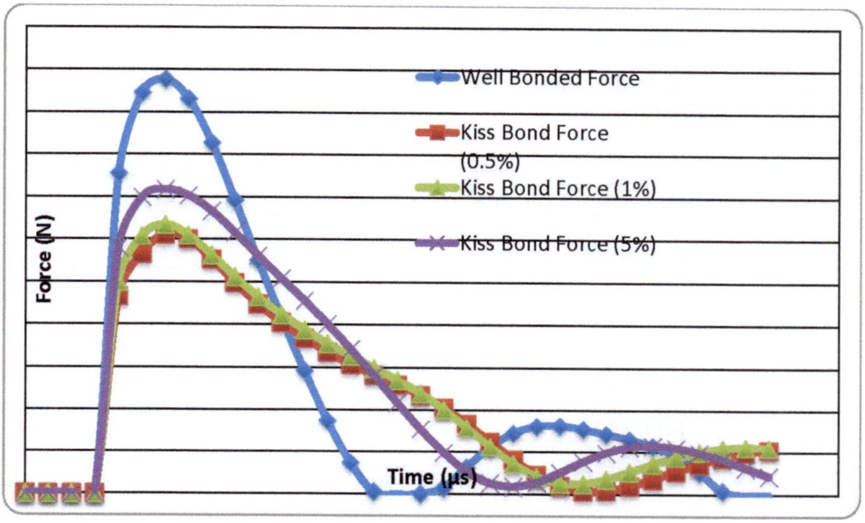

Figure 12. Predicted Force vs Time Curves for Well bonded and Kiss Bonded Models.

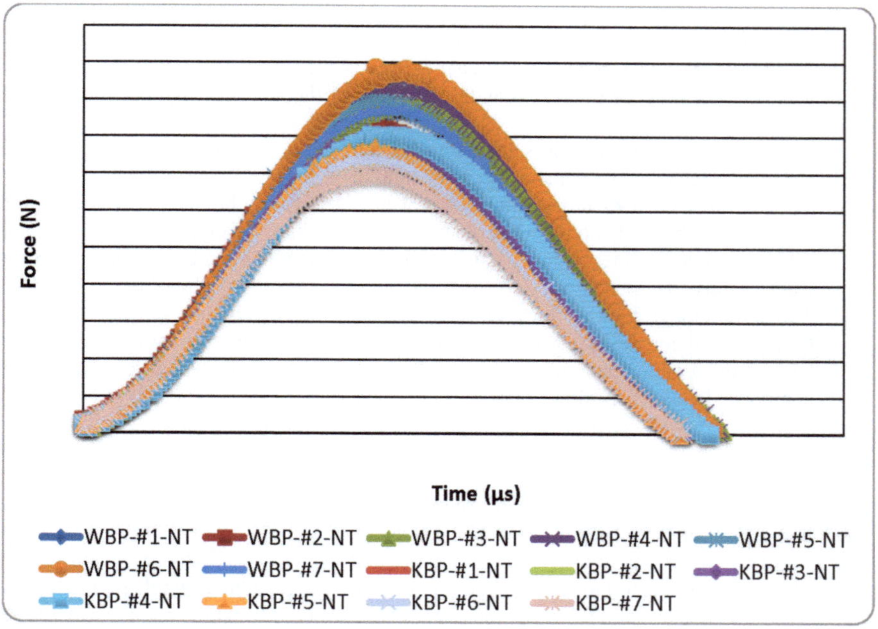

Figure 13. Experimental Force vs Time Curves for Well bonded and Kiss Bonded Panels.

Table 2. **Summary of Predicted Maximum Force Values for Well Bonded and Kiss Bonded Models.**

Panel Description	Max Force (N) at 60μs	% Change from Well Bonded Panel
Well Bonded Panel	9.77	NA
Kiss Bond Panel (5%)	7.17	26.58
Kiss Bond Panel (1%)	6.34	35.15
Kiss Bond Panel (0.1%)	6.11	37.48

Conclusions

Concluding Remarks

The following is a summary of the conclusions from the FEA work completed.

- FEA models correctly predict a reduction in measured percussion force (-26.58% to -37.48%) when a kiss bond (5%, 1%, and 0.5%) is introduced
- The percussion force decreased as the modulus of the adhesive used to simulate a kiss bond decreased.
- The models indicate that less load transfer into the second panel results in the force reduction predicted and observed experimentally.
- The von Mises and Z-component stresses were significantly higher on the back laminate for the well bonded model when compared to the kiss bonded models. This stress difference may be measured by using a load cell on the back of the panel (transmitted force) during the impact.
- The FEA model predicts very close to experimental percussion force measurements. Changes in damping by the kiss bond and the method of how the kiss bond is modeled could have a significant role in accurately predicting the behavior of kiss bonds within composite laminate specimens.

Path Forward

The following is a summary of changes to the model which will be explored to more accurately predict percussion impact events in composite panels.

- The modeling method to create a kiss bond will be varied. One example of a new method is to reduce the modulus in the top layer of elements only within the adhesive layer to better approximate contamination at the bonding surface.
- Structural damping coefficient values will be added into the model to better predict the response of the lower modulus kiss bond and through the thickness of the carbon fiber epoxy laminates.

- An explicit non linear solution (NASTRAN SOL 700) will be used to model the percussion event and compared to the current implicit (NASTRAN SOL 400) solution. The current model run time is long, on the order of 36 hours using 10 μs time increment for 0.32 ms, and the explicit solution may provide similar outputs at a shorter run time.

Acknowledgements

I would like to thank Doug Malcolm from MSC Software Corporation for providing guidance on implicit non linear analysis using PATRAN/NASTRAN software packages and Aren Ebrahimi a UCI undergraduate researcher for help in modeling the percussion rod. I would also like to thank Northrop Grumman Aerospace Systems for contributing $12K from the University Industrial Affiliates program "Evaluation of Bond Strength between Composite Bonded Structures" which helped fund this project.

References

1. Barnes TA, Pashby IR. "Joining techniques for aluminium spaceframes used in automobiles Part II- adhesive bonding and mechanical fasteners." *J Mater Process Technol* 99 (2000): 72-79.
2. Yang S, Lan G, and Gibson RF. "Nondestructive detection of weak joints in adhesively bonded composite structures." *Compos Struct* 51 (2001): 63-71.
3. Piehl MJ, et al. "Efficient Certification of Bonded Primary Structure." SAMPE Conference Proceedings: Education & Green Sky - Materials Technology for a Better World. Long Beach, CA. May 6-9, 2013. Society for the Advancement of Material and Process Engineering.
4. Richard Bossi, et al. "Laser Bond Testing." *Materials Evaluation* (2009): 819-827.
5. Wingfield JRJ. "Treatment of composite surfaces for adhesive bonding." *Int J. Adhes Adhes* 13 (1993): 151-156.
6. Nagy PB. "Ultrasonic detection of kissing bonds at adhesive interfaces." *J. Adhes Sci. Technol* 5 (1991): 619-630.
7. Cawley P and Adams RD. "The mechanics of the coin-tap method of non-destructive testing." *Journal of Sound And Vibration* 122 (1988): 299-316.
8. Poveromo, Scott and Earthman, James. "Development of Percussion Diagnostics in Evaluating 'Kiss' Bonds Between Composite Laminates." SAMPE Conference Proceedings: Education & Green Sky - Materials Technology for a Better World. Long Beach, CA. May 6-9, 2013. Society for the Advancement of Material and Process Engineering.
9. Poveromo, Scott and Earthman, James. "Development of Percussion Diagnostics in Evaluating 'Kiss' Bonds Between Composite Laminates." TMS Conference Proceedings: Advanced Composites for Aerospace, Marine, and Land Applications. Edited by Tomoko Sano, T.S. Srivatsan and Michael W. Peretti. San Diego, CA February 16th-20th, 2014. The Minerals, Metals & Materials Society.
10. Poveromo, Scott and Earthman, James. "Analysis of Kiss Bonds Between Composite Laminates." *Journal of Materials* 66 (2014): 970-978.
11. Stanley BD, Bustemante L and Earthman JC. "Novel instrumentation for rapid assessment of internal damage in composite materials." *Nondestructive Evaluation and Materials Properties III* (1997): 97-100.
12. United States Patent # 6120466. "System and method for quantitative measurements of energy damping capacity." September 19th, 2000.
13. Shiloh K et al. "Aging Characterization of adhesives and bonded joints by non-destructive damping measurements." *J Intel Mat Syst. Str* 10 (1999): 353-362.
14. Srivatsan TS et al. "Electromagnetic measurement of damping capacity to detect damage in adhesively bonded material." *Materials Evaluation* 47 (1989): 564-569.

Advanced Composites for Aerospace, Marine, and Land Applications II
Edited by: T. Sano and T.S. Srivatsan
TMS (The Minerals, Metals & Materials Society), 2015

OXIDATION BEHAVIOR CHARACTERIZATION OF ZIRCONIUM DIBORIDE COMPOSITES AT ABOVE 2000°C

Ziyuan Zhou, Xianghe Peng and **Zhen Wei**

College of Aerospace Engineering
Chongqing University
174 Shazheng Street, Shapingba,
Chongqing, 400044, China

Keywords: Oxidation, Zirconium diboride composites, High temperature

Abstract

Because the highest temperature for most traditional testing systems is limited to below 2000°C, an apparatus capable of testing zirconium diboride based composites, the key members of ultra-high temperature ceramics, was built and applied based on rapid Joule heating in the ultra-high temperature range (above 2000°C). A novel method was introduced to measure the real-time temperature distribution along the length of a ribbon shaped specimen (ZrB_2-20 vol% SiC), which could be heated up to failure with the apparatus. The influence of oxidation on the temperature measurement was evaluated, which suggested that the bubble formation can significantly decrease the measuring temperature. A liquid phase would form at ~2300°C, which might be induced by the eutectic reaction of the ZrB_2 and SiC.

1. Introduction

ZrB_2-base composites, the key members of a family of materials known as ultra-high temperature ceramics (UHTCs), are recognized as the most potential candidates for a variety of high temperature structural applications due to their unique combination of extremely high melting temperature, high thermal and electrical conductivities, low density, and excellent mechanical properties and chemical stability [1-3]. Due to the relatively poor oxidation resistance of nominally phase-pure ZrB_2 at temperatures above ~1200°C, SiC is often added as a second phase which has been proved to enhance the oxidation resistance significantly, especially at temperatures above 1600°C by many researchers.[1, 4-7] The improved oxidation resistance is attributed to the formation of a stable oxygen-isolated borosilicate glass layer on the surface of the oxidized ceramics. Unfortunately, only a limited experimental work involved oxidation behavior of ZrB_2-SiC composites at above 2000°C. The high temperature furnace with traditional heating element may fail at the ultra-high temperature range under oxidized atmosphere. Other new methods have been developed for further investigation, such as plasma torch heating [8, 9] oxyacetylene torch heating,[10-12] high frequency induction heating,[13, 14] and rapid Joule heating of specimen itself [15-19] It is found that the rapid Joule heating has more merits such as fast heating and cooling rate, lower power input, lower equipment cost and unlimited highest heating temperature. Moreover, if the center of the specimen is enough thin or the current supply is enough large that the specimen can be heated up to the melting point. However, compared with the high temperature furnace, the temperature distribution, especially along the length of the self-heated specimen, is not uniform. The shape of the specimen and the features of the electrode (thermophysical properties or shape) can significantly affect the temperature distribution. Previous work did not pay too much attention on characterizing the difference. Most of them only focus on the highest temperature point.[17, 19] In the recent work, Gangireddy et al.[18] tried to characterized the temperature distribution. However, the limited fixed testing points along the length of the specimen they selected may cannot express the explicit information.

In this work, an apparatus capable of testing ZrB_2-base composites in the temperature regime above 2000°C by Joule heating was built and applied. A novel method was introduced to dynamically measure the temperature distribution along the length of a ribbon shaped specimen which can be heated up to fail in air with this apparatus. The surface condition of the specimen during oxidation can be recorded by an industrial camera equipped with filtering lens. This work may provide a more comprehensive understanding of the oxidation behavior of the material, especially at above 2000°C.

2. Experimental procedure

The ZrB_2-based composite (ZrB_2-20 vol% SiC) block fabricated by hot pressing was provided by Harbin Institute of Technology. The initial block was manufactured into several little specimens with special shape for testing using both electro-discharge machine and diamond wheel. More specifically, the specimen was a ribbon with relatively thicker ends. Firstly, due to the appropriate electrical conductivity of the material, the wire electro-discharge machine could be used to cut the block into several little bars (38 mm long, 2 mm wide, and 4 mm thick). Secondly, the center section of the bar was abrasively machined away using diamond wheel to create a ribbon-shaped geometry, resulting in 4 mm-long thick ends at both sides and a 30 mm-long and 1.5 mm-thick region of reduced area in the center. Both of the thick ends of the specimens were clamped using the upper and lower electrodes located at the corresponding sides of the heating equipment. The electrodes which also displayed as fixtures in this equipment were made of tungsten-copper alloy (70 vol% W). It was considered as a suitable material applied in this testing case as it is capable of high thermal and electrical conductivities as well as excellent high temperature strength.

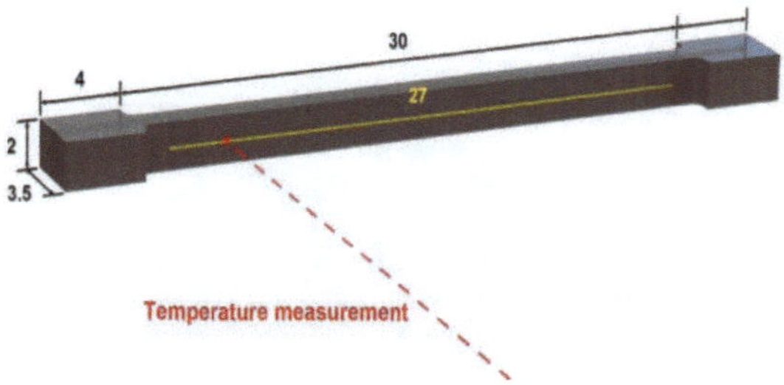

Figure 1. The shape of the ribbon specimen and the length of temperature measurement in the center of the ribbon (27 mm).

The samples were heated by Joule heating using a DC power supply. The electrical current was supplied through the thick ends where the electrical resistance is relatively low to the center of the specimen. Thus, the thin central area was rapidly heated to a very high temperature, while the end sections remained cold. A Hall current sensor was adopted to measure the value of current. In the preceding introduction, it is known that the temperature distribution of specimen is not uniform, especially along the length of the specimen. If the influence of the nonuniformity is expected to be deeply studied, the first problem is to measure the temperature distribution. In the temperature measurement part of the apparatus, a novel method was introduced to measure the real-time temperature distribution along the length of a ribbon shaped specimen. A digital 2- color pyrometer with fiber optic (IMPAC ISR 12-LO, LumaSense Technologies, Germany) for non-contact temperature measurement is adopted. Taking advantage of the fast speed of the response time (2 ms) of the pyrometer, the temperature distribution along the length of the specimen can be measured dynamically by the scanning of the focusable optical head of the pyrometer. The head was installed on a slider which was driven by a stepping motor and was capable of shuttling back and forth along the linear rail with high velocity.

Oxidation time, surface temperature of the specimen, displacement of the optical head and the current applied were recorded simultaneously using a data acquisition card (USB-6008, National Instrument, Austin, USA) to ensure that the corresponding information of the location and the temperature can be obtain at an arbitrary specific oxidation time. Additionally, a camera also used to capture the surface condition of the specimen at the same time.

3. Results and Discussion

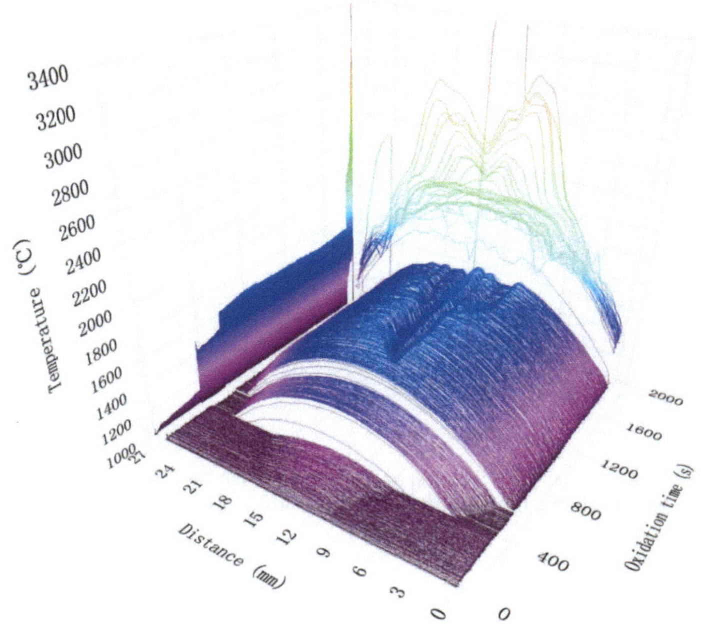

Figure 2. Real-time dynamic temperature distribution measurement along the length of the specimen.

Using the self-developed Joule heating apparatus, the ZrB2-base composite specimen can be heated up to 2000°C or above. Fig. 2 shows the overall temperature distribution of the specimen during heating procedure. It can be found that the heating rate is very fast, which can reach about 100°C/s. There are several stages existed in the graph where each stage corresponds to a specific electrical current value, which is 100 A, 115 A, 130 A, and 140 A, respectively. The temperature distribution could be recognized as quasi-steady and the curves changed smoothly at the initial two stages where the highest temperature is below ~1600°C. And the temperature distribution curve showed an arch-like shape without fluctuation. However, at the subsequent stages where the currents applied were large enough that the highest temperatures could exceed ~1600°C, the center of the curve begins to fluctuate and the overall shape of curve was seemed to begin to turn to a trapezoid-liked one which is approximately consistent with the result given by Gangireddy et al.[18]

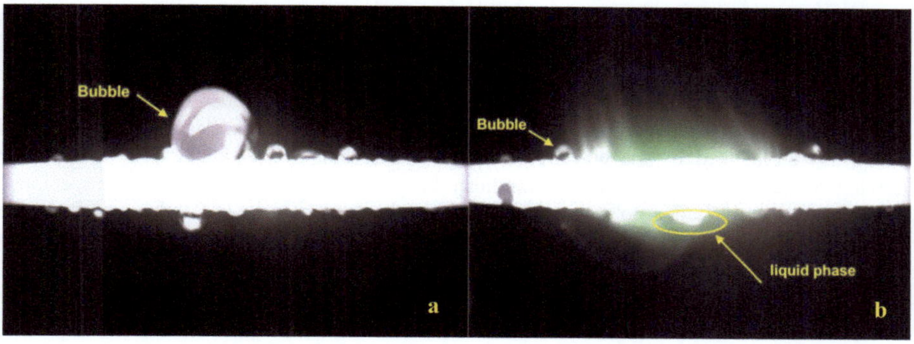

Figure 3. **The morphology of the surface condition of ZrB2-SiC specimen oxidized at over 2000 °C. The bubble burst phenomenon (a) and the flowing of the liquid phase (b).**

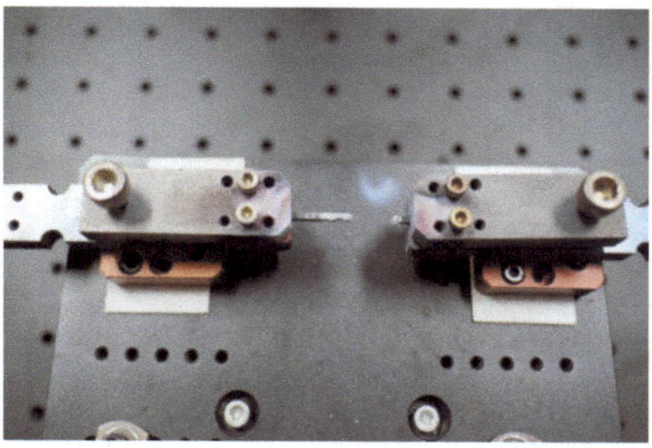

Figure 4 **Heating up to failure of the specimen using the apparatus and the morphology of the failed specimen with a ~9 mm air gap.**

Additionally, the morphology of the surface during the oxidation shown in the video recorded by the camera indicated that the fluctuation of the temperature curve is caused by the effect of formation and collapse of gas bubbles during oxidation. Gangireddy et al.[20] also observed the interesting bubble phenomenon, and they revealed that the onset temperature of the bubble formation and the number of bubbles are depend on the temperature of the oxidation, which can be further evidenced by Fig. 2. Form the shape of the curve, it is obvious that the bubble generated at the third stage where the highest temperature is nearly 1600°C can last for a long time (several minutes), and the number of the bubbles was limited. In contrast, when the current supply increased to 140 A, the highest temperature exceeded 2200°C and the bubbles generated at this stage grew and ruptured drastically resulting in a chaotic variation trend of the measuring temperature. The formation and burst of the bubbles showed a boiling-like phenomena overall, as illustrated in Fig 3a. Moreover, at the final stage, it should be noticed that the temperature distribution cannot reach a stable state as it does at previous stages, which was illustrated by Fig. 2. Meng et al.[17] reported that the material (ZrB_2 -20 % SiC) would fail when the center temperature was higher than 2300°C. In present work, we found in Fig 3b that the a liquid phase would form and subsequently flow down under the influence of the gravity at nearly and above 2300°C. Based on the composite phase diagram[21], the liquid phase may be induced by the eutectic reaction of ZrB_2 and SiC at about 2270°C. And the flowing of material will significantly diminish the cross-section of the center of the specimen, which consequently leaded to a drastic increase of the temperature. The failure mode of the specimen was observed as an sudden explosion which took place out of control and resulted in an air gap in the specimen, as shown in Fig 4.

4. Conclusions

In conclusion, the oxidation behavior of ZrB_2-SiC composite was investigated up to the temperature of material failure using an self-developed apparatus based on Joule heating. A novel method was proposed to measure the real-time temperature distribution along the length of a ribbon shaped specimen. The temperature distribution graph obtained by this method contains many essential information such as heating rate, bubble burst effect, etc. The influence of oxidation on the temperature measurement was evaluated, which suggested that the bubble burst event can significantly decrease the measuring temperature. The flowing of the liquid phase produced due to eutectic reaction of ZrB_2 and SiC was observed, which can significantly diminish the cross-section of the center of the specimen.

Acknowledgements

The authors gratefully acknowledge the financial support from National Natural Science Foundation of China (11272364 and 11332013).

References

1. Opeka, M.M., et al., "Mechanical, Thermal, and Oxidation Properties of Refractory Hafnium and Zirconium Compounds." ***Journal of the European Ceramic Society***, 1999. 19(13–14): p. 2405-2414.

2. Fahrenholtz, W.G., et al., "Refractory Diborides of Zirconium and Hafnium." ***Journal of the American Ceramic Society***, 2007. 90(5): p. 1347–1364.

3. Guo, S.-Q., "Densification of ZrB_2-Based Composites and Their Mechanical and Physical Properties: A Review." ***Journal of the European Ceramic Society***, 2009. 29(6): p. 995–1011.

4. Monteverde, F. and A. Bellosi "Oxidation of ZrB_2-Based Ceramics in Dry Air." ***Journal of The Electrochemical Society***, 2003. 150(11): p. B552–B559.

5. Opeka, M.M., I.G. Talmy, and J.A. Zaykoski, "Oxidation-based Materials Selection for 2000°C + Hypersonic Aerosurfaces: Theoretical Considerations and Historical Experience." ***Journal of Materials Science***, 2004. 39(19): p. 5887-5904.

6. Rezaie, A., W.G. Fahrenholtz, and G.E. Hilmas, "Oxidation of Zirconium Diboride–Silicon Carbide at 1500°C at A Low Partial Pressure of Oxygen." ***Journal of the American Ceramic Society***, 2006. 89(10): p. 3240-3245.

7. Han, W.B., et al., "High-temperature oidation at 1900 degrees C of ZrB_2-xSiC ultrahigh-temperature ceramic composites." ***Journal of the American Ceramic Society***, 2008. 91(10): p. 3328-3334.

8. Monteverde, F. and R. Savino, "Stability of ultra-high-temperature ZrB_2–SiC ceramics under simulated atmospheric re-entry conditions." ***Journal of the European Ceramic Society***, 2007. 27(16): p. 4797-4805.

9. Hu, P., et al., "Effect of SiC Content on the Ablation and Oxidation Behavior of ZrB_2-Based Ultra High Temperature Ceramic Composites." ***Materials***, 2013. 6(5): p. 1730-1744.

10. Han, J.C., et al., "Oxidation-resistant ZrB_2-SiC composites at 2200 degrees C." ***Composites Science and Technology***, 2008. 68(3-4): p. 799-806.

11. Tang, S.F., et al., "Ablation behaviors of ultra-high temperature ceramic composites." *Materials Science and Engineering a-Structural Materials Properties Microstructure and Processing*, 2007. 465(1-2): p. 1-7.

12. Jin, X.X., et al., "Ablation behavior of ZrB_2-SiC sharp leading edges." ***Journal of Alloys and Compounds***, 2013. 566: p. 125-130.

13. Li, X.Y., et al., "A technique for ultrahigh temperature oxidation studies of ZrB2-SiC." ***Materials Letters***, 2008. 62(17-18): p. 2848-2850.

14. Neuman, E.W., et al., "Case study: Building an ultra-high-temperature mechanical testing system." ***American Ceramic Society Bulletin***, 2013. 92(1): p. 36-38.

15. Karlsdottir, S.N. and J.W. Halloran, "Rapid Oxidation Characterization of Ultra-High Temperature Ceramics." ***Journal of the American Ceramic Society***, 2007. 90(10): p. 3233-3238.

16. Gangireddy, S., J.W. Halloran, and Z.N. Wing, "Non-Contact Mechanical Property Measurements at Ultrahigh Temperatures." ***Journal of the European Ceramic Society***, 2010. 30(11): p. 2183-2189.

17. Meng, S., et al., "Mechanisms of material failure for fast heating up at the center of ultra high temperature ceramic." ***Solid State Sciences***, 2010. 12(4): p. 527-531.

18. Gangireddy, S., J.W. Halloran, and Z.N. Wing, "Flexural Creep of Zirconium Diboride-Silicon Carbide up to 2200°C in Minutes with Non-Contact Electromagnetic Testing." ***Journal of the European Ceramic Society***, 2013. 33(15–16): p. 2901-2908.

19. Carney, C.M., T.A. Parthasarathy, and M.K. Cinibulk, "Separating Test Artifacts from Material Behavior in the Oxidation Studies of HfB2–SiC at 2000°C and Above." ***International Journal of Applied Ceramic Technology***, 2013. 10(2): p. 293-300.

20. Gangireddy, S., et al., "In situ microscopy observation of liquid flow, zirconia growth, and CO bubble formation during high temperature oxidation of zirconium diboride–silicon carbide." ***Journal of the European Ceramic Society***, 2010. 30(11): p. 2365-2374.

21. Kaufman, L., "Calculation of Multicomponent Refractory Composite Phase Diagrams." 1986.

ADVANCED COMPOSITES and SYNTACTIC FOAMS

Advanced Composites for Aerospace, Marine, and Land Applications II
Edited by: T. Sano and T.S. Srivatsan
TMS (The Minerals, Metals & Materials Society), 2015

DEVELOPMENT OF LIGHTWEIGHT CARBON NANOFIBER REINFORCED SYNTACTIC FOAM COMPOSITES

Steven Eric Zeltmann, Ronald Poveda, and Nikhil Gupta

Composite Materials and Mechanics Laboratory
Mechanical and Aerospace Engineering Department
New York University Polytechnic School of Engineering
Brooklyn, NY 11201 USA

Keywords: Nanocomposite, Coefficient of Thermal Expansion, Dielectric Constant

Abstract

Syntactic foams are a class of hollow particle filled composites which have found many applications in the marine sector due to their outstanding specific compressive properties and tailorable buoyancy. A limitation of these composites is that the tensile and flexural properties are severely limited due to the addition of hollow particles. Carbon nanofibers reinforcement of the matrix has been explored as a means of improving the tensile properties of such composites without sacrificing light weight. The nanofibers are dispersed between the hollow particles and do not affect the packing efficiency of the filler. The unique electrical and thermal properties of CNFs also present opportunities to tailor these properties of the composite beyond the range traditionally achievable by only particles. Thermal and electrical properties of CNF syntactic foams will be examined with a focus on the potential applications of these composites.

1. Introduction

Syntactic foams, a class of hollow particle reinforced composites, mainly find use in applications where buoyancy and good stiffness under hydrostatic compression are the main parameters [1]. Tailorability of the thermal and electrical properties of syntactic foams based on volume fraction and wall thickness of the hollow filler have been investigated, aiding in the use of such materials where these properties are also a design parameter [2, 3].

Under tensile loading, the interface between the particle and the matrix is not able to support much loading. This leads to lower tensile strength for syntactic foams. Reinforcement of syntactic foams with carbon nanofibers (CNF) in the matrix is of interest because of the possibility of increasing the tensile strength of the composite without affecting the maximum packing fraction of hollow filler particles. CNF can impart excellent mechanical [4], thermal expansion [5, 6], and electrical [7-11] to polymer matrix composites. In addition, a number of studies are available on the mechanical properties of CNF-reinforced syntactic foams [12, 13].

In the present work, CNF/syntactic foams are evaluated for thermal and electrical properties. The linear coefficient of thermal expansion of various compositions of CNF/syntactic foams and CNF/epoxy composites is determined through thermo-mechanical analysis. Electrical properties, namely AC impedance and dielectric constant, are determined for the composites using an electrochemical workstation.

2. Materials and Methods

The composites used in this study are fabricated using an epoxy matrix. Epoxy resin DER 332 (Dow Chemical, Midland, MI) is used with triethylenetetramine hardener (Huntsman Co., The Woodlands, TX) forms the matrix system. The hollow filler is glass microballoons (GMB) (3M, St. Paul, MN) of 220 and 460 kg/m^3 true particle density. Carbon nanofibers of type PR-24-XT- PS supplied by Pyrograf Products (Cedarville, OH) are used as the nano-scale reinforcement.
CNF/epoxy composites are fabricated with 1wt. %, 2wt. %, and 5wt. % CNF content. CNF/syntactic foams containing 30% and 50% GMB by volume are fabricated for each GMB density at 1wt. % and 2 wt. % CNF. Because the viscosity of the mixture increases greatly with CNF content, GMB crushing is deemed to be excessive when fabricating composites with 50% GMB and high fractions of CNF. Therefore, only composites contacting 30 vol. % GMB are fabricated at 2wt. % and 5wt. % CNF. The composite is fabricated by first dispersing CNF in the epoxy resin using high-shear mechanical mixing for 30 min. The GMB are then stirred into the mixture by hand for 15 min., the hardener is added, and this is stirred for 5 additional minutes. The composite is cast into molds and cured at room temperature for 24 hours, followed by 2 hours at 90°C. The composites are named according to the pattern Nxx-yyy-zz, where 'xx' is the weight percentage of CNF, 'yyy' is the microballoon density, and 'zz' is the microballoon volume percentage.

Thermo-mechanical analysis of the composites is conducted using a Q400 TMA (TA Instruments, New Castle, DE). Rectangular prism specimens of nominal dimensions

$4{\times}4{\times}12\ mm^3$ were used for testing. The temperature was ramped at a rate of 3°C/min up to 90°C, similar to previous studies on plain syntactic foams [3]. The CTE value is obtained as the slope of the thermal strain-temperature curve.

Electrical characterization of the composites was performed using a CH Instruments 700D (Austin, TX) electrochemical potentiostat. Specimens of nominal dimensions $18{\times}14{\times}1\ mm^3$ were placed between gold electrodes and impedance and dielectric constant were measured in the range 10^{-2}-10^6 Hz with an applied voltage of 500 mV. The impedance, denoted by *Z,* was obtained using the measurement of the electrical resistance *R* and the capacitive reactance [14] X_c as

$$Z = R^2 + X_c^2 \qquad (1)$$

Both the resistance and the reactance, which represent the real and imaginary parts of the impedance, respectively, can also be utilized to calculate the phase angle, φ

$$\Phi = \tan^{-1} [X_c/R] \qquad (2)$$

The capacitance C is calculated by

$$C = [1/ 2\Pi f X_c] \qquad (3)$$

where f is the frequency. The dielectric constant is calculated by

$$\varepsilon = [C\, t / \varepsilon_0 A] \qquad (4)$$

where t is the specimen thickness, A is the area, and ε_0 is the permittivity of free space. The value of ε_o is taken as $8.854{\times}10^{-12}$ F/m.

3. Results and Discussion

3.1 Thermo-Mechanical Analysis

In all experiments, no non-linearity in the thermal strain-temperature curve was observed within the temperature range tested. The calculated values of the coefficient of thermal expansion of all the compositions tested in this work are presented in Table 1. The CTE of the neat resin was measured to be $65.4{\times}10^{-6}$/°C, and increase of CNF content was observed to monotonically decrease the CTE of the composite. The lowest CTE observed in the CNF/epoxy composites was $60.1{\times}10^{-6}$/°C, which represents an 8% decrease with respect to the neat epoxy. However, it is noted that the CTE of the N1 composite is higher than that of the neat resin, which is likely due to the presence of voids and defects resulting from the fabrication process and the dispersion of CNF.

For the CNF/syntactic foams, the CTE of the composite was observed to decrease with increasing GMB weight fraction and GMB wall thickness. Such decreases can be attributed to the low coefficient of thermal expansion of glass, 4×10^{-6}/°C [3], compared to the neat resin. Maximum reduction in the CTE of the composite with respect to neat resin was observed in the N1-460-50 composite. The CTE of this composition was measured to be 40.3×10^{-6}/°C, which is a decrease of 38% with respect to the neat resin. Thus, by increasing the volume fraction of microballoons in the composite, the composite density and the CTE can be simultaneously reduced. Further, by simultaneously manipulating the volume fraction and wall thickness of the hollow filler, the CTE can be adjusted while maintaining some other design parameter constant. This ability will be important in helping such materials find applications. The effect of the CNF on the CTE of the composite is not as significant as the effect of the GMB. For a given volume fraction and wall thickness of GMB, there is no significant variation in the composite CTE with CNF weight fraction.

3.2 Electrical Properties

The measured electrical properties of CNF/epoxy composites over the full range of frequencies are presented in Figure 1 and Figure 2, which show impedance and dielectric constant, respectively. For all compositions, the impedance and dielectric constant decreases with increasing frequency. The impedance of the CNF composites is lower than that of the neat matrix at all frequencies tested. The dielectric constant of the CNF/epoxy composites is higher than that of the neat matrix at all frequencies and shows a greater sensitivity to frequency. The electrical properties of these composites at the frequency of 10^5 Hz are shown in Table 2. At this frequency, the dielectric constants of the CNF/epoxy composites are all higher than that of the matrix by 2.6-3.4 times. The dielectric constant increases monotonically with CNF content at this frequency. The phase angle of the complex impedance is largely unaffected by CNF content.

The measured dielectric constant of the CNF/syntactic foams tested is shown in Figure 3 as a function of frequency, and the impedance and dielectric constant at the frequency of 10^5 Hz are shown in Table 3. For fixed GMB wall thickness and volume fraction, dielectric constant is observed to increase with CNF content. For fixed CNF content and wall thickness, dielectric constant is observed to decrease with increasing GMB content. The CNF/syntactic foam composites have higher dielectric constant than the neat matrix, except for the N1-220-50 composite. The CNF/syntactic foams also retain lower impedance than the neat matrix, again excepting the N1-220-50 composite.

3.3 Potential Applications

The benefits in thermal and electrical properties that are achieved in syntactic foams by the addition of nanofiber reinforcement will help in making these composites attractive in new applications. As noted above, the impedance of the syntactic foam composites are generally better than the neat resin, and thus the conductivity of the composite is higher. Applications where static charge buildup is a concern may benefit from using thin coating of CNF/syntactic foams as a lightweight conductive medium for charge dissipation. At nanofiber loadings higher than those studied in this work the conductivity benefit may be even greater [15]. The tailorability of the thermal properties is also useful in that it can allow matching of the CTE of the composite to other components of a structure. By selecting a composition that minimizes CTE dimensional stability can be maximized, which is of great importance in applications such as artificial satellites and space mirrors which experience large temperature variations and must retain a desired shape.

4. Conclusions

In this work, the thermal and electrical properties of CNF/epoxy and CNF/syntactic foam composites are evaluated using thermo-mechanical and electrochemical analysis, respectively. In the thermal study, the CTE of these composites was found to decrease with increasing CNF content, GMB content, and GMB wall thickness. Thus, it is possible to realize simultaneous decrease in composite density and CTE, which is useful for many applications requiring low weight and good dimensional stability. In the electrical study, the impedance of these composites was found to decrease with increasing CNF content, and in most cases the composite impedance is lower than that of the neat matrix. The dielectric constant of the composites was found to increase with increasing CNF content, and all of the compositions except the N1-220-50 composite showed higher dielectric constant than the neat matrix. Because both the dielectric constant and the CTE are modifiable by means of CNF fraction, GMB content, and GMB wall size, these properties of the composite are highly tailorable, which is beneficial in finding applications for this advanced class of materials?

Acknowledgments

This work is supported by the Office of Naval Research through the grant N00014-10-1-0988 and the National Science Foundation GK-12 Fellows grant 0741714 (PI: Dr. V. Kapila). The authors thank 3M, MN for providing microballoons and relevant technical information. Dr. Vasanth Chakravarthy Shunmugasamy is thanked for assistance with manuscript preparation.

Table 1. **Measured CTE of CNF/epoxy and CNF/syntactic foam composites.**

Composite type	Density(kg/m^3) Measured	Density(kg/m^3) Theoretical	Experimental CTE ($\times10^{-6}/°C$)
N1	1168.8	1164.7	68.0±3.1
N1-220-30	918.8	883.5	51.2±2.5
N1-220-50	724.9	694.0	41.9±3.4
N1-460-30	909.7	955.5	46.0±1.7
N1-460-50	835.5	814.0	40.3±4.7
N2	1212.2	1169.5	67.0±3.7
N2-220-30	827.5	881.7	52.8±4.9
N2-460-30	949.4	957.7	42.0±4.1
N5	1198.8	1184.0	60.1±4.6
N5-220-30	886.5	896.0	52.1±3.6
N5-460-30	937.1	969.0	42.5±5.1

Table 2. **Electrical properties of CNF/epoxy composites at representative frequency of 10^5Hz.**

Composite type	Resistance ($\Omega\times10^3$)	Capacitance ($F\times10^{-11}$)	Reactance ($\Omega\times10^3$)	Phase angle (°)	Dielectric constant (ε)
Epoxy	4.6	1.1	-151.1	-88.3	4.8
N1	2.2	2.6	-64.8	-88.0	12.4
N2	2.8	3.1	-51.5	-86.9	13.8
N5	1.9	3.1	-52.0	-87.9	15.8

Table 3. **Electrical properties of CNF/syntactic foams at representative frequency of 10^5Hz.**

Composite type	Resistance ($\Omega\times10^3$)	Capacitance ($F\times10^{-11}$)	Reactance ($\Omega\times10^3$)	Phase angle (°)	Dielectric constant (ε)
N1-220-30	2.7	1.7	-95.2	-88.4	7.7
N2-220-30	3.1	2.0	-80.6	-87.8	9.5
N5-220-30	6.3	2.5	-65.3	-84.4	12.1
N1-220-50	4.7	0.72	-255.7	-88.9	3.5
N1-460-30	1.5	1.5	-104.7	-89.2	7.1
N2-460-30	2.9	1.9	-86.1	-88.0	8.4
N5-460-30	3.6	3.3	-49.0	-85.8	14.1
N1-460-50	3.9	1.1	-142.5	-88.4	5.4

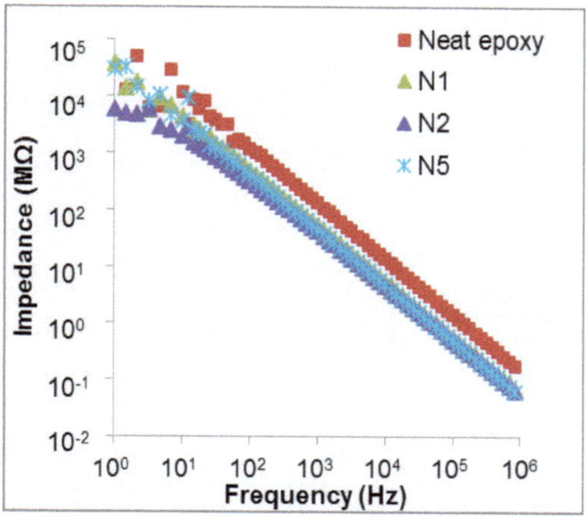

Figure 1. Impedance of CNF/epoxy composites as a function of frequency.

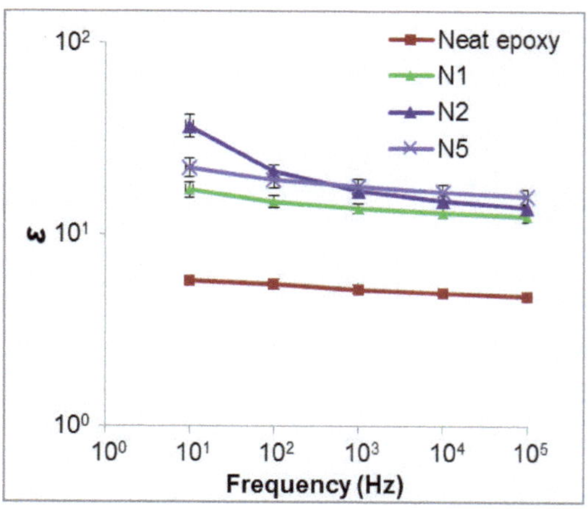

Figure 2. Dielectric constant of CNF/epoxy composites as a function of frequency.

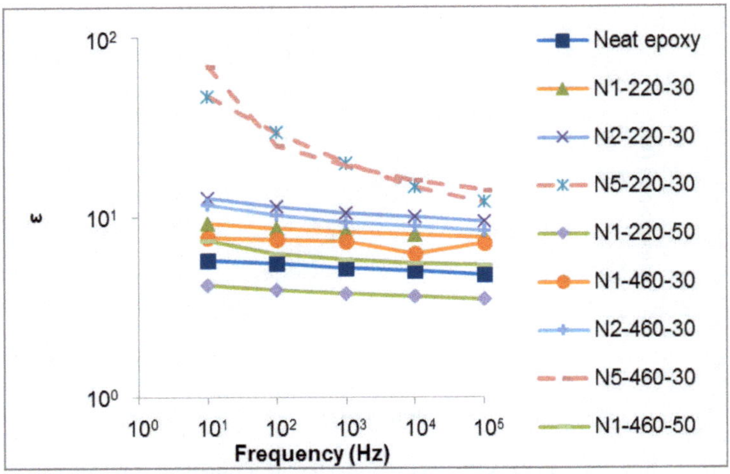

Figure 3. **Dielectric constant of CNF/syntactic foams as function of frequency.**

References

1. Gupta, N., et al., *Applications of polymer matrix syntactic foams*. **JOM: Journal of The Minerals, Metals, and Materials Society**, 2014. **66**(2): p. in press, DOI: 10.1007/s11837- 013-0796-8.

2. Shunmugasamy, V.C., D. Pinisetty, and N. Gupta, *Electrical properties of hollow glass particle filled vinyl ester matrix syntactic foams*. **Journal of Materials Science**, 2014. **49**(1): p. DOI: 10.1007/s10853-013-7691-0.

3. Shunmugasamy, V., D. Pinisetty, and N. Gupta, *Thermal expansion behavior of hollow glass particle/vinyl ester composites.* **Journal of Materials Science**, 2012. **47**(14): p. 5596- 5604.

4. Al-Saleh, M.H. and U. Sundararaj, *Review of the mechanical properties of carbon nanofiber/polymer composites.* Composites Part A: **Applied Science and Manufacturing**, 2011. **42**(12): p. 2126-2142.

5. Chen, Y.-M. and J.-M. Ting, *Ultra high thermal conductivity polymer composites.* Carbon, 2002. **40**(3): p. 359-362.

6. Poveda, R., S. Achar, and N. Gupta, *Thermal expansion of carbon nanofiber-reinforced multiscale polymer composites.* **JOM**, 2012. **64**(10): p. 1148-1157.

7. Chawla, S., M. Naraghi, and A. Davoudi, *Effect of twist and porosity on the electrical conductivity of carbon nanofiber yarns.* **Nanotechnology**, 2013. **24**(25): p. 255708.

8. Al-Saleh, M.H. and U. Sundararaj, *A review of vapor grown carbon nanofiber/polymer conductive composites.* **Carbon**, 2009. **47**(1): p. 2-22.

9. Al-Saleh, M.H., G.A. Gelves, and U. Sundararaj, *Carbon nanofiber/polyethylene nanocomposite: Processing behavior, microstructure and electrical properties.* **Materials & Design**, 2013. **52**(0): p. 128-133.

10. Bal, S., *Experimental study of mechanical and electrical properties of carbon nanofiber/epoxy composites.* **Materials & Design**, 2010. **31**(5): p. 2406-2413.

11. Al-Saleh, M.H. and W.H. Saadeh, *Hybrids of conductive polymer nanocomposites.*, **Materials & Design**, 2013. **52**(0): p. 1071-1076.

12. Poveda, R. and N. Gupta, *Carbon-nanofiber-reinforced syntactic foams: compressive properties and strain rate sensitivity.* **JOM**, 2014. **66**(1): p. 66-77.

13. Poveda, R.L., S. Achar, and N. Gupta, *Viscoelastic properties of carbon nanofiber reinforced multiscale syntactic foam.* **Composites Part B: Engineering**, 2014. **58**(0): p. 208- 216.

14. Arfin, T. and N. Yadav, *Impedance characteristics and electrical double-layer capacitance of composite polystyrene–cobalt–arsenate membrane.* **Journal of Industrial and Engineering Chemistry**, 2013. **19**(1): p. 256-262.

15. Poveda, R.L. and N. Gupta, *Electrical properties of carbon nanofiber reinforced multiscale polymer composites.* **Materials & Design**, 2014. **56**(0): p. 416-422.

Advanced Composites for Aerospace, Marine, and Land Applications II
Edited by: T. Sano and T.S. Srivatsan
TMS (The Minerals, Metals & Materials Society), 2015

EVALUATION OF SYNTACTIC FOAM FOR ENERGY ABSORPTION AT LOW TO MODERATE LOADING RATES

T.R. Walter, J. Sietins, P. Moy

The United States Army Research Laboratory
4600 Deer Creek Loop
APG, Maryland 21005

Keywords: Syntactic Foam, Energy Absorption, Strain Rate

Abstract

Many structural applications require the use of lightweight energy absorbing materials for the increase in mass efficiency. Sandwich composite structures have been implemented to fill this need in both aerospace and marine applications. To maximize the structural performance the core material should have excellent compressive stiffness and strength as well as good shear strength. Recently syntactic foams have been developed to offer a simple cost effective method to produce a number of complex molded shapes with high relative energy and stiffness. This study compared the mechanical properties of traditional foams and recently developed syntactic foams to determine the energy absorbing characteristics at low and moderate rates of loading. Test results were fit to a phenomenological model to further study the mechanical behavior. Results from this study will demonstrate the need for the development of new, efficient cellular material for energy absorption and protection applications.

1. Introduction

Foam materials offer several advantages when used in the design of structures. Although foams have very low strength compared to other structural materials, they may be utilized to add stiffness, increase strength and enhance the ability of a structure to absorb energy without adding significant mass due to their low density. Sandwich construction using foam core materials is a well established design technique, and many industries incorporate sandwich design to increase stiffness while maintaining or reducing mass. One example of the importance of structural foams is the application to the auto-industry. Using foam to fill hollow structural components of automobiles allows the use of lighter construction materials while maintaining or even increasing crash test performance. This is critical to the design and manufacture of fuel efficient vehicles while maintaining occupant safety. Using similar principles it may be possible to incorporate structural foams into both the design of military vehicles and personal protection equipment to help mitigate energy and protect soldiers during impact and blast.

Foams are produced from a variety of materials including polymer, metal and ceramic and are often classified by their internal structure such as open-cell or closed-cell. Many polymer foams are produced using a liquid polymer above the melt temperature or in an uncured state with the addition of a blowing-agent to produce the cellular structure. In some cases the reaction to produce the polymer itself can release the blowing agent as in the case of polyurethane foams. Foams are also classified by their mechanical response such as brittle, plastic, or elastic [1] with most rigid, high-strength foams having large permanent deformations and softer elastic foams having near complete recoverability. Other examples of foam material include syntactic foam. These foams are composite materials which include hollow micro-spheres or micro-balloons, blended with a matrix material and may be cast into any variety of shapes. This process allows the production of complex shapes using advanced matrix materials without the need for complex blowing agents.

Polymer foams have three traditional regions of deformation during compression. The initial linear elastic region terminates at the compressive collapse stress which is followed by a long plateau region and then densification. The collapse stress is sometimes clearly identified by a small drop in stress before the plateau. This has been attributed to a large instability caused by the collapse which results in non-uniform propagation of damage during the plateau. Research has shown [1] that the value of stress required to initiate collapse is highly dependent on the relative density and constitutive properties (e.g. modulus and compressive strength of the solid polymer). The densification region is characterized by the sharp increase in stress which is due to the compaction or the bulk material. The beginning of this region, defined as the densification strain, is highly dependent on the porosity or relative density of the material. The plateau region is responsible for most of the energy absorbing capabilities and is dependent on the collapse stress and densification strain. The stress-strain curve may be evaluated to determine the work per unit volume at any point by integrating the response. While this energy is an important property for the foam it may be more informative to examine the efficiency [2] which is defined as the energy at a point along the stress strain curve divided by the stress at that instance along the response:

$$E = \frac{\int \sigma d\varepsilon}{\sigma} \tag{1}$$

A material with a slowly increasing plateau region may absorb more energy but will have lower efficiency. Also, as the foam response enters the densification region, the efficiency will show a significantly drop.

Several models have been proposed to capture the response of foams. Gibson and Ashby [1] have proposed a simple yet effective model which separates the three regions into a piecewise linear response. The parameters are highly dependent on the density and solid material properties. Another model originally proposed by Rusch [3, 4] and later modified by Avalle et al [5] was developed to capture the compressive response however it is unable to capture the tensile response. Liu and Subhash [6-8] developed a phenomenological model which is continuous in both tension and compression. This model is shown below,

$$\sigma = A\frac{e^{\alpha\varepsilon}-1}{B+e^{\beta\varepsilon}} + e^{C}(e^{\gamma\varepsilon}-1) \tag{2}$$

Contrary to common practice, compression is considered positive while tension is negative. The strain is typically percent engineering strain although other forms are valid. The fist term of the model captures the linear and plateau region while the second term captures densification. The parameters of the first term have been shown to correlate well with common engineering parameters. Collapse stress correlates to 'A', the modulus is dependent on 'α' while 'β' controls the slope of the plateau region. The tensile strength is related to the ratio A/B and therefore dependent on 'B'. The parameters 'C' and 'γ' control the onset and rate of densification respectively.

Compression tests are easy to perform and provide great insight into the mechanical behavior of foam. However, due to the structural nature of the material it is difficult to understand the mechanisms involved during the initial collapse and densification of the material. X-Ray computed tomography (CT) uses a series of x-ray images to capture detailed internal structure of the material. By performing in-situ compression tests during CT scanning, the internal deformation mechanisms may be recorded.

In this paper six different foams are evaluated. These foams include syntactic foams produced with glass microspheres and traditional rigid PVC foam. Compression tests were performed at low and moderate rates. The results were fit using the Liu-Subhash (LS) model to determine the rate effects on the model parameters while also providing insight into the energy absorption characteristics of the materials. In-situ compression tests were also performed using CT to observe the deformation mechanisms during testing. The results are compared and potential applications are discussed.

2. Materials

Six different polymer foam materials were used in this study. These foams include both syntactic and traditional polymer foams of various densities. A summary of the foams are presented in Table 1 and discussed in detail below. Scanning electron microscopy (SEM) and optical images of the foam structures are shown in Figure 1.

Table 1 Summary of foam materials

Material	Density (kg/m^3)	Matrix	% Balloon
UL2	*352*	*Silicone*	*62*
ALT2	*335*	*Silicone*	*66*
HEL830	*319*	*Epoxy*	*79*
H35	*38*	*PVC*	-
H45	*48*	*PVC*	-
HP60	*65*	*PVC*	-

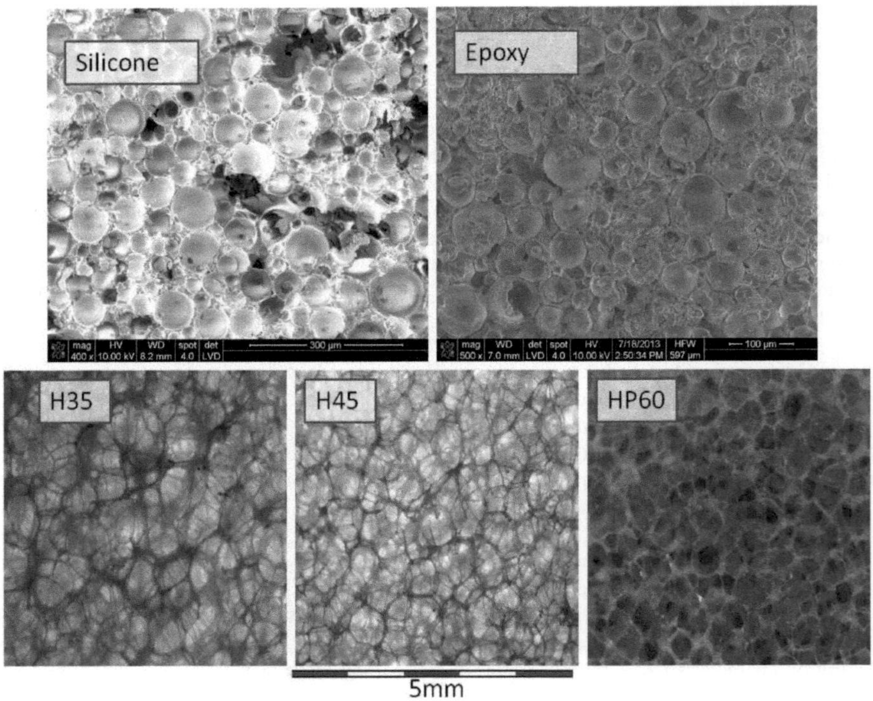

Figure 1 Micrographs of polymer foam structures. Examples of the syntactic foams (top) were imaged using SEM while optical images were used for PVC foam (bottom)

2.1 Trelleborg syntactic foam

Three syntactic foams were tested during this study. These included two silicone based foams ALT2 and UL2 with densities of 335 kg/m^3 and 352 kg/m^3 respectively and one epoxy foam HEL830 with a density of 319 kg/m^3. The density is controlled by the percent loading of glass micro-balloons. The spheres used to produce the foam are Trelleborg Eccosphere SI-130. The panels were custom made and supplied by Trelleborg AB, Sweden in sheets approximately 19 mm thick. Specimens were cored out of the sheets to produce 19 mm diameter cylinders.

2.2 DIAB Divinycell

Divinycell is a polymer foam produced by DIAB Group, Desoto TX in several different grades and densities ranging from 35kg/m^3 to 250kg/m^3. The bulk polymer is a PVC blend containing Polyurea and Polyamide produced using a thermal process that allows a blowing agent to expand the foam into sheets. Although the foam is produced in sheets, the final product is thermal formable and can be molded into simple shapes. The foams used in this study include H35, H45 and HP60 with nominal densities of 38, 45, and 65 kg/m^3 respectfully. The HP grade is a higher performance series of foam but is not available in lower densities. Although the Divincell foams used in this study have much lower densities than the syntactic foams, the manufacture data sheets indicate strengths similar to the Trelleborg silicone based foam. The foam was provided in large sheets approximately 12.7 mm thick. These sheets were sectioned into small test coupons; for traditional compression tests these coupons had a square cross section of 25.4 mm x 25.4 mm, while 13 mm diameter cylinders were cut for computed tomography testing. Due to limitations in the maximum stroke during in-situ CT tests; these specimens were also trimmed to approximately 7 mm to achieve densification strains.

3. Test procedures

3.1 Mechanical Testing

Compression tests were performed on each material to acquire the entire compressive response including densification. Tests were performed on a electro-mechanical test frame (Instron 1122) at strain rates of 0.01 /s and 0.1 /s. Higher rate tests were performed using a servo-hydraulic test frame (Instron 1331) at rates of 1.0 /s and 10 /s. Due to the brittle nature of the epoxy based syntactic foam, an acrylic confinement cell was used to extend the test duration into densification. Strain was determined in each case using machine displacement.

3.2 Micro-CT Procedure

A Bruker Skyscan 1172 x-ray micro-computed tomography (Micro-CT) scanner was used to obtain three-dimensional views of the HP60, H35, and H45 foam microstructures. The scans were conducted with a 60KeV voltage, 167uA current, and a resolution of 4.5 microns/pixel. In-situ compression tests were conducted on the foams with cylindrical sample sizes of 13 mm in diameter and approximately 7 mm in height. Samples were loaded at a rate of 0.2 mm/min and scans were conducted prior to any loading and at 0.5, 3, and 5.3 mm displacements. Prior to each scan after loading, samples were held for 20 minutes in order to stabilize due to relaxation. Micro-CT scans were conducted only on the DIAB foams due to force limitations of the in-situ loading stage for the syntactic foams and the inability to achieve adequate resolution which is limited by the distance between the specimen and x-ray source.

4. Results and discussion

4.1 Mechanical

Representative curves of the compressive response for each foam are shown in Figure 2. These graphs compare the response of the foam at four rates of loading. From these graphs several observations can be made. Although the density of the silicone foam is much higher than the Divinycell, the collapse strengths are very similar at quasi-static rates. The ATL2 foam strength correlates well with H35 and UL2 correlates to HP60. The plateau region of the UL2 foam is initially flat but quickly begins to harden as the foam deforms. The ALT2 shows a hardening response immediately after the collapse stress. This behavior is in contrast to the response of the Divinycell foam which has a very constant flat plateau region. The effect of this behavior will be further examined later by determining the efficiency of the foam. As the loading rate increases the response also increases. It was noted that the rate of increase is much higher in the silicone foam. The UL2 foam is clearly stronger then the HP60 foam at a strain rate of 10 /s. This is likely due to the visco-elastic nature of the silicone matrix.

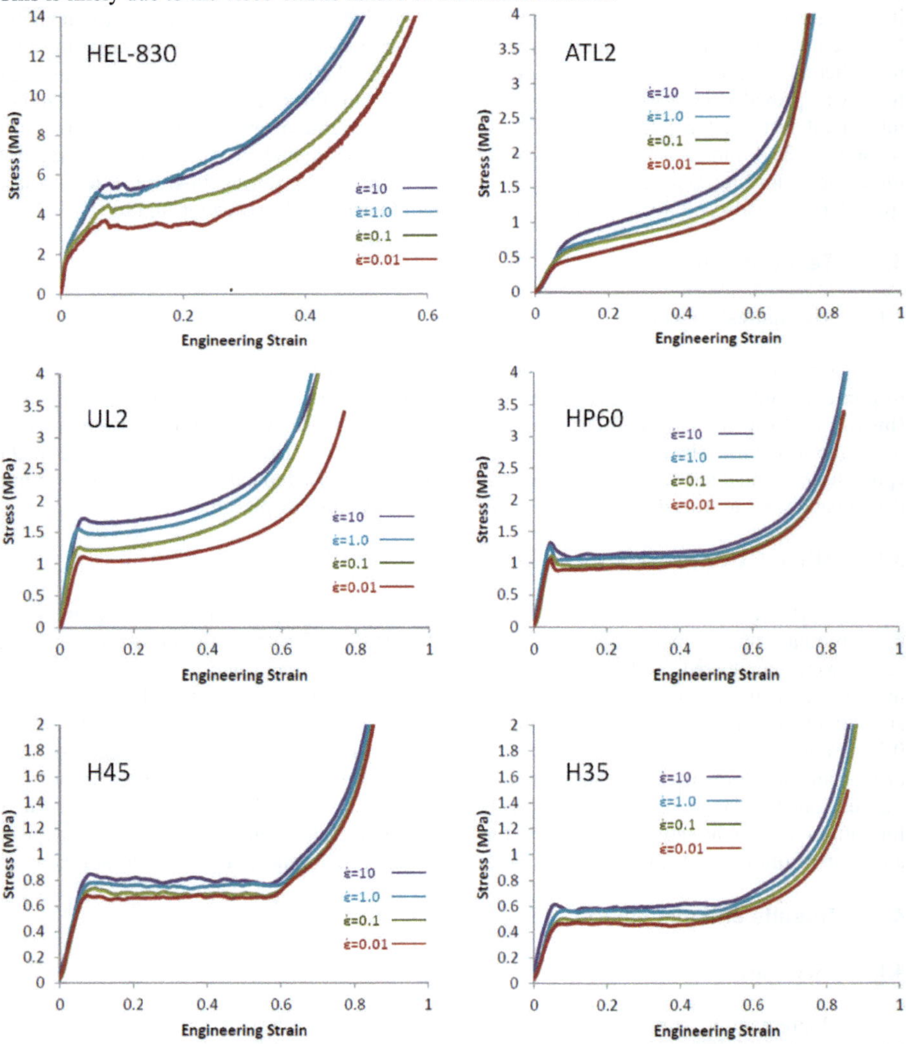

Figure 2 Representative stress-strain curves for each foam material at four strain rates

As mentioned above a confinement cell was used in the testing of the epoxy foam due to the brittle nature of the foam. The use of confinement prohibits the damaged material from being ejected from the specimen, allowing the test to continue into the densification region. Even with confinement, the response shows signs of brittle failure with a jagged plateau. The response shows two distinct linear regions before collapse. This behavior is interesting and is a sign of an initial damage mechanism which is later superseded by the collapse of the foam structure. Further analysis is necessary to explain this behavior and will not be explored in the current work. Like the previous foams the epoxy foam also shows clear rate dependency on its compressive response. The strength and plateau are both much higher than the other foams tested in this study.

Post-mortem observations of the specimens reveal large permanent deformation. This is most significant in the epoxy foam which has no recovery. The Divinycell foams show slight elastic recovery but have little remaining strength before densification. Although the silicone foams appear to recover once load is removed, further testing revealed it to have lost most of its strength. Scanning electron microscopy was used to examine the damaged syntactic foams and example images are shown in Figure 3. It is clear that the micro-balloons have considerable damage resulting in total loss of most contributing strength. This indicates that the loss of strength in the silicone syntactic foam is likely due to the fracture of the micro-balloons.

Figure 3 SEM micrographs of damaged syntactic specimens showing fractured micro-balloons. Image on left is silicone foam, image on right is epoxy foam.

4.2 Modeling

A MatLab program was used to perform a nonlinear regression analysis using the LS-model (Eq.2) to fit each set of data. For each set of tests, model parameters were determined which represent the average response and are shown in Table 2. As previously discussed the LS-model is designed to fit both compression and tension data, and since this study is focused on the compression response the 'B' parameter is set to a value of 1. This results in an assumption that the collapse strength is equal to the max tensile strength. The model parameters provide a smooth response which will be used to represent the average response of the foam at each strain rate. The model can be numerically integrated and further analyzed to determine the energy absorbing characteristics of the foam materials.

Table 2: LS-model parameter averages for polymeric foam at various strain rates

	H35				H45				HP60			
$\dot{\varepsilon}$ 1/s	0.01	0.1	1	10	0.01	0.1	1	10	0.01	0.1	1	10
A	0.47	0.48	0.52	0.55	0.67	0.68	0.79	0.83	0.91	1.00	1.08	1.18
α	0.52	0.47	0.55	0.63	0.44	0.45	0.41	0.49	0.75	0.62	0.80	0.78
β	0.52	0.47	0.54	0.63	0.44	0.44	0.41	0.49	0.75	0.62	0.80	0.78
C	-7.42	-9.27	-9.52	-8.97	-9.14	-10.92	-9.46	-8.42	-7.44	-7.34	-7.06	-6.16
γ	0.09	0.11	0.11	0.11	0.11	0.13	0.12	0.11	0.10	0.10	0.09	0.08
	HEL830				UL2				ALT2			
$\dot{\varepsilon}$ 1/s	0.01	0.1	1	10	0.01	0.1	1	10	0.01	0.1	1	10
A	3.71	4.29	4.85	5.39	0.92	1.15	1.38	1.60	0.46	0.55	0.60	0.76
α	0.67	0.69	0.70	0.50	0.54	0.69	0.78	0.52	0.37	0.37	0.35	0.25
β	0.67	0.70	0.72	0.52	0.53	0.69	0.78	0.52	0.36	0.35	0.33	0.24
C	-1.41	-0.66	-0.59	0.28	-8.05	-6.88	-4.95	-4.81	-11.56	-10.26	-10.12	-8.76
γ	0.07	0.06	0.06	0.05	0.12	0.11	0.08	0.08	0.16	0.15	0.14	0.13

4.3 Energy and Efficiency

The absorbed energy for each foam material and loading rate was computed using the LS-model and average model parameters by integrating the response. Figure 4 shows the results from this analysis. It is clear that the epoxy foam absorbs the most energy before densification. However the higher stress required to initiate collapse results in much higher forces to absorb lower levels of energy. The Divinycell PVC foams are capable of absorbing much higher energy at low stress making them more desirable to prevent injury during impact. Slight rate sensitivity was observed from the analysis however the overall trends do not change with strain rate.

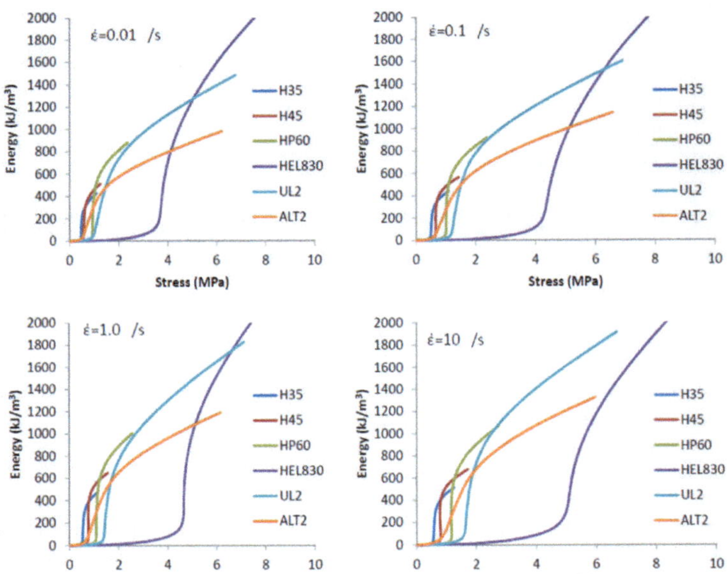

Figure 4 Energy versus stress plots for each strain rate.

From the energy determined using the LS-model, the efficiency was calculated as discussed previously. Efficiency is an indicator of the foams ability to absorb energy without increasing the stress or force to deform the material. Figure 5 shows the efficiency versus strain for each foam and rate of loading. It was observed that the Divinycell foam achieved much higher peak efficiency and maintained a higher efficiency through its deformation. This is due to the level plateau stress these foams achieve which is similar to the behavior of an ideal energy absorber. Since the PVC foams also have much lower relative density, this high efficiency is carried over a larger range of deformation. From these results it is clear that although the epoxy syntactic foam absorbs higher energy it does so at the cost of efficiency. The Divinycell foam has higher efficiency, lower density, and high relative strength. It was also noted that the efficiency was not highly rate dependent.

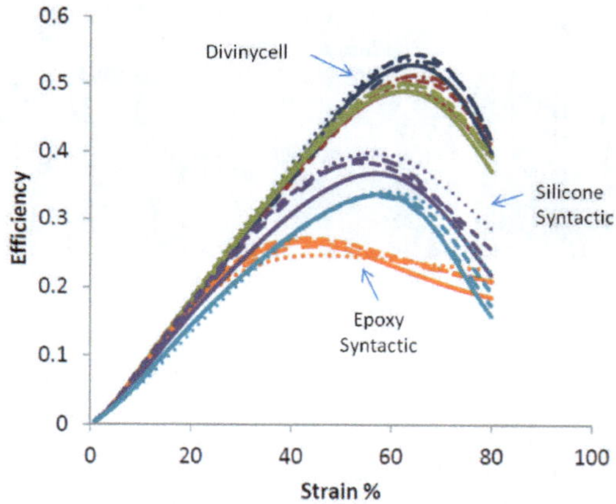

Figure 5: Efficiency plots determined using LS-Model parameters for each Foam and strain rate. Low strain rates are shown with solid lines while the highest strain rates are shown with dots. No clear trend was observed regarding strain rate

4.4 Micro-CT Results

After sample reconstruction, all three viewing planes were observed in DataViewer. Images in the thickness direction for each foam and scanning load are shown in Figure 6. Prior to any loading, the struts of the foam appear straight with no observable deformations, as expected. At 0.5 mm displacement, several cell walls began to collapse and buckle. At 3 mm displacement, it was observed that some of the collapsed and buckled cell walls are starting to come into contact with one another, indicating the onset of the densification stage. Numerous cells walls that previously collapsed were in contact at 5.3 mm displacement.

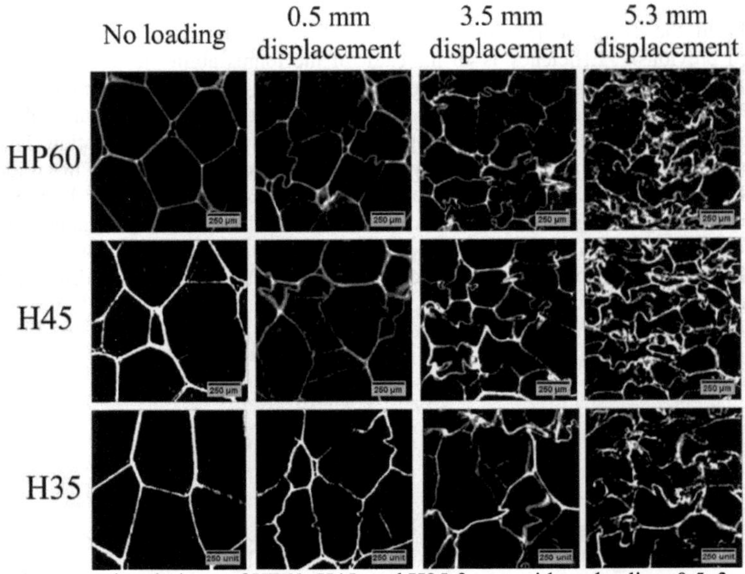

Figure 6. Cross-sectional views of HP60, H45, and H35 foams with no loading, 0.5, 3, and 5.3 mm displacement

Stress-strain responses for the in-situ micro-CT compression tests are shown in Figure 7 (left), with the red markers indicating the positions the compression loading was stopped before a twenty minute hold followed by the x-ray scan. The response of the foam closely resembles the quasi-static loading presented above. CTan was used to quantify the percent porosities at varied stages of loading for each foam and was plotted as a function of strain as shown in Figure 7 (right). This figure shows approximately the same percent porosity reduction for each foam as a function of the strain. It is important to note that the percent porosity calculation is highly dependent on the thresholding value when converting the 2D image slices to binary images for quantitative analysis. This is due to the density of the foam increasing under each additional compressive load. Therefore, the x-ray attenuation through the sample changes and more x-rays are absorbed within the foams at higher densities. Because the same thresholding value could not be selected for comparative analysis due to the density changes, the automatic thresholding feature was used to remain unbiased in the selection of the thresholding value. The true value of the percent porosity may be slightly higher or lower with the selection of a different value. A thresholding sensitivity study found that with a change in thresholding value of 2, the resulting percent porosity changed by less than 0.5%.

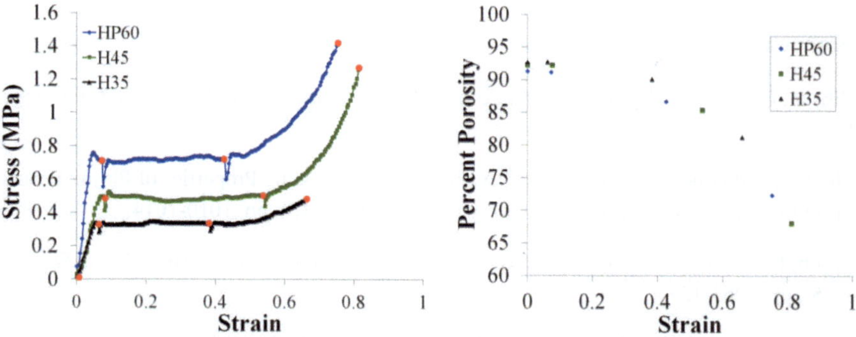

Figure 7. Stress-strain responses for HP60, H45, and H35 foams during in-situ compression loading in the Micro-CT (left) and quantified percent porosity as a function of percent strain for HP60, H45, and H35 foams (right)

5. Conclusions

Syntactic and traditional polymer foams were tested under compression loading to determine the stress strain response. Epoxy syntactic foam demonstrated a high collapse strength but at much higher density than traditional foam. Silicone syntactic foam performed as well as traditional PVC foam but also at much higher density. The syntactic foams failed to achieve high efficiency compared to Divinycell foams. The PVC foams showed considerable strength to weight ratio and demonstrated an ideal energy absorption behavior which is desirable for impact and shock mitigation. While epoxy foam may absorb higher amounts of energy it achieves this at a cost of much higher stress. The higher stresses required to initiate collapse can result in higher accelerations during impact and possibly injury. It is important to choose the appropriate material which has low strength to prevent subsequent injury yet have the capacity to absorb the energy of impact. Neither syntactic foam nor traditional PVC foam demonstrated recoverable strength. Using SEM imaging it was possible to identify the fracture of the glass micro-balloons used to produce the syntactic foam. The fracture of these balloons contributes to the permanent loss in strength. X-ray Computed Tomography was used to examine the internal deformation for the PVC foam. These images reveal the buckling of the cell walls during the beginning of the plateau region. This buckling continues until interaction from the cell walls result in the stiffening response during densification. This micro-buckling resulted in local plastic deformation in the foam struts. Although the energy absorbing properties of Divinycell show high efficiency, the lack of recoverable strength limits their applications. The need for light-weight, high-strength and recoverable energy absorbers is critical in the development of protection systems to mitigate repeated impact and blast.

References

[1] Gibson, L.J., and Ashby, M.F., 1999, *Cellular Solids: Structure and Properties*, Cambridge University Press, Cambridge.

[2] Miltz, J., and Gruenbaum, G., 1981, "Evaluation of Cushioning Properties of Plastic Foams From Compression Measurements," Polym. Eng. Sci., 21(**15**), pp. 1010-1014.

[3] Rusch, K.C., 1970 "Load-Compression Behavior of Brittle Foams," J. Appl. Polym Sci. 14(**5**), pp. 1263-1276.

[4] Rusch, K.C., 1969 "Load-Compression Behavior of Flexible Foams," J. Appl. Polym. Sci. 13(**11**), pp. 2297-2311.

[5] Avalle, M., Belingardi, G. and Ibba, A., 2007, "Mechanical Models of Cellular Solids: Parameters Identification From Experimental Tests," Int. J. Impact Eng., 34(**1**), pp. 3-27.

[6] Subhash, G., and Liu, Q., 2004, "Crushability Maps for Structural Polymeric Foams in Uniaxial Loading Under Rigid Confinement," Exp. Mech., 44(**3**), pp. 289-294.

[7] Liu, Q., Subhash, G. and Gao, X.-L., 2005, "A Parametric Study on Crushability of Open-Cell Structural Polymeric Foams," J. Porous Mater., 12(**3**), pp. 233-248.

[8] Walter, T.R., Richards, A.W., and Subhash, G., 2009, "A Unified Phenomenological Model for Tensile and Compressive Response of Polymeric Foams," J. Eng. Mater. Technol., 131(**1**), 011009-2.

Advanced Composites for Aerospace, Marine, and Land Applications II
Edited by: T. Sano and T.S. Srivatsan
TMS (The Minerals, Metals & Materials Society), 2015

STUDY OF MICROSTRUCTURE AND MECHANICAL PROPERTIES OF PARTICULATE REINFORCED ALUMINUM MATRIX COMPOSITE FOAM

Suresh Kumar and **O.P. Pandey***

Metallurgical Research Lab
School of Physics and Materials Science
Thapar University, Patiala-147004, India

Abstract

Metal foams cellular metals have gained an important role in the field of metallurgy, though barely a few decades old. Aluminum composite foam exhibit unique properties such as light weight, blast palliation, sound absorption, high energy absorption, and flame resistance. In the present investigation the effect of variation in the amount of $CaCO_3$ as blowing agent on the microstructure and wear behavior of LM13 alloy foams has been studied. The blowing agent was blended in highly viscous semi-solid melt by stirring process. The process parameters that influence the formation of bubbles like the melt temperature, size and amount of blowing agent and its distribution has been optimized to get uniform size foams. The distribution behavior of blowing agent is influenced by the melt viscosity and stirring speed. For packaging application, the dry sliding wear behavior of the prepared foam was investigated by using a pin on disc method at applied loads of 9.8, 19.6 and 29.4 N at room temperature. The results indicate that the wear rate is dependent on the cell size and cell wall thickness of the foam.

KEYWORDS: Metal foam; Temperature; Wear mechanism; Blowing agent.

*** Corresponding author**

O.P. Pandey
Senior Professor
School of Physics and Materials Science
Thapar University, Patiala (Punjab), INDIA, 147004.
Email: oppandey@thapar.edu, Tel: +91 1752393883; fax: +91 1752393005

1. Introduction

Metallic foams, especially closed-cell aluminum foams offer a unique combination of properties such as low density, high stiffness, strength, and energy absorption capability. Among the various possible applications, some are common to other types of foams [1-3]. Aluminum foams are mostly used for packaging, energy absorption and dissipation for passive safety systems. Closed cell aluminum foams have superior performance in terms of energy absorption with respect to polymeric foams, even considering the greater weight per volume unit [3]. There are various foaming methods which include metal casting, powder metallurgy and metal deposition. The casting method for the preparation of aluminum foam allows the production of high porosity foams which is suitable for commercial applications because of its low cost compared with other methods [4]. The physical and mechanical properties of aluminum foams are not only controlled by their composition, but are also affected significantly by cell structure [5]. Normally different hydride agents are used as foaming material [6]. However, health hazards causing eye, skin, and respiratory tract irritation due to high flammability of hydrides and higher cost, their use are not recommended [7]. In the present work, we developed the Al foam by employing the foaming process proposed by Yang et al. [8, 9] by controlling of the viscosity and concentration of blowing agent to minimize the imperfections. Aluminum alloy is foamed by incorporating the blowing agent having different particle size into the molten metal. The blowing agent gets decomposed in molten aluminum and generates gas bubbles therein. The manufacturing of a foamed aluminum includes two important considerations. Firstly, the blowing agent must be able to generate the desired amount of gas bubbles of desired sizes during the processing. Secondly, the molten aluminum must possess a certain viscosity, which also gets affected by the addition of ceramic particles so that the generated gas bubbles are retained in the aluminum matrix during solidification. The microstructure and wear behavior is controlled by the process variables such as foaming temperature, impeller design, impeller speed, size and volume fraction of the blowing agent particles during the foaming process. However, to the best of our knowledge, no one has done a systematic study of these parameters which influence the cell structure of aluminum foams. The present study is aimed to investigate the effect of variation in concentration of blowing agent on the microstructure and wear behavior of LM13 alloy foam.

2. Experimental details

2.1 Materials

In the present work, commercial grade Al-Si alloy (LM13) of near eutectic composition (Al-12Si-3Mg) is used as a foaming metal matrix material. LM13 alloy was obtained in the form of ingots from the Emmes Metals Pvt. Ltd. Mumbai, India. LM13 alloy was employed for the foam making with the addition of blowing agent, $CaCO_3$ in different concentration (1.5wt. %, 2.0wt. % and 2.5wt %). $CaCO_3$ powder having particle size in the range of 05-20μm was used for foaming process. Before charging, the blowing agent ($CaCO_3$) was preheated at 300 °C to remove moisture and other volatile substances.

2.2 Preparation and characterization of foam specimen

The LM13 alloy foam was developed by stir casting route. For it required quantity of LM13 alloy was taken in a graphite crucible and melted in an electric furnace. No extraneous viscosity enhancing agents like calcium, magnesium or SiC were added in melt. The temperature of melt was raised to 750 °C. This molten metal was stirred in open air using a graphite impeller at a rotation speed of 630 rpm to create the vortex. Stirring molten aluminum in open air raises its viscosity by a factor of 1.3 to 1.8 [8]. After the formation of vortex in the melt, the blowing agent particles were charged inside the vortex at the rate of 15-20 g/min into the melt during stirring with the help of funnel kept on top of vortex. After charging the blowing agent stirring speed was increased to 660 rpm to get uniform distribution of blowing agent. When the gas bubbles inside the melt start coming up the impeller was quickly removed from the crucible containing the mixture. The crucible was kept in the furnace maintaining the temperature at 850°C for 10 minutes to allow the decomposition of $CaCO_3$ which releases the gas bubbles. After that the crucible containing the foamed melt was removed from the furnace and cooled in air. Specimens of the dimensions $15\times15\times20$ mm^3 for the wear test and microstructural analysis were prepared by cutting the foam with the help of diamond cutter. To study the microstructure of the specimens, samples first ground on fine size SiC paper having grit size 400 to 800 and then polished with 0.5μm diamond polishing paste and etched by Keller's reagent. Dry sliding wear tests were carried out using a conventional pin-on-disc testing machine (wear and friction monitor, TR-20CH-400, Ducom Instruments, Bangalore, India) at room temperature. Wear of the foam was conducted against an EN32 steel disc with a hardness of 65HRC by making pin samples of 12×12 mm^2 size. The pins were in a conformal contact with the steel disc rotating at a speed of 1.6m/sec. The test was carried out for distance up to 3000m under the different applied loads of 9.8N, 19.6N and 29.4N. The wear properties were measured in terms of volume loss of the materials.

3. Result and Discussion

3.1 Optimization of blowing agent content

The optimization of blowing agent concentration in the melt aluminum alloy is followed out by reviewing the macrostructure of the foam samples. An optimum amount of blowing agent is required because low amount may lead to insufficient decomposition and large amount of blowing agent may lead to wastage of material. Blowing agent ($CaCO_3$) having 1.5wt%, 2wt% and 2.5wt% concentration was chosen for study because using higher amount leads to undecomposed $CaCO_3$ in the casting. The amount of blowing agent also plays an important role in deciding the pore size of LM13 alloy foam. One way for foaming melts is to add a blowing agent directly to the melt instead of injecting gas into it. Heat causes the blowing agent to decompose and release gases, which then propels the foaming process [11]. After the addition of blowing agent into the melt, it starts expanding slowly and gradually fills the foaming vessel. The foaming takes place at constant pressure. The decomposition reaction of $CaCO_3$ takes place on a definite boundary between the molten aluminum. This boundary moves towards the center of the compact but remains spherical in shape. The rate of the decomposition reaction of $CaCO_3$ is controlled by the transfer of heat (molten temperature LM13 alloy) to this reaction boundary.

After cooling the vessel below the melting point of the alloy, the liquid foam turns into solid aluminum foam and can be taken out of the mold for further investigation. Developed metal foams are reviewed at macro level by taking photographs of foam samples after cutting and grinding. Data obtained from these photographs and micrographs is summarized in table 1 and 2. Table 1 indicates density, foaming efficiency, average pore size and cell wall thickness of metal foam produced at 850°C with variation of blowing agent concentration. An empirical relationship exists between average cell diameter and the viscosity of the melt [12].

Table 1: Density, foaming efficiency, average pore size and cell wall thickness of metal foam produced at 850°C

Foam	Wt.% of $CaCO_3$ with particle size (05-20μm)	Density (gm/cm^3)	Foaming Efficiency (%)	Average pore size (mm)	Cell Wall Thickness (mm)
Sample-1	1.5	0.95	64.8	2.0	1.5
Sample-2	2.0	0.65	76	2.2	1.0
Sample-3	2.5	0.51	81	2.3	0.8

The viscosity of molten aluminum can also be enhanced by bubbling gas mixtures through the melt. Metal alloys can be foamed by mixing the foaming agent in the molten mass that releases gas when heated. The foaming agent $CaCO_3$ begins to decompose into CaO and gaseous CO_2, when heated from 850°C to 900°C. Large volumes of CO_2 gas produced in molten alloy create bubbles that can lead to closed cell foam. The structure of aluminum foam with 1.5wt. % blowing agent $CaCO_3$ is shown in fig.1 (a). The optimum temperature for foaming of aluminum lies above the liquids temperature of the used alloy (850°C). However, the melting temperature of aluminum alloy (660°C) is not sufficient for the proper decomposition of $CaCO_3$ which was used as a foaming agent in this case. At lower foaming temperatures the foaming agent releases less amount of CO_2. The CO_2 gas is inadequate for generating uniform pore size foam at 850°C (fig. 1(a)). With increase in the amount of blowing agent from 1.5wt. % to 2wt. %, an increment in the pore size as well as volume fraction of pores (Fig. 1b). This increment is due to large production of CO_2 gas with uniform pore size in molten mass. The molten mass at 850°C containing 2.5wt. % foaming agent releases large amount of CO_2 gas. This gas expands in the liquid metal and creates large size bubble structure. These bubbles are interconnected to each other as shown in fig.1(c). After decomposition the bubbles so formed tries to come out at the surface. Consequently, most of the CO_2 bubbles in the melt meet each other and acquires large area. Under the influence of temperature these unviable bubbles burst on reaching the melt surface. Few stable bubbles ascended to the top surface of the aluminum melt and formed an initial hull of two or three layers of roughly ellipsoidal foam cells [13]. The ensuing foaming caused the shape of the foam cells below the hull to be changed from ellipsoidal to increasingly compressed ones. Foam can be optimized on the basis of various parameters like surface morphology, cell size and cell wall thickness. Figure 1(a) shows the closely packed bubbles like structure of the LM13 alloy foam developed by the addition of 1.5 wt. % $CaCO_3$. Foam with a very fine pore structure with some signs of collapse is shown on the some portion in this image. This is due to diffusion of gas from one cell to another which is responsible for the drainage and coarsening of the foam during solidification. Varuzan et al. [14] also observed pores with different diameters that have different internal pressures because pressure is proportional to the inverse of the pore radius, so a smaller pore with a higher pressure disappears after giving its gas to a larger pore. In addition, the membranes between two adjacent cells can rupture spontaneously.

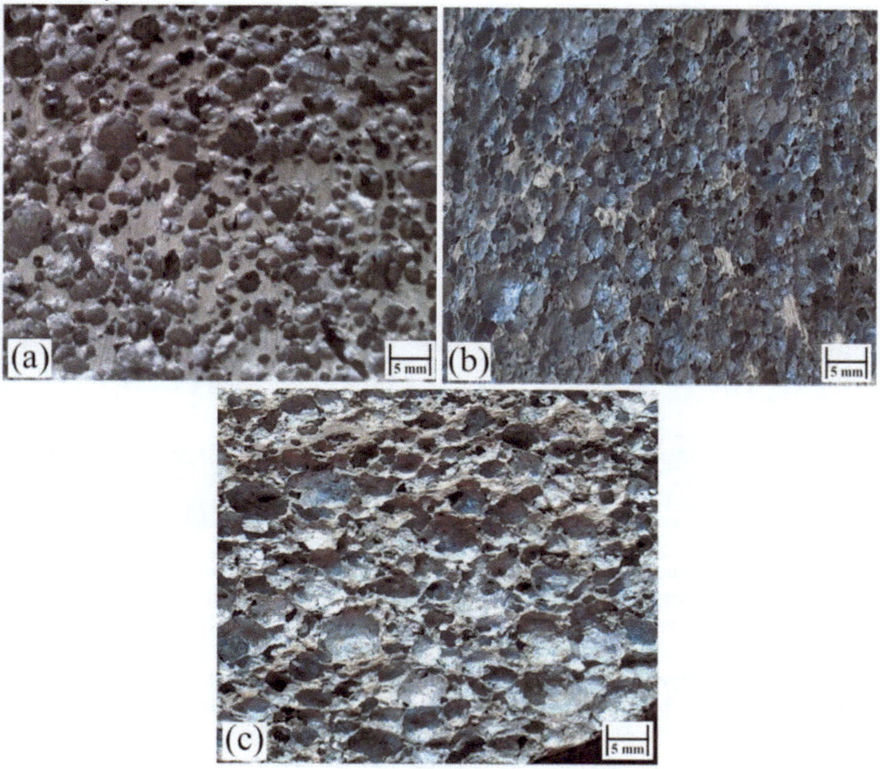

Figure 1. Macroscopic image of aluminum foam with (a) 1.5wt. %, (b) 2wt. % (c) 2.5wt. % blowing agent $CaCO_3$ at 850°C.

Due to drainage and coarsening, there are very large and irregular pores in the middle. Collapsed sections can be seen at the top of the fig.1a. However, on the left bottom part of the foam in fig.1b shows larger number of interconnections between pores, which is due to higher solidification role of foam. The spiral growth inside the pores demonstrate that the gas is released which is a good sign of foaming agent. At 850°C, CO_2 gas expansion make larger size bubble like structure, where pores are interconnected to each other as shown in fig.1(c). The cell size of the aluminum foam got increased with increasing rate of CO_2 release. The 1.5wt. % $CaCO_3$ foaming agent results in a much smaller cell size in comparison to the larger cell size produced at 2wt. % and 2.5wt. % $CaCO_3$ foaming agent as can be seen in fig. 2 (a-c). Consequently most of the air bubbles formed in the melt was large and unstable, and they burst on reaching at the melt surface [15]. This was due to increased size and insufficient strength of the foams under the increasing weight of the aluminum foam accumulated on top of the aluminum melt. The result was a closely packed bubble like structure of the foam. Therefore, it can be concluded that the increase in amount of foaming agent from 1.5wt. % to 2.5wt. % during the preparation of the aluminum foams has a great effect on the aluminum cell size. Figure 2 (a-c) also indicates the relation of the cell wall thickness and node size versus cell size of the aluminum foams prepared with increasing amount of foaming agent from 1.5wt.% to 2.5wt.%. Figure 2(a & b) shows the increment in amount of foaming agent from 1.5wt.% to 2wt.%, leads to decrease in cell wall thickness and the node size are decreased with the increase in cell size, which is in agreement with Simone and Gibson's study [15] where only the data for mid span of the cell wall is reported.

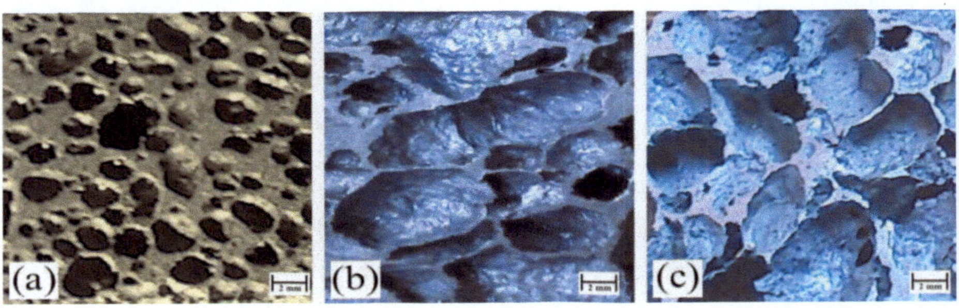

Figure 2. Macroscopic image of Aluminium foam showing variation in thickness of cell wall with variation in amount of blowing agent $CaCO_3$ at 850°C (a) 1.5wt. %, (b) 2wt. % and (c) 2.5wt. %.

Furthermore, the change in the amount of $CaCO_3$ from 2 wt. % to 2.5wt. % (fig 2b & 2c), seems to have less effect on the cell wall thickness and node size of the aluminum cells. The effect of the higher rate of gas release at 2.5 wt. % $CaCO_3$ on cell wall thickness and node size are shown in fig.2 (c) where the cell wall thickness and node size are decreased with increasing cell size. The thinner cell wall and node size in contrast with lower rate of gas release at 1.5wt.% $CaCO_3$ observed in fig. 2 (a) as compared to the higher rate of gas released at 2wt.% and 2.5wt.% $CaCO_3$. This experimental phenomenon comes from the effect of rate of gas release on the size of the closed cell foams.

3.2 Optimization of Foaming Temperature

Foaming temperature of the melt also affects the density, foaming efficiency, average pore size and cell wall thickness of metal foam. To get the optimize foaming temperature, we develop the melt foam with 2 wt. % of $CaCO_3$ at different foaming temperatures, as 2 wt. % $CaCO_3$ resulted optimize concentration for good surface morphology in terms of cell size and cell wall thickness. It is observed that less variation in the porosity occurs when the temperature is at 850°C. The cell diameter increases steeply above 850°C. Foaming temperature directly affects the melt surface tension and the foaming agent decomposition. The relationship of melt surface tension and temperature range from 850 to 875°C is calculated as [16]:

$$\sigma = 842 - 0.204\,(T - T_m)$$

where σ is the surface tension of melt, T is the experimental temperature and T_m is the melting point. Surface tension is known to decrease with increasing foaming temperature. Therefore, the stability of the bubble is reduced with increasing temperature. The foaming agent may decompose causing quick release of gas at high temperature [17] and make the experiment too difficult to control. The viscosity decreases and the stability of bubbles reduces at a certain temperature causing increased trend of the bubble mergence and the capacity of the escaped gas, as a result, non-uniform cells can be obtained. At very high temperature foaming agent ($CaCO_3$) disappears due to very fast decomposition, hence very low amount of $CaCO_3$ is available for the mixing in the molten metal. Therefore, the addition of foaming agent should be as much as possible at a relatively low temperature. According to the present investigation, the foaming process ought to be controlled between 850°C and 900°C. Effect of foaming temperature related to density, foaming efficiency, average pore size and cell wall thickness of the foam metal is shown in table 2.

Table 2. Density, foaming efficiency, average pore size and cell wall thickness of metal foam produced with 2 wt. % of $CaCO_3$.

Foam	Foaming temperature (°C)	Density (gm/cm³)	Foaming efficiency (%)	Average pore size (mm)	Cell wall thickness (mm)
Sample-4	875	0.55	79.6	2.0	1.2
Sample-5	900	0.86	68	1.8	1.4

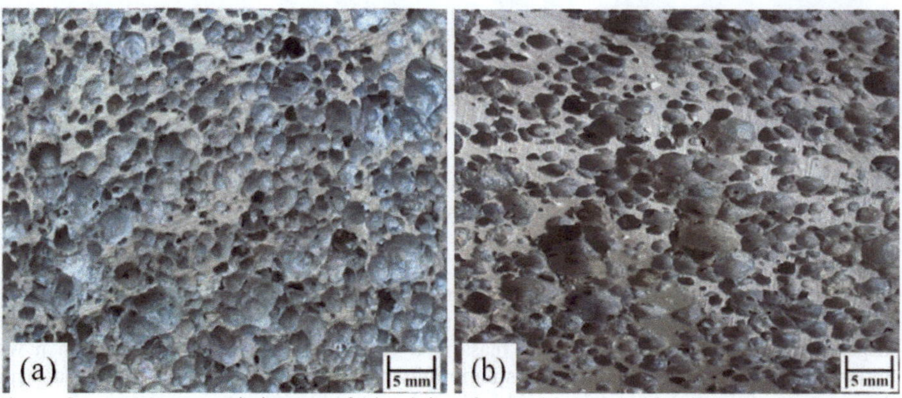

Figure 3. Macroscopic images of Aluminium foam with 2wt. % blowing agent $CaCO_3$ at (a) 875°C and (b) 900°C.

Figure 3 (a & b) shows the macroscopic images of the aluminum foam with 2 wt.% $CaCO_3$ at different foaming temperatures 875°C and 900°C respectively. The $CaCO_3$ with 2wt. % used as foaming agent decomposes at 850°C to CaO and CO_2 gas. This gas expends the liquid metal and creates bubble like structure. After its formation it tries to come out at the surface. As already mentioned pores with different diameters have different internal pressures because pressure is proportional to the inverse of the pore radius. This explains the signs of collapse observed on very fine pore structures that can be seen on some portion in figure 1a. This is also due to diffusion of gas from one cell to another which is responsible for the drainage and coarsening of the foam during solidification. In addition, the membranes between two adjacent cells can rupture spontaneously. Due to drainage and coarsening, there are very large and irregular pores in the middle. Increasing the temperature from 875°C to 900°C, the cell size in the foam is increased as shown in fig. 3 (a & b) respectively. Number of interconnected pores increases with increasing temperature from 850°C to 875°C. However, at 900°C, decrement pore size is observed as shown in fig. 3 (b).

The effect of the foaming temperature on the cell size of the LM13 alloy foam is shown in fig. 4 (a, b and c). Cell size of the aluminum foam is decreased with increasing temperature from 850°C to 900°C with 2wt.% $CaCO_3$ used as foaming agent. At 850°C with 2wt. % foaming agent (fig. 2b) results in a larger cell size compared with other temperature (875°C and 900°C). At 875°C, gas bubbles were concentrated at the central part and there was increasing density near the chiller surface (fig. 3a).

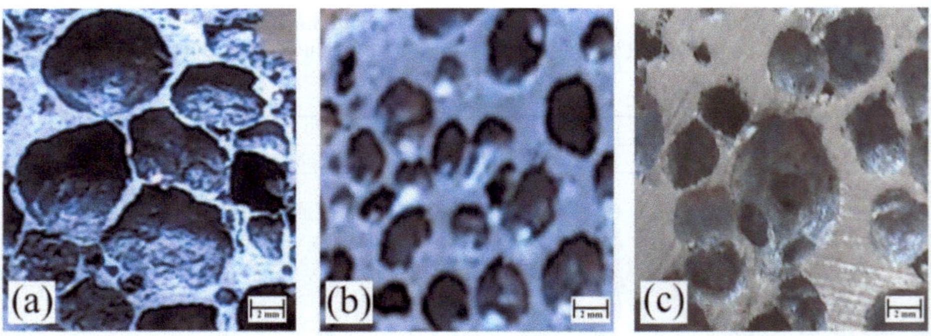

Figure 4. Macroscopic image of 2wt.% $CaCO_3$ showing variation in cell wall thickness and node size of the aluminum foams processed at (a) 850°C, (b) 875°C and (c) 900°C.

The non-uniformity of distribution and undesirable large size (fig. 3b) of some cells has been treated in several ways. Firstly, by using high speed mixing particles of the foaming agent that dispersed throughout the molten metal mass in a very short time. Without exception it was observed that the uniform mixing results uniformity in foam morphology. Secondly, increasing the viscosity of the molten metal can aid in the subsequent blowing step, preventing the escape of bubbles. However, the introduction of a viscosity increasing agent into a molten metal mass is more applicable. Gases produced in molten metal may also acts as a viscosity-increasing agent. The effect of the temperature on the cell wall thickness of the LM13 alloy foam is shown in fig. 4 (a, b and c). Cell wall thickness and node size of the aluminum foams are increased with the increase in temperature, and the change of the melting temperature has less effect on the cell wall thickness and node size of the aluminum cells. At 850°C with 2wt. % foaming agent cell wall thickness and node size of the aluminum foams is shown in fig. 4a. At this foaming temperature, alloy slurry is of lower viscosity. As a consequence, the easier drawing of liquid material to cell wall intersections by surface tension causes the thinner cell walls. The change of temperature from 850°C to 875°C, cell wall thickness and node size of the aluminum foams is increased due to the increase in the temperature, which leads to decrease in viscosity [15]. At 875°C temperature, lower rate of gas release of CO_2 into the melt is completely mixed and create bubbles on the surface of the melt, forming liquid foam. The cell wall thickness of the sample foam is decreased with the increasing the temperature from 875°C to 900°C, as shown in fig. 4 (b & c). The increase in the thickness of the cell wall at certain temperature implies that the viscosity of the LM13 alloy melt is becoming higher. The air trapped in the liquid foam upon forming at the increased temperatures is responsible for the cell size reduction due to its more expansion which makes the cell wall larger. However, the increase in the cell wall thickness is caused mainly by increasing the viscosity of the LM13 alloy melt.

In all cases the experimental results clearly indicate that the foaming efficiency is increasing with respect to increasing the amount of blowing agent. Generally, foamable precursors with a lower porosity resulted in foam samples with a higher apparent density and a lower foaming efficiency. Under the same foaming conditions (foaming agent, temperature), the average pore size of the foam samples was influenced by the density of foaming precursors and the initial size of the foaming particles [18]. Figure 4 shows the effects of temperature on the foam cell wall thickness and node size. The porosity continues to decrease with increasing temperature from 850°C to 900°C. The stirring process aims to make the foaming agent particles disperse uniformly, so it is critical to obtain suitable stirring time. Too short or too long stirring time is not suitable to fabricate the uniform size of cell of aluminum foam. Holding is a process of gas growing which affect the pore shape [19]. In the initial period, the gas begins to grow, and the pore is circular shape. When the volume of gas in the melt reaches a certain proportion, the bubbles get crowded. Under the pressure inside the bubble, the melt among the bubbles turns into thin film. According to Laplace's law, the additional surface pressure of the big bubbles is relatively small, when it could not offer enough tension force to keep a big bubble in the sphere, the bubble gives way to the pressure from the melt and becomes a polyhedron, which forms the cellular structure. Based on above analysis, the critical temperature of 875°C, and effective foaming agent (2wt. %) are appropriate for the formation of circular foams.

3.3 Wear behavior

3.3.1 Effect of amount of blowing agent on wear behavior of LM13 alloy foam

Wear rate against sliding distance of LM13 alloy foam with different applied loads and amount of blowing agent ($CaCO_3$) at 850°C are shown in fig 5. The wear rate of LM13 alloy foam measured with different loads and with 1.5wt%, 2wt% and 2.5 wt.% amount of blowing agent is shown in Fig.5 (a, b & c) respectively. It is observed in Fig.5 that wear behavior of the LM13 alloy foam is similar for all loads at room temperature. However, the wear rate of all LM13 alloy foams increases with the increase in applied load. This increment in the wear rate may be due to the increase in the actual contact area between LM13 alloy foam samples and counter disc at higher load. Initially the wear rate is higher which corresponds to the run in wear state. However, the steady state wears approaches at nearly 2000 meter sliding distance. Run-in wear of LM13 alloy foam (sample-3) is higher as compared to sample (2 and 3) at all loads as shown in Fig.5 (a, b & c). Wear rate depends on the cell size, cell wall and node of the foam material. With increasing the amount of the blowing agent in the LM13 alloy, cell size in the foam increases (fig. 1). Increment in the cell size is due to the fast decomposition rate of the blowing agent. This increment in the cell size of foam is also responsible for the decrement in the cell wall thickness. Thus decrement in the cell wall thickness leads to lower real contact area of the foam samples for the wear test. Figure 5 reveals that foam sample having lower cell wall thickness (lower real contact area) indicates high wear rate. This is due to that load transfer per unit area for the foam sample containing lower contact area is high, which leads to higher wear rate of the material. Foam samples having higher cell wall thickness introduce higher real contact area, which leads to lower load transfer per unit area and hence lower wear rate.

3.3.2 Effect of foaming temperature on wear behavior of LM13 alloy foam

Graphs plotted in figure 6 show wear behavior measured at different load of the LM13 foams having different foaming temperatures. Figure 6 (a & b) show the wear behavior of the LM13 foam with foaming temperature 875 °C and 900 °C. Figure 6 also indicates that the wear rate of all the foam samples increases with increasing the applied load. However, wear rate of LM13 foam decreases with the increment in the foaming temperature. Foam sample having higher foaming temperature (fig. 6a) shows lower wear rate at all the applied load in comparison to other foam samples (fig. 5 and 6 b). This decrement in the wear rate is due to the presence of larger real contact area of the sample (fig. 3a & 4b). However, higher viscosity of the foam sample with higher foaming temperature (table 2) is also responsible for the lower wear rate of the sample. Wear rate data collected of LM13 alloy foam samples at 1000m and 2000m sliding distance at different loads is given in table-3.

Table 3. Wear rate data

Sliding distance	**1000m**					**2000m**				
Sample	**Sample-1**	**Sample-2**	**Sample-3**	**Sample-4**	**Sample-5**	**Sample-1**	**Sample-2**	**Sample-3**	**Sample-4**	**Samp 5**
1kg	11.43	10.08	9.41	8.77	10.17	5.30	6.05	5.74	4.44	4.7
2kg	13.70	14.02	15.17	13.20	12.97	7.34	8.48	8.59	7.36	6.6
3kg	15.16	16.99	17.10	15.12	15.83	9.24	9.95	10.05	8.62	8.3

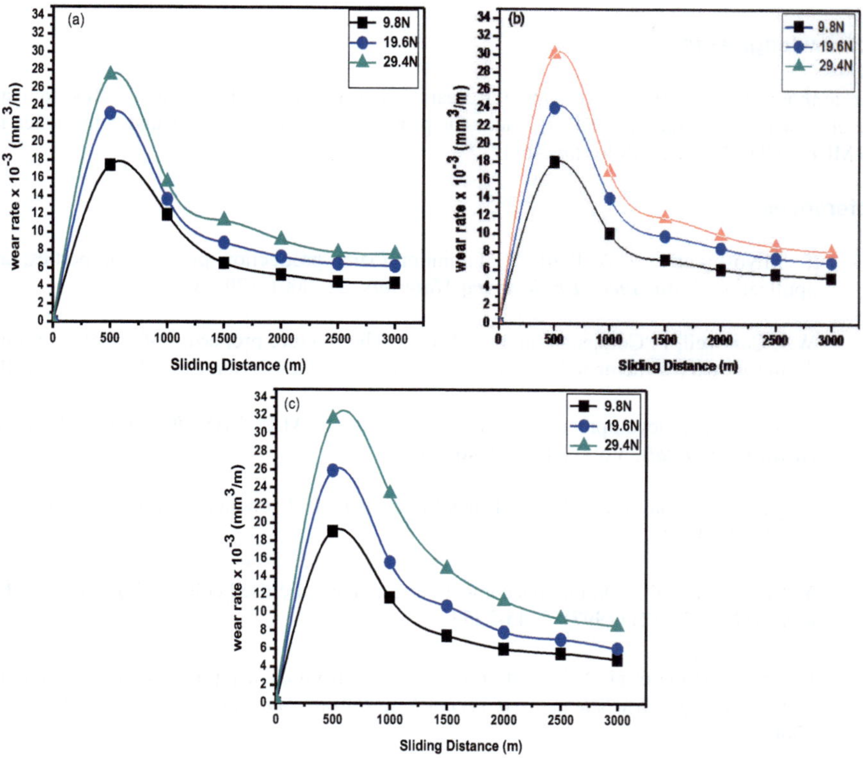

Figure 5. Wear rate against sliding distance of aluminum foam with (a) 1.5wt.%, (b) 2wt.%, and (c) 2.5wt.% blowing agent $CaCO_3$ at 850°C.

4. Conclusions

Proper control of the process parameters, such as foaming agent, experimental temperature and stirring and holding time, leads to the production of LM13 alloy foam with a uniform cell structure and high porosity. The average pore size and porosity change little with 1.5wt. %-2.5wt. % CaCO3, while the average pore size and porosity decrease with increasing foaming temperature. The experimental findings showed that the apparent density of the foam samples is inversely proportional to the density of the foaming precursor.Wear rate of all foam sample increases with applied load. LM13 alloy foam having higher amount of blowing agent evidences higher wear rate. LM13 alloy foam having higher foaming temperature shows lower wear rate.

Acknowledgments

The authors are thankful to Armament Research Board (ARMREB), Defence Research and Development Organization (DRDO), India for providing financial support under the letter no. ARMREB CDSW/2012/148 for this study.

References

1. T. Miyoshi et al., "ALPORAS aluminum foam: production process, properties, and applications," ***Advanced Engineering Materials***, 2 (2000), 179-183.

2. W.J. Cantwell, P. Compston, and G. Reyes, "The fracture properties of novel aluminium foam sandwich structures," ***Journal of Materials Science Letters***, 19 (2000), 2205-2208.

3. M. Itoh et al., Development of Foamed Aluminium "ALPORAS, ***Bulletin of the Japan Institute of Metals,*** Vol. 26 (1987) No. 4 P 311-313.

4. Shigeru Akiyama et al., "Foamed metal and method of producing same," U.S. Patent 4 (713) (1987), 277.

5. Y. Sugimura et al., "On the mechanical performance of closed cell Al alloy foams," ***Acta materialia,*** 45, (12) (1997), 5245-5259.

6. F von Zeppelina et al., "Desorption of hydrogen from blowing agents used for foaming metals," *Composites Science and Technology-Porous Materials*, 63, (16) (2003), 2293-2300.

7. L. E. G. Cambronero et al., "Manufacturing of Al–Mg–Si alloy foam using calcium carbonate as foaming agent," ***Journal of Materials Processing Technology***, 209, (4) (2009), 1803-1809.

8. C.C. Yang, et al., US Patent 5632319 (1997).

9. H. Nakae, and C.C. Yang, "Mechanism of Pore Formation" In: 134th Japan Foundry Engineering Society Conference, Tokyo Japan, (1999), 37, in Japanese.

10. I. Duarte, and J. Banhart, "A study of aluminium foam formation-kinetics and microstructure," ***Acta materialia***, 48, (9) (2000), 2349-2362.

11. A.W.D. Hills, "The mechanism of the thermal decomposition of calcium carbonate," Chemical Engineering Science, 23, (4) (1968), 297-320.

12. J. Banhart, and J. Baumeister, "Deformation characteristics of metal foams," ***Journal of materials science,*** 33 (6) (1998), 1431-1440.

13. V. Shapovalov, "Porous and cellular materials for structural applications," vol. 521, ed. D.S. Schwartz et al. (Warrendale, PA: MRS, 1998), p 281.

14. Varuzan Kevorkijan et al., "Influence of the foaming precursor's composition and density on the foaming efficiency, microstructure development and mechanical properties of aluminum foams," *MTAEC9 Materials and technology*, 45, (2) (2011), 95-103.

15. A. E. Simone and Lorna J. Gibson, Aluminum foams produced by liquid-state processes, ***Acta materialia,*** 46, (9) (1998), 3109-3123.

16. Biljana Matijasevic and John Banhart, "Improvement of aluminium foam technology by tailoring of blowing agent," ***Scripta Materialia,*** 54, (4) (2006), 503-508.

17. Zhen-lun Song et al., "Evolution of foamed aluminum structure in foaming process," ***Materials Science and Engineering A***, 298 (1) (2001), 137-143.

18. Bo Young Hur, Soo Han Park, and Arai Hiroshi, "Viscosity and surface tension of Al and effects of additional element," ***Materials Science Forum,*** 439, (2003), 51-56.

19. A.E. Markaki, and T.W. Clyne., "The effect of cell wall micostructure on the deformation and fracture of aluminium-based foams," ***Acta Materialia,*** 49, (2001), 1677-86.

Advanced Composites for Aerospace, Marine, and Land Applications II
Edited by: T. Sano and T.S. Srivatsan
TMS (The Minerals, Metals & Materials Society), 2015

ATMOSPHERIC PLASMA TREATMENT OF NYLON 6,6 FOR IMPROVED INTERFACIAL ADHESION IN THERMOPLASTIC COMPOSITES

A.A. Bujanda[1], C.Wu[1], J.D. Demaree[1], E.J. Robinette[1]
A. Weerasooriya[1], and D. Flanagan[1]

[1]The United States Army Research Laboratory
4600 Deer Creek Loop
APG, Maryland 21005, USA

Keywords: Atmospheric Plasma, Dielectric Barrier Discharge, Thermoplastic, Adhesion,

Abstract

The physio-chemical modifications of polyamide (Nylon 6,6) obtained from exposure to an atmospheric dielectric barrier discharge were examined. Specifically, the surface energy, surface chemical composition, and interfacial adhesive strength were studied using water contact angle goniometry, x-ray photoelectron spectroscopy, and single joint lap shear testing. Nylon substrates were plasma treated with both O_2 and water vapor (H_2O) in He for 30, 60, and 180 seconds, resulting in the functionalization of the surface via the addition of reactive chemical groups such as –OH that change the energy, composition, and reactivity of the surface. The studies revealed that He/H_2O plasma treatments were more effective at functionalizing the surface than He/O_2 plasma treatments, resulting in a more than 40% decrease in the water contact angle. XPS results show a significant increase in the amount of surface oxygen after treatment, and lap shear experiments show an almost ~300% increase in interfacial adhesive strength.

1. Introduction

Thermosetting resins have long been preferred as matrix materials in polymer matrix composites because, uncured, they remain in a liquid state at room temperature, making processing fast and convenient. However, once cured, many thermosetting resins often exhibit low impact resistance and low damage tolerance. To mitigate the low impact resistance, thermoplastics such as nylon 6,6 are now increasingly being integrated as matrix materials in polymer matrix composites. This is a departure from typical applications where the thermoplastic materials were used as reinforcements in conventional thermosetting-matrix composites [1]. However, the more widespread use of thermoplastic composites is limited by inherently poor interfacial adhesion between the matrix and the reinforcing material [1]. The adhesive properties of the fiber-matrix interface are largely governed by the physical and chemical interactions of the fiber-matrix interface. It is, therefore, important to understand and optimize the interface between thermoplastic resins and the adhesive materials present at composite interfaces.

Plasma treatment of polymeric materials aims to modify their surface and near-surface properties for the purposes of adhesive bonding, wettablility, biocompatibility, and durability among others. [2-3] One of the main advantages of plasma treatments is that the chemical and morphological effects are limited to very shallow depths (10 nm or less) and therefore leaves the bulk properties such as mechanical strength unaffected. [4] Plasma treatments, specifically atmospheric discharges, have proven to be a very effective alternative to conventional surface modification techniques such as wet chemistry methodologies. Depending on plasma parameters such as gas composition, treatment time, power settings, etc, atmospheric plasma can modify polymeric surfaces by forming oxygenated chemical groups at the surface, allowing for enhanced surface chemical reactivity. [5] Atmospheric plasma processes provide a uniform modification, require short processing times, are industrially scalable, and are typically environmentally friendly. [4,6] In this paper, we discuss the physiochemical modification of Nylon 6,6 polyamide exposed to atmospheric pressure dielectric barrier discharges. We attempt to create an enhanced interfacial adhesive bond by creating a functionalized surface at the Nylon-adhesive interface.

2. Experimental Procedure

Atmospheric dielectric barrier discharge (DBD) plasma treatments were employed to modify the physio-chemical surface properties of Nylon 6,6 substrates. The goal is to find the optimal treatment conditions for attaining the greatest enhancements in both the surface energy and accessible hydroxyl groups (-OH) on the substrate surface. A custom-built atmospheric glow DBD plasma system was utilized to treat the Nylon 6,6 substrates. The system consists of a symmetric L-shape high voltage (HV) electrode, encased in plastic plates with customized gas manifold above it, a motorized stage as the ground electrode moving horizontally with a speed of ~ 0.9 inch/s in the direction perpendicular to the HV electrode. The distance between electrodes was fixed at 2 mm. The complete setup of the system and images of the DBD electrodes are illustrated in Fig. 1. The system was equipped with an AC microsecond-pulsed power supply that generated pulses with a width of 30 μs at a duty cycle of 8%. The pulses had a 20 kV peak voltage and 300 W peak power. Helium (He, Praxair, 99.999%) with a constant flow rate of 100 cm^3/s (6 standard liter per minute) was used as the carrier gas with the reacting gases, either water vapor (H_2O) or oxygen (O_2, air products, 99.999%) with a 5% volumetric ratio to He, respectively. For He/ H_2O plasma, the H_2O vapor was provided via blowing the He through a vial containing HPLC-grade distilled water as shown in Fig. 1(a). Three plasma exposures, 30 sec, 60 sec, and 180 sec, respectively, were employed for both the He/O_2 and He/H_2O plasmas, respectively.

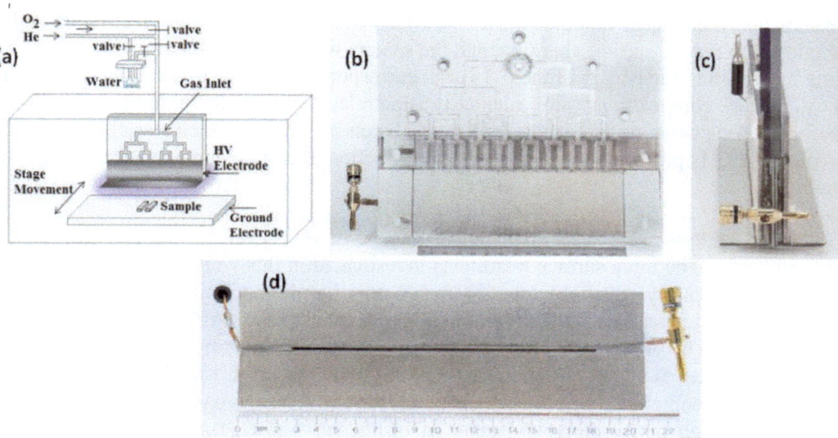

Figure 1. **The DBD plasma system: (a) the schematic diagram showing the entire setup. (b) ~ (d) are images for DBD electrodes: (b) side view showing the customized gas manifold with split canals between the two pieces of HV electrodes, (c) end view showing the L-shape of HV electrodes, and (d) bottom view showing the narrow gas outlet (~1.5 mm).**

Nylon 6,6 films (Goodfellow, USA) were obtained in 0.05mm and 5mm thicknesses. The 50μm films were used for the plasma treatments and chemical analyses, and the thicker 5 mm films were used as adherends for the lap shear evaluation. The films were cut into 2.5 mm × 2.5 mm squares in and the thicker films were cut into 5 mm × 30 mm strips for use in lap shear testing.

The hydrophilicity/hydrophobicity of the untreated and treated substrates was examined by measuring the static water contact angle (WCA) [1] using an automated goniometer/tensionmeter (Model 290 with DROPimage Advanced v2.6) equipped with a CCD camera. A drop of the HPLC-grade distilled water, with ~ 2.5 μL volume, was administered on the substrate to create a real-time image of two-dimensional drop profile. The tangent angles were then measured on the left and right sides with the vertex at the three-phase lines formed between the substrate and the drop profile and their average was referred to as WCA. For each substrate sample, at least five water drops were positioned across the long side and the grand average from their WCAs was calculated. An aging study was also performed to examine any possible change in the wettability for plasma-treated surfaces by repeatedly measuring the WCA after different durations (one and six days, respectively) after plasma treatments.

Near-surface compositional measurements were conducted by x-ray photoelectron spectroscopy (XPS), using a Kratos Axis Ultra system equipped with a hemispherical analyzer. The sample was irradiated with a 150 W monochromatic Al Kα (1486.7 eV) beam, and both magnetic and electrostatic lenses were used to select photoelectrons from a 1 × 2 mm area of the nylon surface with a take-off angle of 90°. The pressure in the XPS chamber was held between 10−8 and 10−9 torr. Elemental high resolution scans for C 1s, N 1s, and O 1s were acquired in the constant analyzer energy mode with a 20.0 eV pass energy. A low energy electron neutralizer was employed to reduce surface charging, and the spectra were adjusted for any residual charging by adjusting the energy of the C 1s peak corresponding to C-C to a binding energy of 285.0 eV. Surface compositions were calculated using relative sensitivity factors supplied by Kratos for the Axis Ultra system, using the Kratos Vision 2 software, which was also used to deconvolute the C 1s peak to determine the relative fraction of carbon bonding states present in the sample.

Lap shear tests were used to evaluate the effect of Nylon 6,6 surface treatment on adhesive strength. The three surface treatments investigated in this work were (i) untreated and cleaned with an isopropyl alcohol wipe, (ii) treated only by plasma, and (iii) treated by plasma followed by a silane treatment with 3-aminopropyltriethoxysilane (APS) (Sigma Aldrich, St. Louis, Missouri). The plasma condition yielding the most surface hydroxyl groups was chosen for the plasma treated samples. Single-lap-joint specimens were fabricated by bonding Nylon-6,6 substrates of geometry 10 cm × 2.5 cm x 0.5 cm with an epoxy adhesive. The epoxy monomer was a high purity diglycidyl ether of bisphenol A (DGEBA), EPON 825, with an epoxy equivalent weight of about 176 g/mol from Momentive (Columbus, OH). The epoxy was cured with a polypropylene oxide based-Jeffamine diamine having a molecular weights of 230 g/mol (Huntsman, The Woodlands, TX). The lap-shear joints were bonded using a custom bonding tooling fixture that ensured proper alignment of the 1.25 cm x 2.5 cm overlap and a uniform bondline thickness of 2.5 mm. Pressure was applied to the adhesive bond area by placing lead weights on top of the fixture. Figure 3 gives images for the bonding procedure using the custom tool. To ensure the adhesive did not flow out of the bonding zone before gelation, the viscosity of the thermosetting epoxy was cured in a mixing container until a noticeable viscosity build-up occurred, which required overnight curing (approximately 16 hours) followed by 145 minute cure at 45°C. The bonded specimens sat overnight prior to being post-cured at 120°C for at least 12 hours to achieve full cure. The lap-shear joints were pulled in tension according to ASTM D1002 at a rate of 1.27 mm/min with an Instron (Model 1123 with a 5500 series controller and Bluehill software) until the specimen failed. The lap-shear strength was calculated as the maximum load divided by the bonded area as measured from the failed specimens. The mode of failure (adhesive, cohesive, or mixed) was determined by visual inspection of the failed specimen.

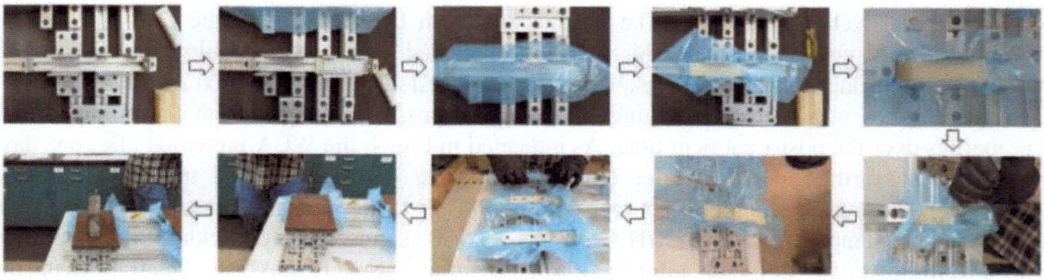

Figure 2. The 80/20 aluminum set up and images showing the procedure of making lap shears.

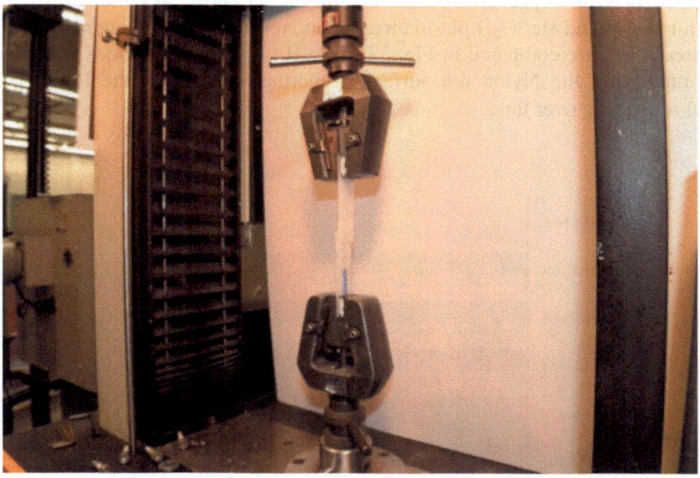

Figure 3. A picture showing a lap joint undergoing the tension test.

3. Results and Discussion

1. WCA Measurements

The results obtained from WCA measurements strongly demonstrate a close relation between the wettability of Nylon 6,6 films and experimental parameters implemented in plasma treatments. Figure 4 shows the WCA results for all Nylon 6,6 films treated by different plasma experimental conditions including the reactive gas mixture and exposure. As expected, the measured WCAs were at least 30% smaller for all treated films than their untreated counterparts. This suggests that the surface wettability was substantially improved through plasma treatment using either He/O_2 or He/H_2O . The enhancement can be attributed to the increase of –OH groups on the substrate surface which can be explained by the XPS results. The WCA also decreased with increasing plasma exposure for both tested plasma gas mixtures. On the other hand, the WCA results from the aging study suggest possible changes on the substrate surface properties over the post-treatment time. As indicated in Fig. 4, the WCA increased after one day, followed by further increases after 6 days of plasma treatment for all treated specimens. Moreover, the treated samples also appear to age faster if the plasma exposure increases. The WCA for the sample treated by He/H_2O plasma for 180 sec increased particularly for more than 40% from 36° to 51° in 6 days after the plasma treatment. Nevertheless, the detail mechanisms for the changes in surface chemistry and physics associated with these observed aging phenomena are not yet fully understood despite that some previous reports had suggested the possibility of oxide formation on the treated surface overnight [1] In conjunction with our previous experiences on oxygenated plasma treatments, we chose 30 s and 180 sec as the optimal exposures for He/O_2 and He/ H_2O plasma treatments, respectively. The aging behaviors between these two treatments are compared in Fig. 5The He/H_2O plasma was more effective than He/O_2 plasma in modifying the Nylon 6,6 surface properties in light of the lower WCA and the relatively slower aging over time,

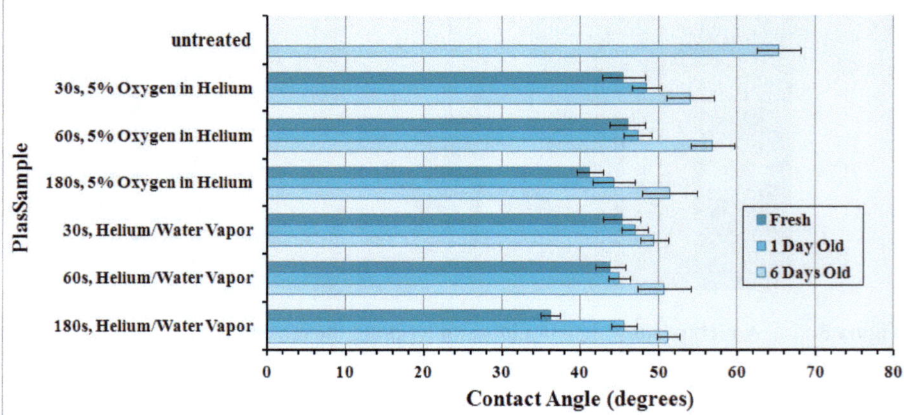

Figure 4. WCA measurements for all plasma treatment and aging conditions.

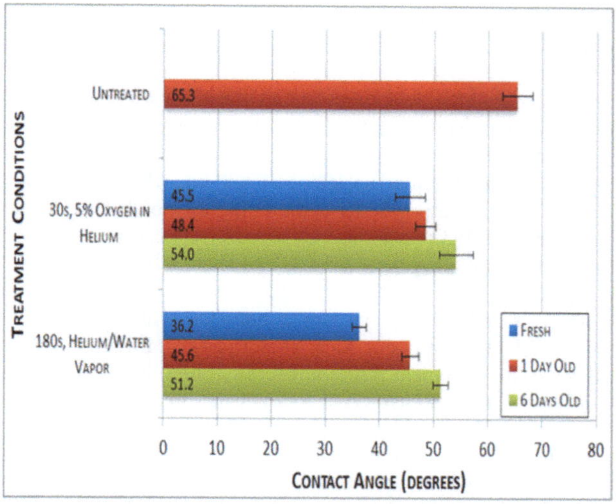

Figure 5. **Aging studies of Nylon 6,6 films treated by He/O_2 plasma for 30 sec and He/H_2O plasma for 180 sec.**

2. XPS Results

XPS was used to examine the nylon surface before and after exposure to both He/O2 and He/H2O plasmas, and the compositional results are compiled in Figure 6. The surface of the control sample was found to have a slight deficit in oxygen and nitrogen relative to the ideal chemical composition of nylon-6 (12.5% O and 12.5% N) and to a published reference sample of spin-coated nylon-6 (13.2% O and 11.9% N). [7] This very slight deficit is likely due to two or three monolayers of surface hydrocarbon contamination, and after even the briefest exposure to either plasma, the composition was very close to ideal nylon-6, indicating that most surface contamination is easily removed with plasma treatment.

At increasing exposure times, however, He/O2 plasma treatment was found to be ineffective at oxidizing the surface. Both the oxygen and nitrogen content measured remained constant at exposure times up to 180 seconds, indicating that while the He/O2 plasma was not significantly damaging the nylon, it was also failing to create new oxygen functionality. Exposure to He/H2O plasma, however, was extremely effective at increasing the amount of oxygen present at the surface, likely due to oxidation reactions between the plasma gases and the polymer. Even at exposure times as long as 480 seconds, the reaction does not appear to saturate, as oxygen content continues to increase with increasing exposure. XPS measurements made using a 30 degree take off angle (not shown here) found nearly identical oxygen contents for each exposure (except the control and the 60 sec samples), indicating that the plasma-modified layer is thicker than the photoelectron escape depth (~6-7 nm) after 180 sec of plasma exposure.

The continued increase in oxygen content observed must be from continued interaction of the He/H2O plasma gases with the polymer, and not merely the growth of a thicker surface oxide.

To confirm this, the C 1s peak was deconvoluted to ascertain the effect of plasma exposure on carbon bonding states (see Figure 7). In untreated nylon, carbon exists in four different states [7], resulting in four distinct C 1s peaks. The peak with the lowest binding energy (285.0 eV) corresponds to the polymer backbone (C-C*-C), while the peak with the highest binding energy (288.1 eV) corresponds to carbon doubly bonded to oxygen (C=O). Two intermediate binding energies (285.6 and 286.4 eV) result from carbon near the double bond (C-C*-C=O) and carbon bonded to nitrogen (N-C*-C). The relative fraction of these carbon bonding states for samples exposed to both He/O2 and He/H2O plasmas is shown in Figure 8. In the case of He/O2 plasma, there is little change in carbon bonding, with the possible exception of a decrease in carbon bound with nitrogen. Most importantly, the amount of carbon bound to oxygen remains constant, in agreement with the overall composition measurement, confirming that He/O2 are not effective at attaching oxygen functionality to nylon.

In the case of He/H2O plasma, the fraction of carbon atoms attached to nitrogen remains fairly constant, in agreement with the observation that the total amount of nitrogen present does not decrease. There is, however, a steady decrease in backbone carbon, and a steady increase in carbon bound to oxygen. We conclude that the primary interaction between the plasma and the polymer is an oxidative attack on the carbon backbone, and attachment of carbon-oxygen functional groups, with little interaction between the amine group and the plasma gases. We do note a slightly greater increase in the amount of carbon in a C-N bonding state (Figure 8) than in the total amount of nitrogen present (Figure 2), which may indicate the formation of C-N-C crosslinks, as noted by others investigating plasma and electron beam modification of polyamides. [3] Nevertheless, the primary effect of the He/H2O plasma exposure on nylon is to attach oxygen functionality, which should lead to favorable interaction mechanisms with epoxy adhesives.

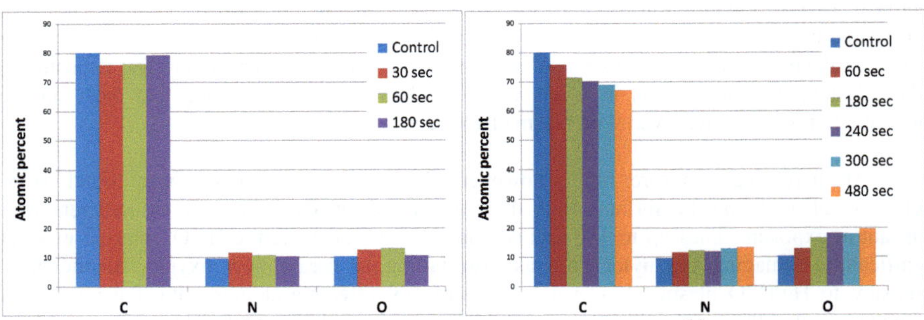

(a) (b)

Figure 6. **XPS surface composition after exposure to**
(a) He/O2 plasma, and
(b) He/H2O plasma.

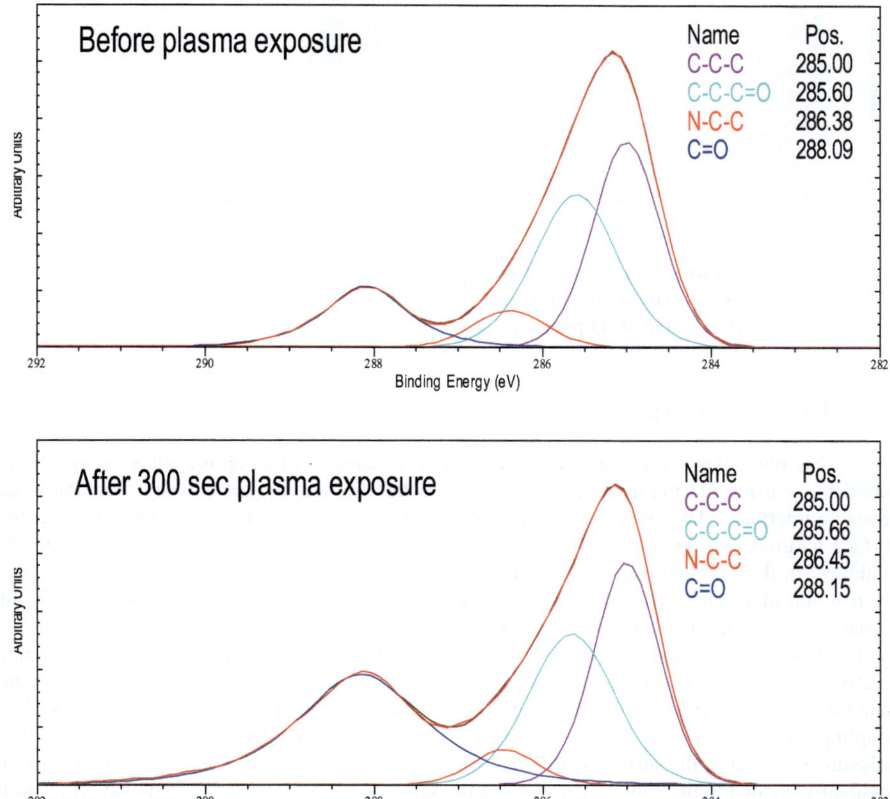

Figure 7. Deconvolution of C 1s peak before and after exposure to 300 sec He/H_2O vapor plasma.

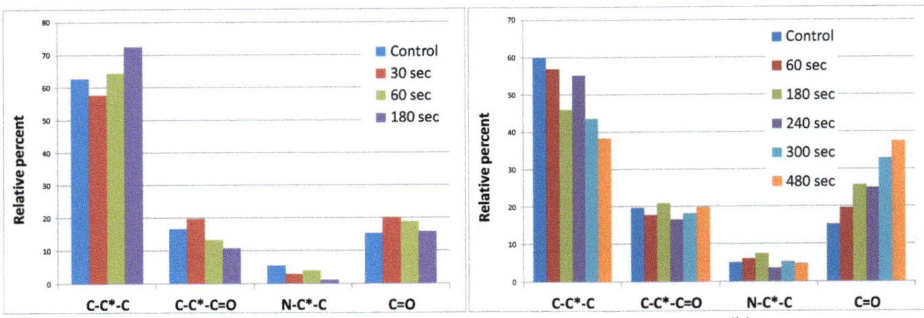

(a) (b)

Figure 8. **Bonding states of C after exposure to**
(a) He/O2 plasma, and
(b) He/H2O plasma.

3. Lap-shear Testing

The objective of this work was improving the adhesion of an epoxy adhesive to a Nylon substrate by using plasma treatment to increase wettability and impart covalent bonding at the adhesive interface. Two approaches were taken to improve adhesion that hinged upon creating surface hydroxyl groups, which was optimized using a 180 sec exposure to a He/H_2O plasma medium. In the first approach, the substrates received only a plasma treatment and relied upon the reaction of surface hydroxyl groups with the epoxy to impart covalent bonding [8]. In this scenario, a competition existed between the adhesive curing chemistry (i.e. epoxy-amine reaction) and the hydroxyl-epoxy reaction occurring at the interface. Typically, the epoxy-amine reactivity rate is higher than a hydroxyl-based reaction. The second approach coupled and amine-functional silane chemistry with the same plasma treatment in the first approach. The APS silane coupling agent condensed (after hydrolysis) with the surface hydroxyl groups to form a polysiloxane layer at the surface with amine-functionality. The amine groups in the polysiloxane covalently bonded to the epoxy monomers in the same epoxy-amine reaction pathway of the bulk adhesive.

The adhesive strength in this work was evaluated with a lap-shear test. Three surface treatments were investigated to compare the surface modification approaches and the results are given in Figure 9 and Table I. The poor adhesion of the epoxy and Nylon was demonstrated for the untreated sample, where the lap-shear strength was 0.49 ± 0.08 MPa. However with plasma treatment only, the lap-shear strength increased roughly 300% to 1.46 ± 0.48 MPa. This result confirms that by adding a high concentration of surface hydroxyl groups, a significant level of covalent bonding can be achieved despite a competitive epoxy-amine reaction occurring in the adhesive bulk. When combining plasma treatment with an amine-functional silane treatment, the lap-shear strength increased another 10% to 1.64 ± 0.47 MPa. By converting the surface hydroxyl groups to amine functionality, we were able to achieve a higher degree of covalent bonding that resulted in increased adhesive strength. It should be noted that increased covalent bonding does not necessarily translate to increased adhesive strength. [9] Future work will explore optimizing the level of covalent bonding at the epoxy-Nylon interface.

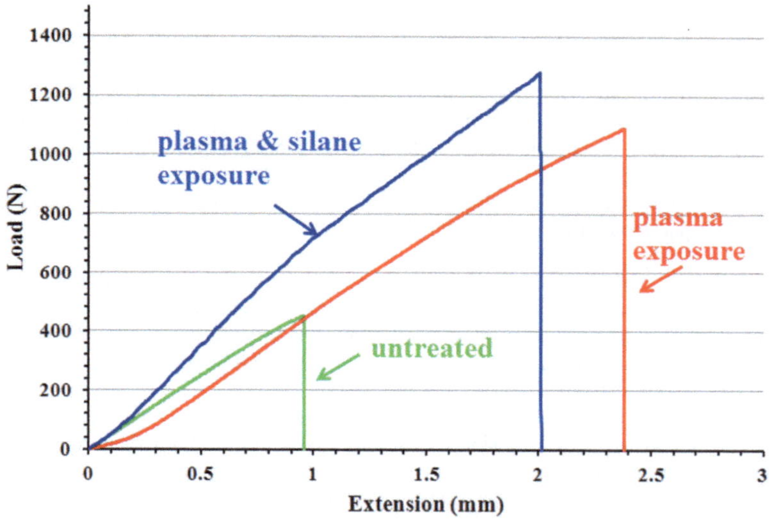

Figure 9. **Representative load-displacement curves for the lap-shear experiments conducted on Nylon with surface treatments of (i) untreated, (ii) plasma treatment only, and (iii) plasma treatment followed by silane treatment.**

Table I. **Adhesive Bond Strength for Differently Treated Nylon 6,6 Substrate Surfaces.**

Treatment condition	Lap-Shear Strength (MPa)	Maximum extension (mm)	Mode Failure
Untreated	0.49 ± 0.09	0.9 ± 0.25	Adhesive
He/H_2O Plasma	1.46 ± 0.48	2.4 ± 0.97	Adhesive
He/H_2O Plasma & silane	1.64 ± 30.47	2.0± 0.72	Adhesive

4. Conclusions

In this study we were able to successfully modify the surfaces of Nylon 6-6 thermoplastic materials using an atmospheric DBD plasma. Both He-O and He/H_2O vapor gas compositions of plasma were able to produce a significantly lower water contact angle and increased wettability when compared to as-received Nylon 6-6 films. At 180 seconds of exposure, the He/H_2O vapor discharge was able to reduce the WCA from 65.3 to 36.2 degrees. Aging studies revealed that the surface wettability diminishes over time with exposure to ambient conditions, and 6 days post-treatment, the WCA of the specimens subjected to a 180 second He/H_2O vapor exposure increased to 51.2 degrees. This aging behavior of the reactive surface places a time-constraint on subsequent manufacturing steps. It is suggested that any further processing of the surfaces exposed to plasma be carried out immediately after exposure. If not, storing the specimens in an inert atmosphere is paramount to preserving surface reactivity.

XPS studies revealed that the He/H_2O vapor plasma was more effective at grafting oxygen-containing groups and removing surface contamination on the surface of the Nylon 6-6 than He/O_2 plasma. This is most noticeable tracking an increasing oxygen content on the surface as exposure times increased, as well as showing an increase in the C=O bonding within the Carbon 1s peak.

Lap shear testing showed that by exposing the surfaces of single joint lap shear specimens to He/H_2O vapor discharges, a nearly 300% increase in adhesive strength could be achieved. Combining the plasma activation with a silane treatment improved the adhesive strength by another 10%. These improvements are a direct result of the increased surface reactivity initiated by plasma exposure.

References

1. D.D. Pappas, A.Bujanda, J.D. Demaree, J.K. Hirvonen, W. Kosik, R. Jensen, and S. McKnight,"Surface modification of polyamide fibers and films using atmospheric plasmas", ***Surf. Coat. Technol***., 201 (2006), 4384-4388.

2. C. Wang, L. Lin, C. Chen, and J. Chen, "Hydrophobicities and antimicrobial activities of silicone polyester/titanium dioxide composites on nylon fabrics after argon plasma treatment", ***J. Polym. Res***., 21 (2014), 408.

3. A.Boulares-Pender, I. Thomas, A. Prager, A. Schulze, "Surface Modification of Polyamide and Poly(vinylidene fluoride) Membranes", ***J. Appl. Polymer Sci***., 128 (2013), 322-331.

4. D. Pappas, "Status and potential of atmospheric potential of atmospheric plasma pricessing of materials", ***J. Vac. Sci. Technol***., 29 (2011), 020801.

5. N. Cui, D. J. Upadhyay, C.A. Anderson, N.M.D. Brown, "Study of the surface modification of a Nylon-6,6 film processed in an atmospheric pressure air dielectric barrier discharge", ***Surf. Coat. Technol***., 192 (2005), 94-100.

6. V. Rodriguez-Santiago, A.A. Bujanda, B.E. Stein, D.D. Pappas, "Atmospheric plasma processing of polymers in helium-water vapor dielectric barrier discharges", ***Plasma Proc. Polym.***, 8 (2011), 631-639.

7. P. Louette, F. Bodino, and J. Pireaux, "Nylon 6 (N6) Reference XPS Reference Core Level and Energy Loss Spectra," ***Surf. Sci. Spectra***, 12 (2005), 12-17.

8. F. Kolar, J. Svitilova, "Kinetics and mechanism of curing epoxy/anhydride systems", ***Acta Geodyn. Geomater***., Vol. 4, No. 3 Issue 147 (2007), 85-92

9. I. Lee, R. Wool, "Thermodynamic analysis of polymer-solid adhesion: sticker and receptor group effects", ***J. Polym. Sci., Part B (Polym. Phys.)***, 40 (2002), 2343-2353.

Advanced Composites for Aerospace, Marine, and Land Applications II
Edited by: T. Sano and T.S. Srivatsan
TMS (The Minerals, Metals & Materials Society), 2015

NEXT GENERATION CAR – EXAMPLE OF FUNCTION INTEGRATION AT THE LIGHT URBAN VEHICLE (LUV) VEHICLE CONCEPT

G. Kopp[1], S. Brückmann[1], M. Kriescher[1], H.E. Friedrich[1]
[1]Institute of Vehicle Concepts
German Aerospace Center (DLR)
Pfaffenwaldring 38-40
70569 Stuttgart, Germany

Keywords: Next Generation Car (NGC), Light Urban Vehicle (LUV), Lightweight Strategies, Functional Integration, Lightweight Design, Sandwich Design, Crash Behavior, Polyurethane Foam, Vehicle Concept, Crash Tests

Abstract

In times of climate change vehicle emissions have to be reduced clearly. One possibility is to reduce the mass of the body in white using lightweight sandwich structures. The department 'Lightweight and Hybrid Design Methods' of the Institute of Vehicle Concepts develops a vehicle body structure by using sandwiches with aluminum top layers and polyurethane foam as core material. For that the foam and the sandwiches were investigated under different load cases, e.g. pressure loading and in-plane tests. In tests with components the high potential of the sandwich materials were shown. On the dynamic component test facility of the institute, vehicle front structures were tested successfully. The results of all investigations regarding sandwich materials, integration of functions (e.g. crash, thermal) in vehicle structures and the concept LUV are developed under the research program of Next Generation Car of the DLR. We will show the development and results of the LUV.

State of the Art

In an age of global climate change, reducing the emission of greenhouse gases and pollutants is essential in order to slow down continuing climate warming. Moreover, the shortage of resource is also an important aspect which has to be considered in the development of new vehicles or vehicle components. Steady population growth and the increasing need for mobility associated with it will in future continue to exacerbate the situation with regard to the emission of greenhouse gases and pollutants. These factors, plus the statutory requirements, allow the conclusion that reductions in the consumption and emission values of passenger cars are essential. The main way of achieving this is by improving the efficiency of vehicles. As well as improving aerodynamics and the power train, increasing efficiency is possible through lightweight design, since a large proportion of fuel consumption is generated by the vehicle mass. The mass-dependent fuel consumption of passenger cars is detailed in Fig. 1.

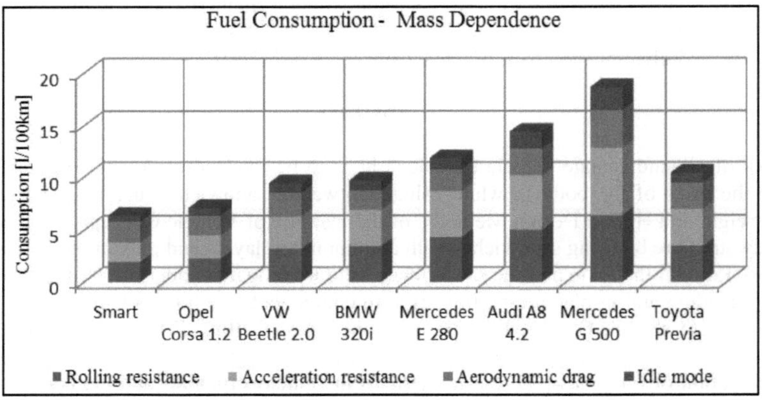

Figure 1: Fuel consumption of different vehicles; mass dependence [1]

The share of consumption due to aerodynamic drag and idling consumption is around 30 to 40% of total consumption, whereas the mass-dependent consumption of a vehicle due to rolling and acceleration resistance is up to 70%. These values are irrespective of the total vehicle mass (cf. Fig. 1). It is therefore evident that by saving on mass, a large proportion of consumption can be influenced and that the type and the mass of vehicle being investigated are irrelevant. In vehicles with an alternative drive, e.g. electrically driven vehicles, a reduction in the mass of the body in white structure is of considerable importance, because electrical energy storage systems are heavy.

There are various lightweight design strategies for reducing vehicle mass, e.g. sandwich construction. These include conditional, conceptual, material and form lightweight construction. These options of reducing vehicle mass, in combination with sandwich construction, open up further potential to reduce mass. New lightweight materials and the combination of materials through hybridization have shown in the past further improvements in lightweight solutions. For the future we need additional possibilities to reduce the weight of vehicle structures. A central possibility is to integrate additional functions. Functions and improved performance like mechanical (e.g. crash functions), thermal (e.g. isolation functions), electric (e.g. wires or sensor functions) or climate functions.

The goal of the LUV project, therefore, is the development of a highly efficient, affordable urban vehicle, with excellent passenger safety. The concept is a 2-seated urban vehicle with an overall weight of 450-500 kg (constraint of the German vehicle class L7e). The maximum speed will be 45 km/h (L7e), with a range of 100 km (project objective). This will result in a prototype car with increased overall efficiency due to very low weight and a highly efficient drive train. Very high passenger safety and low manufacturing cost, especially regarding the initial investment cost, are equally important goals. To reach these goals the solution structure has to be a combination of materials and integration of functions e.g. crash and thermal functions.

	Stiffness	Bending strenght	Mass
Only layers 	1,0	1,0	1,0
Core thickness t 	7,0	3,5	1,03
Core thickness 3t 	37,0	9,2	1,06

Figure 2: Comparison of stiffness, bending strength and mass of a sheet and two sandwich configurations [3]

One example of a sandwich structure can be constructed from cover layers (e.g. aluminum), adhesive and core material (e.g. polyurethane foam). The cover layers of a sandwich composite mainly absorb tensile and compressive forces. Forces at panel level are distributed two-dimensionally onto the core, so that where the design is optimal no local weakening is created. Furthermore, the cover layer material should possess a high Young's modulus, good impact resistance, high surface quality and high wear resistance [2]. The core material serves to absorb thrust and to guarantee the spacing of the cover layers. Moreover, in most cases it has a low density, possesses high shearing strength, a high degree of stiffness in the direction of the cover layer and often a good thermal insulation effect [2]. Because of the thickness of a sandwich structure, the geometrical moment of inertia of the structure is high compared with an orthodox sheet metal structure with approximately the same mass. Consequently, when the structure thickness is quadrupled, stiffness and bending strength increase by a factor of 37 and 9.2 respectively, whereas the mass only increases by around 6% (cf. Fig. 2).

Investigation of generic polyurethane foam samples

The aim of the investigations was the applicability of sandwich materials in vehicle areas affected by cases of crash load. The analytical design of sandwich elements is only possible if foam behavior is known. For this reason, different models were investigated in order to calculate the Young's modulus and the yield strength of the foams and then compared with the manufacturer data. The comparison produces considerable differences. This connection becomes clear in Figs. 3 and 4. Especially in high density areas (200 kg/m³), a difference of over 70% in the Young's modulus and over 100% in the yield strength can be seen.

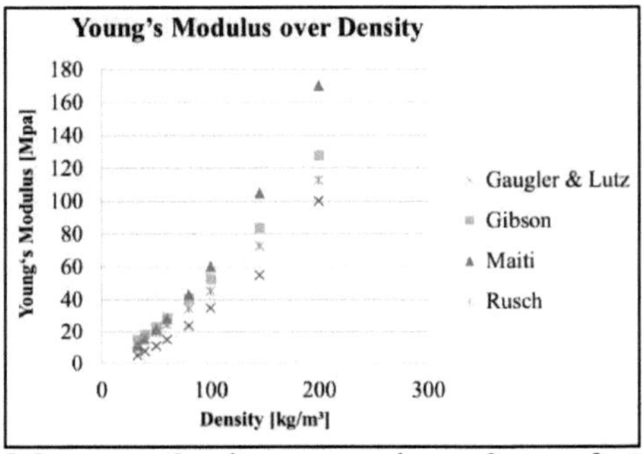

Figure 3: Young's modulus over density: comparison of manufacturer data (Gaugler & Lutz) with analytically calculated values

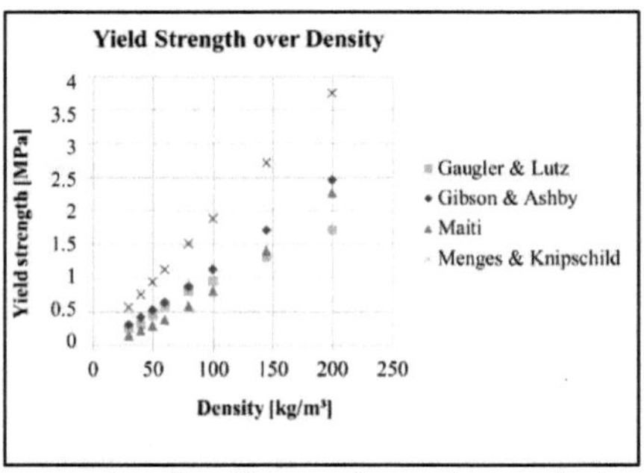

Figure 4: Yield strength over density: Comparison of manufacturer data with analytically calculated values

In the case of a foam density of 100 kg/m³, the manufacturer data for Young's modulus agrees with the calculated values according to Menges & Knipschild. In the case of a foam density above 100 kg/m³, all calculated values are above the manufacturer data quoted here. Below this foam density, the calculated values according to Menges & Knipschild are lower than those of the manufacturer. A similar connection can be seen for yield strength. Here, in the lower density range, the calculation values according to Maiti are below, and in the upper range above the manufacturer data. A comparison of the various calculation models with the manufacturer parameters allows the conclusion that generic tests on different polyurethane foams must be carried out. The foams are stressed for compression, tension and shear strain. The parameters of Young's modulus, shear modulus, compression strength, tensile strength and shear resistance were determined. In addition, force-displacement diagrams of the test piece series were prepared, in order to be able to add them to FE models.

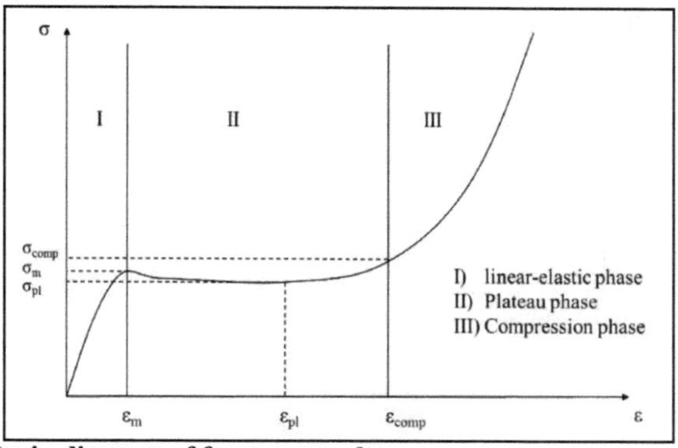

Figure 5: Stress-strain diagram of foam, example

All stress-strain diagrams produced, which were determined during the static compression tests, possess the shape shown by way of example in Fig. 5. At the end of the linear-elastic phase a local stress peak occurs in the gradient, followed by a stress drop to the plateau phase. This stress drop indicates that the first cell walls of the foam are collapsed. In the plateau phase, the approximate plasticization stress of the foam is read off. In this phase further foam cells collapse continuously. After this follows the compression phase, in which the foam is compressed to the final value of 85% compression (cf. DIN EN ISO 844 [4]).
The results of the compression and tensile tests are shown in Fig. 6. Shown in the left-hand diagram is the tensile strength, in the middle diagram the secant modulus and in the right-hand diagram the elastic compression limit.

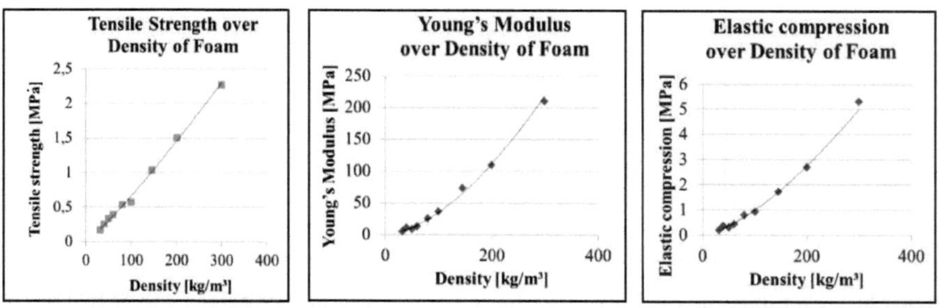

Figure 6: Results: Top right: Young's modulus, left middle: elastic compression, bottom right: tensile strength; overall density of foam

The tensile test was carried out in accordance with DIN 53292 [5]. Whereas the curve progressions of the secant modulus and of the elastic compression limit exhibit an exponential gradient, the tensile strength of the foams increases in linear fashion with the density. The physical test data generated is required for extrapolating the tests to generic sandwich elements. On the basis of the material parameters determined, the cover layer and core thicknesses can be calculated and a selection made as to which foam density is appropriate in the sandwich composite.

Generic sandwich elements

Using generic sandwich elements, the influence of cover layer thickness, foam density and core thickness on the failure behavior of the test pieces was investigated. In Fig. 7, three different failure modes are shown. On the left, global buckling can be seen. In the middle picture, buckling of the cover layers can be seen, which leads to the detachment of large areas of the cover layers from the foam core. On the right, cover layer wrinkling has occurred. In this case, the test piece remains vertical and does not buckle. Small wrinkles form almost symmetrical to the sandwich center axis. In the area of the wrinkles, the cover layers become detached from the core. Local buckling behavior and wrinkling is the preferred failure type for energy absorption in sandwich elements stressed in the in-plane direction.

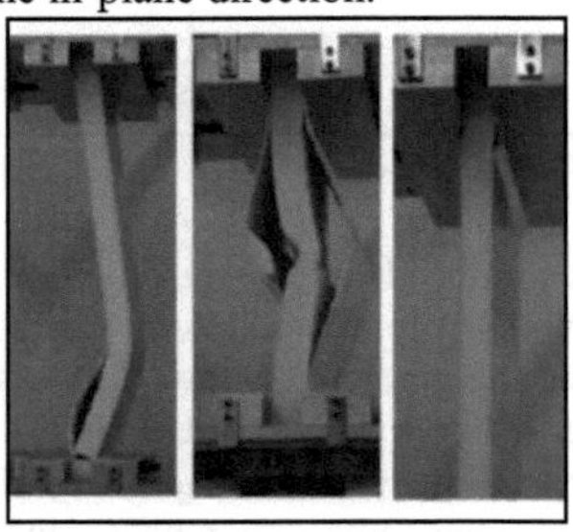

Figure 7: Various types of failure in sandwich elements; left: global buckling of the test piece, center: cover layers detach and buckle, right: wrinkling and buckling behavior of the cover layers

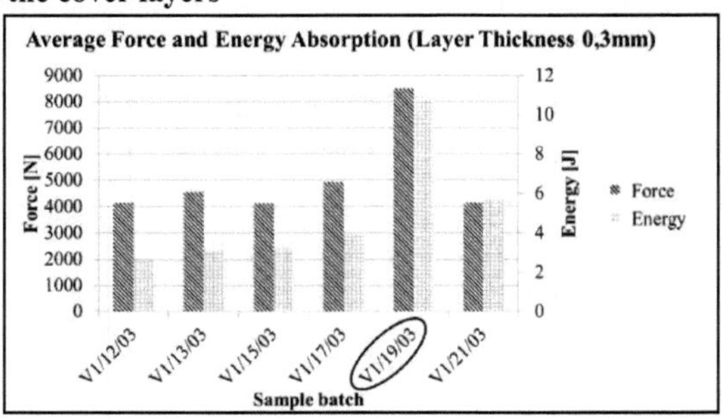

Figure 8: Comparison of the failure forces and energies of various sandwich test pieces with a cover layer thickness of 0.3mm

In Fig. 8, a comparison is shown of various sandwich elements. The force level at the point of failure and the absorbed energy with respect to a defined distance are compared. All test pieces have a cover layer thickness of 0.3 mm. The foam density continued to be maintained. The core thickness was changed. It can be seen that all test pieces except for V1/19/03 are situated roughly on the same force and energy absorption level. The failure types of these test pieces were comparable to the global failure mechanisms shown in Fig. 7. However, test piece V1/19/03 fails on cover layer buckling behavior. This proves that cover layer wrinkling absorbs energy more effectively in comparison with the other failure options.

Fig. 9 shows an example of a so-called failure mode map. Failure behavior can be predicted with the aid of a failure mode map. Failure is dependent on the core density (foam density) and the core thickness. In this simplified failure mode map, a distinction is made between global failure, shear failure and cover layer wrinkling. Detailed differentiation of the various global failure mechanisms is not necessary at this point, as the aim is cover layer wrinkling. The available failure mode maps are not applicable to the material combination being investigated here. Customized failure mode maps must therefore be drawn up. If the material behavior changes depending on the test speed, further differentiation is necessary.

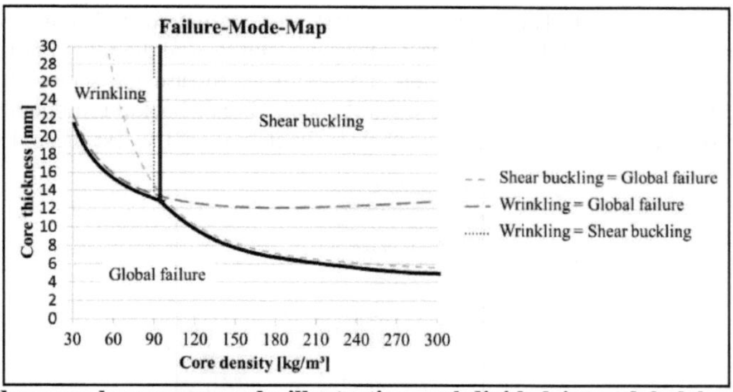

Figure 9: Failure mode map, sample illustration, subdivided into global buckling, cover layer wrinkling and shear failure

Triggering mechanisms

In order to adjust the force level at the time of initial failure, so-called triggering mechanisms are employed. The reduction of peak force at the beginning of failure is important, for example, in keeping interior acceleration in a vehicle as low as possible. This results in lower deceleration, which acts on the occupants and means that the risk of injury can therefore be decreased. In metallic constructions, beading and nicking are used for triggering purposes. Beads are also conceivable in sandwich elements. In addition, the possibility exists of weakening sandwich components in a targeted manner by means of holes or slots (cf. Fig. 10).

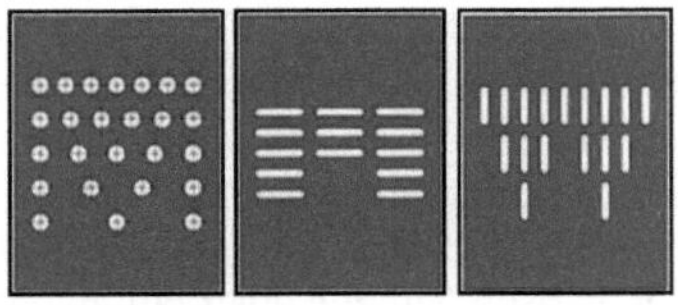

Figure 10: Various types of triggering; left: hole pattern, center: horizontal slot, right: vertical slot

The use of external triggering mechanisms is also conceivable. These include, for example, wedge-shaped elements, which weaken the sandwich composite in the event of a desired failure at the front and thus induce wrinkling of the cover layers. The effects of functioning triggering mechanisms are shown in Fig. 11. Pictured above is a two-row triggering mechanism with horizontal slots. Below left, a variant with holes and, below right, a design with vertical slots can be seen. In the case of the two-row triggering mechanism with horizontal slots, symmetrical wrinkles form. The apex of the wrinkles is in the top row of slots, the bottom end in the second row of slots. This can be accounted for by the weakening of the cover layers and easier buckling made possible as a result. The wrinkles in a test piece triggered with holes form completely symmetrically, both toward the sandwich center plane and toward the row of holes. In the case of a triggering mechanism with vertical slots, no symmetry toward the sandwich center plane is discernible. Moreover, the partitions between the slots do not buckle reproducibly. The triggering mechanism with vertical slots is nevertheless functional, as the test piece does not fail elsewhere.

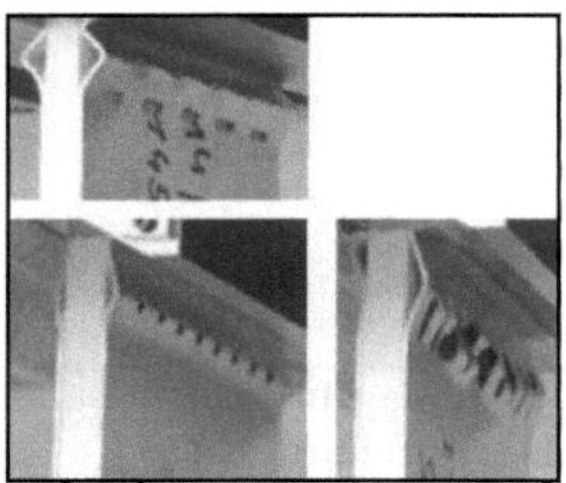

Figure 10: Different triggering mechanisms; top: two-row slot triggering horizontally, below left: one-row hole triggering, below right: slot triggering vertically

Geometric influence on failure behavior

The findings from the preliminary observations were incorporated in component tests. Fig. 12 shows two different test pieces. Pictured on the left is a box with a square cross section which was triggered with a row of holes, as hole triggering mechanisms proved to be the most appropriate. Repeated wrinkling can be seen. As predicted, in the actual course of the test the triggering mechanism worked, i.e. the initial failure occurred on the triggering mechanism and spread out from there in the form of wrinkles in line with the progress of the deformation. Pictured on the right is a cruciform structure, which was tested untriggered. During the test, a rotary motion became apparent in the test piece, so that it became twisted. Nevertheless, wrinkles formed and no bending of the test piece was detected. The preliminary tests are being translated into the development of a complete vehicle body.

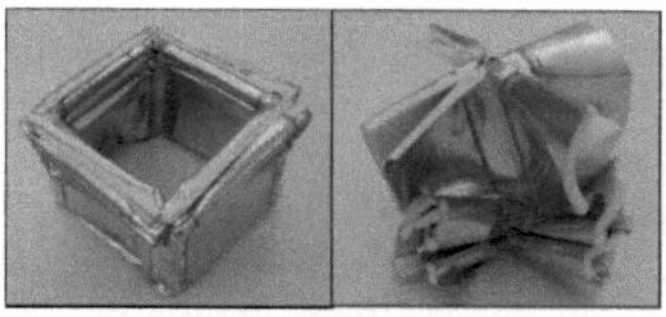

Figure 11: Compression tests on test pieces with geometric variance; left: triggered box structure, right cruciform structure

Vehicle concept in sandwich design

The Institute of Vehicle Concepts is developing a new kind of vehicle concept using sandwich construction. The virtual vehicle structure is shown in Fig. 13 right side. The front end and rear structures of the vehicle are constructed from sandwich panels and are comparable to the test pieces shown in Fig. 13 left side. Triggering mechanisms in the form of hole patterns are also used. The vehicle is designed to be driven by a battery fuel cell system. The aim is a very light and secure vehicle structure. First structural components were tested on the facilities of the institute. Fig. 13 (right side) shows the front structure of the LUV (Light Urban Vehicle) before and after crash test. The deformed structure shows the wrinkling of the aluminum layers. The crash speed was 9.5 m/s and 38 kJ of energy were absorbed bei deformation. The displacement lenght is 350 mm.

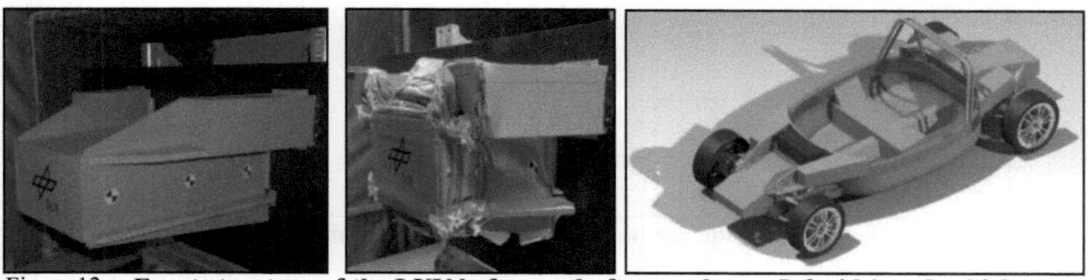

Figure 12: **Front structure of the LUV before and after crash test (left side) and vehicle concept in a new kind of sandwich construction (right side)**

Results of crash simulations

To date, 5 different crash load cases have been simulated, which all has shown excellent crash behaviour:

- US-NCAP full frontal crash, 56 km/h
- Euro-NCAP ODB crash, 64 km/h
- FMVSS 208 Frontal crash at 30° angle, rigid wall 48 km/h
- Euro-NCAP side-impact, 950 kg, 50 km/h
- Euro-NCAP pole-impact, 29 km/h

The front structure can easily absorb the kinetic energy of a US-NCAP full frontal crash (Figure 14), even without the use of an additional Crashbox, which is still under development. The mean deceleration of the structure is around 30 g. The deformation of the passenger compartment is small, at around 70 mm.

With the Euro NCAP-crash, there is a larger deformation of the impacted side, which is expected. The energy absorption of the front structure is still sufficient, however. The mean deceleration is slightly lower, compared to the US-NCAP full frontal test, at around 25 g. The deformation of the suspension parts is very important, since contact of the wheel with the passenger compartment must be prevented, in order to avoid the necessity for heavy reinforcement of the compartment. Therefore, the geometry of the front double wishbone suspension is designed in such a way, that the wheel rotates away from and to the side of the passenger compartment. The maximum Intrusion is higher, at 102 mm, and a slight distortion of the passenger compartment is visible. In contrast to a passenger compartment made of hollow structures, the metal monocoque structure can tolerate a certain deformation without buckling or collapse. This is due to the support of the core material, which prevents early buckling of the metal skin.

The Euro-NCAP side impact and pole impact result in a very low deformation, compared to the state of the art. In these load cases, the foam filled ring structure is especially effective in the distribution of deformation over a wide area. The maximum intrusion of the metal monocoque compared to a state of the art car body is lower in the side impact and pole crash and only slightly higher in the Euro-NCAP-ODB crash (Figure 18).

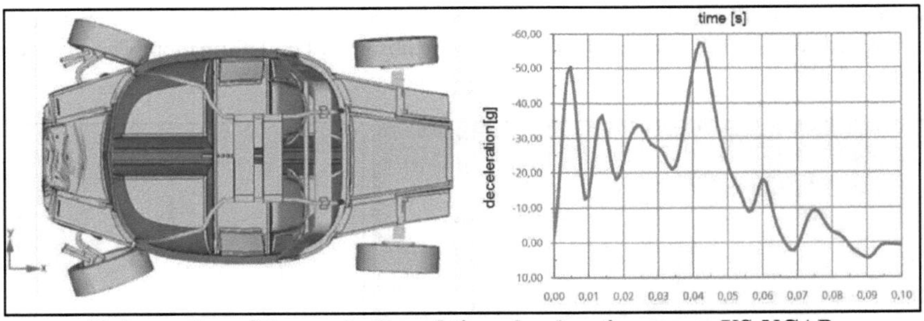

Figure 13: **Deformation of the car body and time-deceleration curve, US-NCAP**

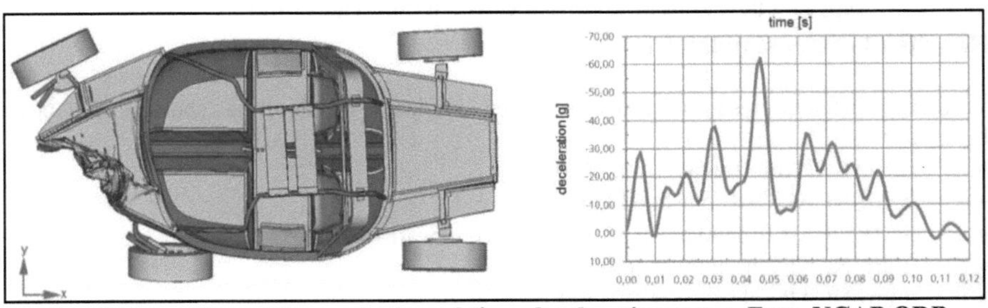

Figure 14: **Deformation of the car body and time-deceleration curve, Euro-NCAP ODB**

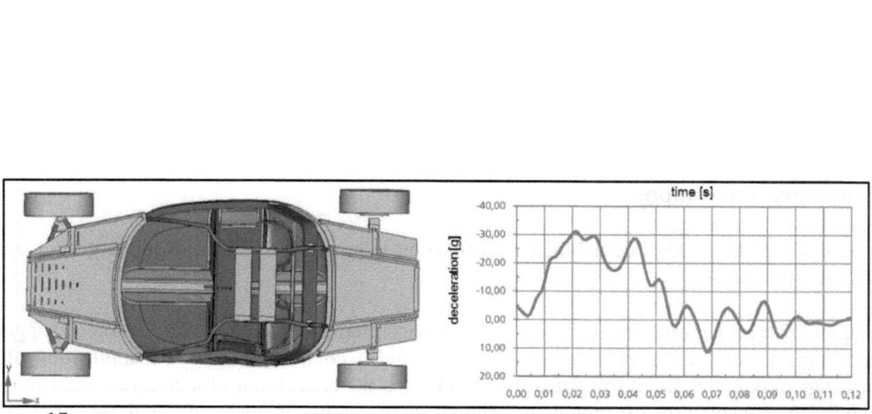

Figure 15: Deformation of the car body and time-deceleration curve, Euro-NCAP side impact

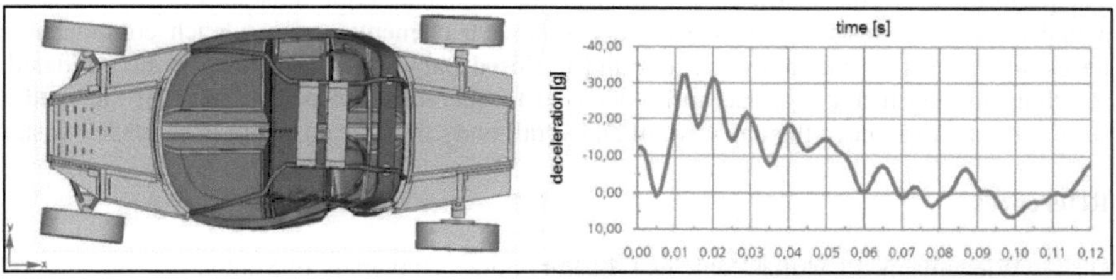

Figure 16: Deformation of the car body and time-deceleration curve, Euro-NCAP pole crash

Load case	Max. Intrusion Metal monocoque structure	Max. Intrusion State of the art body in white
Euro-NCAP frontal ODB	102	84
Euro-NCAP side impact	155	211
Euro-NCAP pole impact	206	381

Figure 17: Maximum Intrusion of the metal-monocoque structure in comparison to a state of the art car body

Summary and outlook

The lightweight strategies combined with function integration could fulfill the target for lighter vehicle structures.

Beginning with generic tests on foam and sandwich elements, via the investigation of triggering mechanisms for the targeted weakening of sandwich components and the application of the findings to box and cruciform structures, through to the conception of a complete vehicle using sandwich construction, it could be shown that sandwich materials and an appropriate construction have the potential to be used in automobile manufacture.

The performance of the sandwich materials and of the construction must be validated in ongoing actual tests. More tests on foam materials, sandwich elements and sandwich components are necessary. Also the simulation process and the used parameters in it have to be validated. A matching joining technology has to be detected to guarantee the performance of the sandwich structures and to achieve the objective of an overall body in white structure in sandwich design.

References

[1] Wiedemann, J.: Motor Vehicles I, Lecture notes, University of Stuttgart

[2] Zenkert, D.: An Introduction to Sandwich Construction. 1st edition. EMAS, London,1997.

[3] Starlinger, A.: Development of Efficient Finite Shell Elements for Analysis of Sandwich Structures under Large Instabilities. VDI-Verlag GmbH, Düsseldorf, 1991.

[4] DIN EN ISO 844: Harte Schaumstoffe – Bestimmung der Druckeigenschaften; Rigid cellular plastics – determination of compression properties; October 2009

[5] DIN 53292: Zugversuch senkrecht zur Deckschichtebene; Testing of sandwiches; tension test in flatwise plane; February 1982

[6] M.Aegerter (Ed.), Handbook Aerogels, Springer, New York 2012

[7] Cabot Aerogels, see http://www.cabot-corp.com/aerogel

Advanced Composites for Aerospace, Marine, and Land Applications II
Edited by: T. Sano and T.S. Srivatsan
TMS (The Minerals, Metals & Materials Society), 2015

THE SYNTHESIS AND PROCESSING OF SELF-HEALING MATERIALS: A LAMELLAR SHAPE MEMORY ALLOY IN COMPOSITE STRUCTURE

Bakr Mohamed Rabeeh[1] and Yasser Fouad[2]

[1] *Professor*
Engineering and Materials Science
German University in Cairo, GUC.

[2] *Associate Professor*
Engineering and Materials Science
German University in Cairo, GUC.

Keywords: Composite, HIPing; Hybrid; interface, interphase; fibrous, laminated, LMP, Self-healing.

Abstract

Nontraditional self-healing alloy composite has been introduced to address the need for self-repairable high-strength structural materials. Cu-Zn/Al laminates are uniaxial hot pressed for either lamellar or particulate composite. A new bridging and stitching technique is introduced with crack mitigation via the synergetic effect of interfacial alloying elements. Low melting phase (LMP) induced structures of Al phase, Cr-Fe phase, and Al-Cu-Zn phase via alloy segregation. Microstructural and mechanical characterization is also established via scanning electron microscopy, energy dispersive x-ray spectroscopy and tensile testing. The interphase kinetic established with micro plasticity, metal flow, low melting phase, LMP and delocalized Al enriched zone and Cu enriched zone. The improvements in performance of self-healing interphase with SMA composite structure are due to heating of low melting phase induce crack closure, and crack mitigation via stitching and bridging mechanisms. Engineering material can be designed with multifunctionalities ranging from the macro- to Nano-scale to optimize performance while minimizing time and the need for extensive repair.

1. Introduction

Aluminum based metal matrix composites are used in aerospace, defense and selected automotive applications such as high performance racing applications [1]. The most commonly used materials as reinforcement are SiC, Al2O3 or B4C into the aluminum matrix. These materials are used to improve elastic modulus, enhanced heat and wear resistance [2]. However, the addition of copper and/or brass in a lamellar structure to aluminum based composite is an objective. A solid state processing, unidirectional pressing, is selected at different parametric study; pressure, temperature and holding time. The most important concept in this technique is that the synergetic effects of parametric study [3]. While applying this technique for lamellar composite distribution, the kinetic of interface/interphase is being considered for mechanical alloying (MA). Though, it is not really „alloying" due to insolubility of the dispersoids in the matrix [4]. Mechanical alloying, as originally developed by Benjamin, was used to produce a combined oxide dispersion strengthening with gamma prime precipitation hardening in a nickel- based super alloy [5]. The purpose of the present studies is to investigate the influence of engineered interface on SMA processing of composite [8-9]. Interface/interphase size, shape and microstructural changes is being controlled and engineered for producing new delocalized constituents of composite reinforcement and their corresponding properties. A shape memory composite which is made of two types of shape memory materials, namely shape memory alloy (SMA) and shape memory hybrid. This composite has repeated instant self-healing function by means of not only shape recovery but also strength recovery (over 80%). The activation of the self-healing function is triggered by joule heating the embedded SMA [10].

2. Methodology

2024 Al alloy foils is introduced with brass foils at 0.5 mm thickness each and a solid state processing technique uniaxial pressing is applied at different temperatures 450, 500, 540 ^{O}C for different holding time (10, 20, and 90 minutes) and press level of 100 Ton. Design architectures as well as ply numbers are being considered for composite processing. The chemical composition of the 2024Al powder is shown in Table 1. As-received 2024 Al foils was arranged in a foil-foil technique and at different architectures 3 plies, 5 plies and seven plies. The constituent's weight fraction is being considered as 2:1, and 3:2 for 3, 5 and 7 plies symmetry arranged lamellar structural composite materials. Chemical and mechanical interfacial bond is established and is controlled via the synergetic effect of interfacial alloying elements Zn [LMP], Fe and Cr.

Table 1. The Chemical Composition of 2024Al foil (Wt. %)

Component	Al	Cr	Fe	Cu	Zn	others
Weight %	balance	Max. 0.1	Max. 0.5	3.8-4.9	Max. 0.25	Max. 0.05

3. Results and Discussion

A self-healing, metal matrix composite along with shape memory alloy is introduced via the engineering of lamellar structure composite materials. Engineering of Brass-Al-Brass foils layup, stacked in a symmetrical 3 plies, and then hot isostatic pressed, in a unidirectional press. Figure 1 presents scanning electron microscopy of 3-ply brass/Al at 560°C and 10 minutes holding time with 50 MPa pressure. After 30 minutes with the same parametric study, brass (Cu-Zn) layer is completely consumed and diffused within Al layers producing delocalized particulates (Cu-enriched) in Al-matrix composite (Fig. 2). Figure 3 presents 5 plies symmetrically unidirectional hipped [Brass/Al/brass] at 520°C for 20 minutes holding time and 50 MPa pressure and produced lamellar metal matrix composite with clear and bonded interface. Figure 4 presents high magnification of interface/interphase in scanning electron microscopy of 5-ply brass/Al/brass at 520°C and 20 minutes holding time with 50 MPa pressure. However, the macrostructures of Brass/Al/Brass are completely different from the parent material structure in the as bonded condition; the results are completely different if we switch the arrangement [Al/brass/Al]. The diffusion kinetics as well as the architecture is different in its sequential and directional form. Figure 5 presents scanning electron microscopy of 5 ply of Brass/Al/Brass symmetrically arranged at 540°C and 20 minutes holding time with 50 MPa pressure. The commencement of LMP diffusion along interface may produce minor delamination. Figure 6 presents effect of temperature as well as holding time on crack closure with bulk interphase diffusion at 540°C, 30 minute holding time and 50 MPa. Lateral and transverse bulk interphase flows induced by inter diffusion mechanism and LMP that proceed at 45 minute [Figure 7] and after 60 minutes it completely transform the final microstructure and macrostructure into new emerging materials symmetrically arranged [Figure 8]. The diffusion kinetic of alloying elements established with energy dispersive X-ray spectroscopy, EDX, at high magnification of the newly developed structure, Figure 8. It reveals that the outer lamellar composite structure, shape memory alloy, with 75.24 wt. % Cu, 18.74 wt. % Zn and 6.02 wt. % Al. A newly developed shape memory hybrid composite structure is intrinsically and extrinsically processed. Alloy segregation along with low melting phase [Zn] induces fibrous structure [Fe, Cr] in ductile aluminum enriched matrix [Al, Cu and Zn] as a leading interphase [Figure 10]. EDX of leading interphase is presented in Figure 11a,b,c and d. Trailing interphase [Figure 12] produces delocalized bulk and fibrous structure [Fe, Cr] in aluminum enriched matric [Al, Fe, Cu, Zn]. EDX of trailing interphase is established and presented in Figure 13a, b, and c. Heat flow and alloying elements segregation direction and low melting phase that produce aluminum metal matrix composite reinforced with delocalized particulates of copper alloy [Figure 1, and 2]. Architectural design and ply number produced either lamellar composite structure [Figure 3] and/or new nontraditional composite structure [Figure 8]. However, delamination is a lamellar structure limitation [Figure 5], the control of heat flow, holding time and pressure is dominant for the commencement of bulk interphase [Figure6]. Transient liquid phase bonding (TLP) and diffusion kinetics produce transverse and lateral material flow [Figure 7]. A lamellar composite structure (outer ply), symmetrically synthesis with micro fibrous interphase and central bulk particulate composite structure is established [Figure 8, and 9]. EDX reveals micro fibrous composite interphase with Fe-Cr fibrous structure in aluminum enriched matrix [Al-Cu-Zn] that

Figure 1. Scanning electron microscopy of 3-ply brass/Al at 560°C and 10 minutes holding time with 50 MPa pressure.

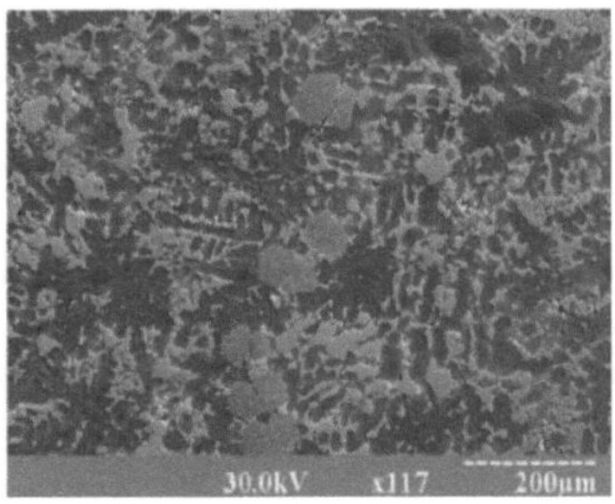

Figure 2. Scanning electron microscopy of 3-ply brass/Al at 560°C and 30 minutes holding time with 50 MPa pressure.

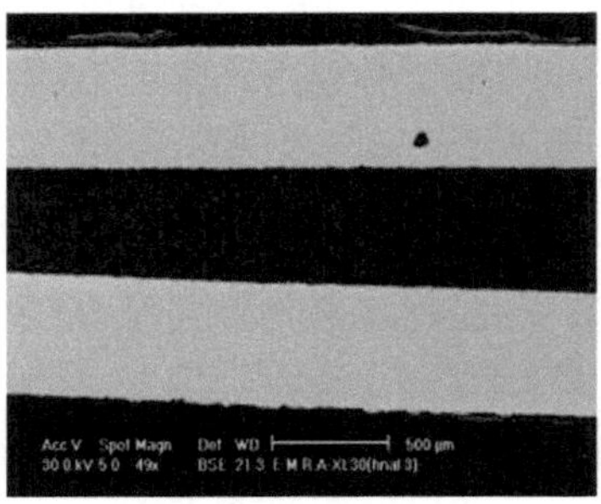

Figure 3. Scanning electron microscopy of 5-ply brass/Al at 520°C and 20 minutes holding time with 50 MPa pressure.

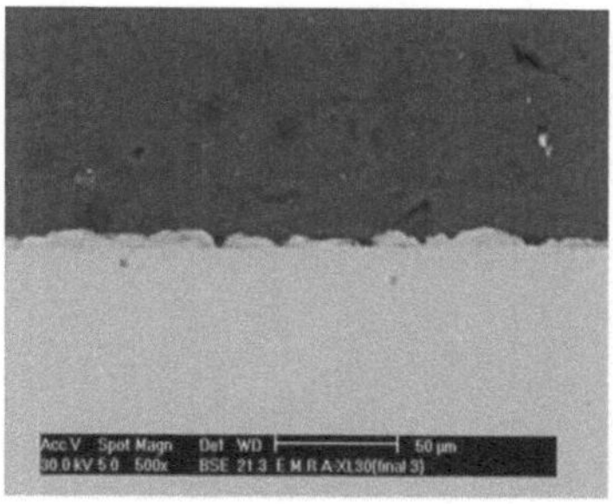

Figure 4. **Interface/Interphase in scanning electron microscopy of 5-ply brass/Al/brass at 520^{0}C and 20 minutes holding time with 50 MPa pressure. [High mag.]**

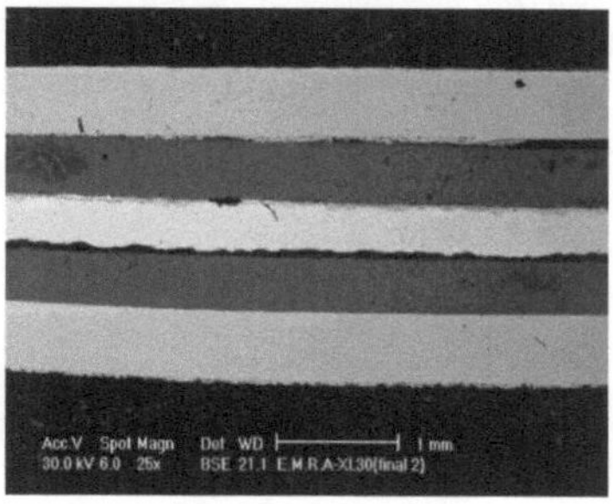

Figure 5. **Scanning electron microscopy of 5 ply of Brass/Al/Brass symmetrically arranged at 540^{0}C and 20 minutes holding time with 50 MPa pressure.**

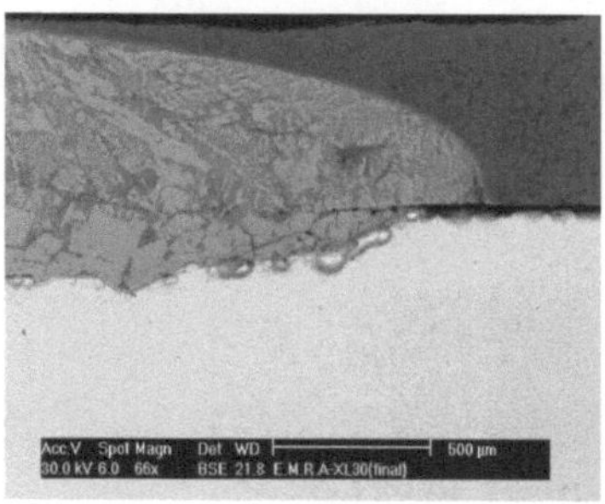

Figure 6. **Scanning electron microscopy of inter ply diffusion in Al/Brass interface at 540^{0}C and 30 minutes holding time with 50 MPa pressure.**

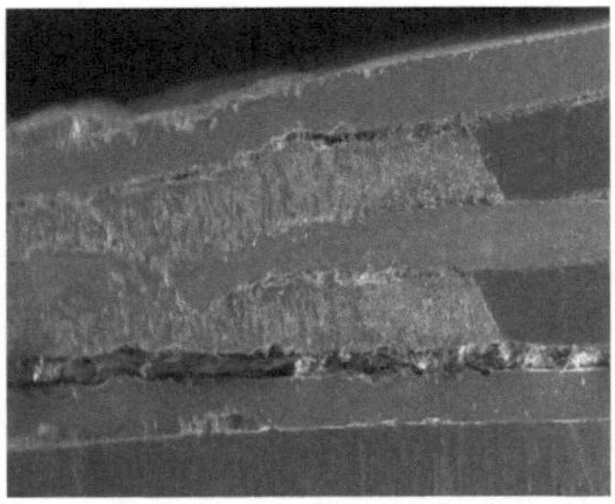

Figure 7. **Scanning electron microscopy of inner 3 ply inter diffusion Brass/Al/brass at 540^{0}C and 45 minutes holding time with 50 MPa pressure.**

Figure 8. **Scanning electron microscopy of final inter ply diffusion in Brass/Al/Brass 5 ply at 540^{0}C and 60 minutes holding time with 50 MPa pressure.**

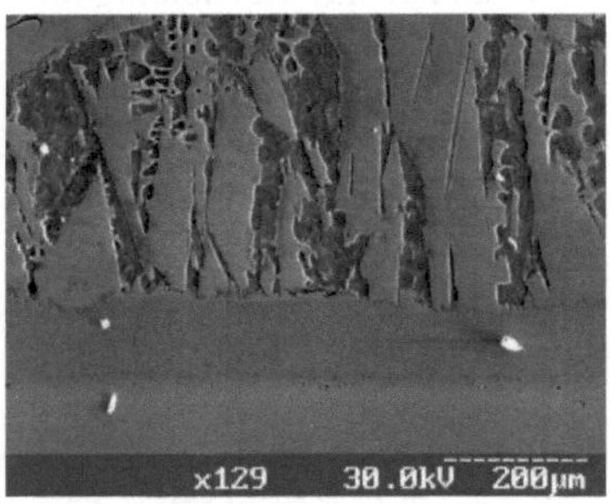

Figure 9. **Scanning electron microscopy of one half 5 ply brass/Al/brass at 540°C and 60 minutes holding time with 50 MPa pressure.[High mag.]**

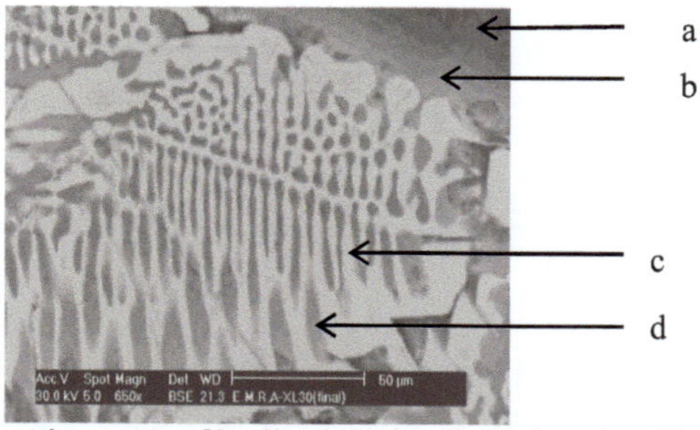

Figure 10. **Scanning electron microscopy of leading interface/Interphase in of 5-ply brass/Al/brass at 540^{0}C and 60 minutes holding time with 50 MPa pressure**

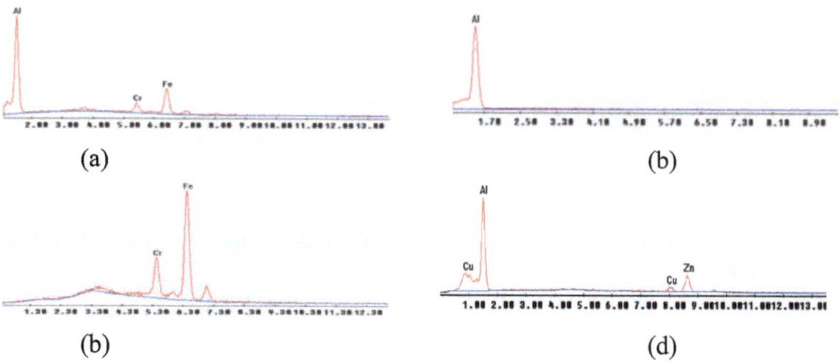

Figure 11. **Energy dispersive X-ray spectroscopy of aluminum foil along interphase fiber (dark).**

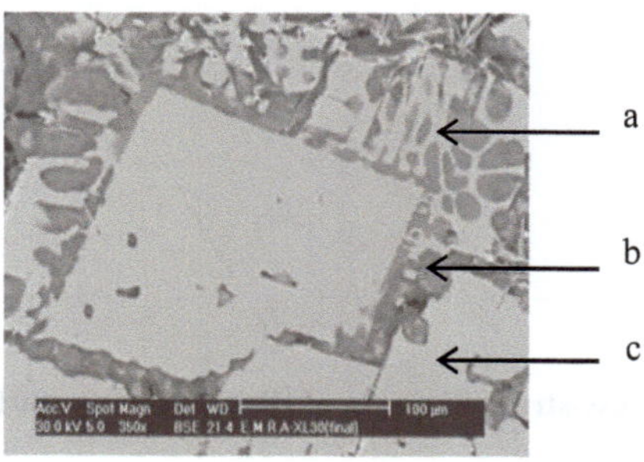

Figure 12. **Scanning electron microscopy of trailing interface/Interphase in of 5-ply brass/Al/brass at 540^{0}C and 60 minutes holding time with 50 MPa pressure**

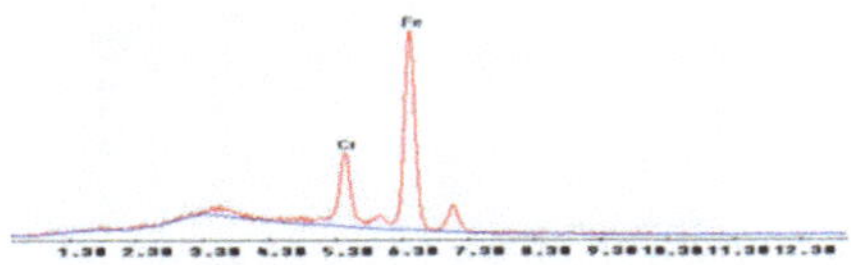

(a)

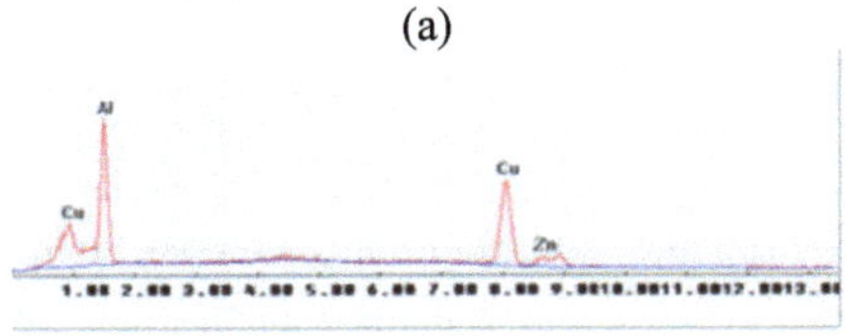

(b)

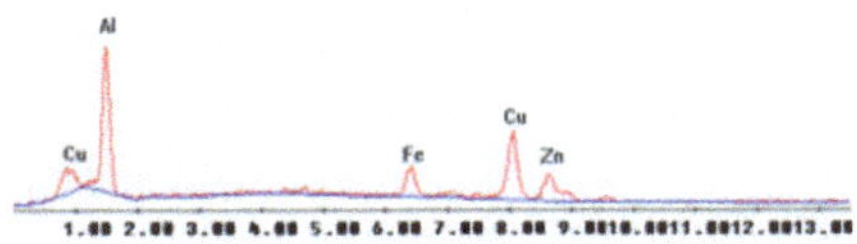

(c)

Figure 13. **Energy dispersive X-ray spectroscopy of aluminum foil along interphase fiber (white).**

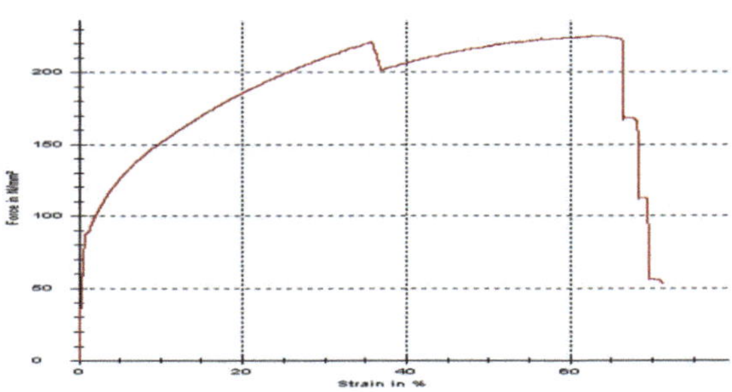

Figure 14. **Stress-strain diagram of [Brass/Aluminum/Brass] 5 plies.**

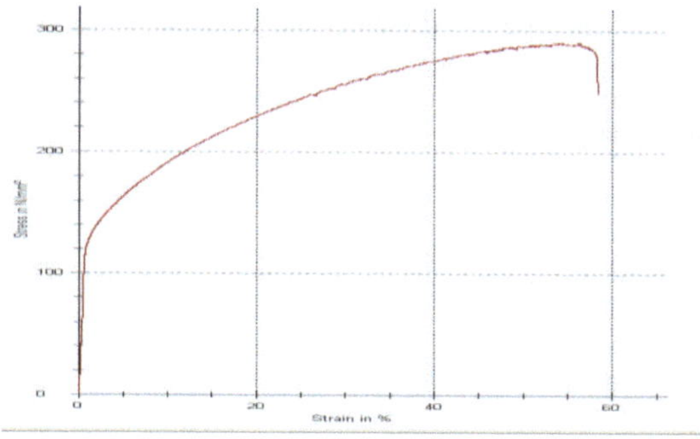

Figure 15. Stress-strain diagram of [Brass/Aluminum/Brass] in SMA with self-healing interphase.

presented in Figure 10. The central composite is dominated by aluminum enriched matrix reinforced by fibrous structure, Fe-Cr, and Cu-Zn enriched delocalized shape.

Mechanical characterization is applied for the two types of composite in a lamellar and in SMA outer lamellar composite structure with self-healing interphase composite. Stress-strain diagram of [Brass/Aluminum/Brass] 5 plies is conducted and presented in Figure 14, with stepped delamination induced fracture. No clear fracture obtained in stress-strain diagram of [Brass/Aluminum/Brass] in SMA hybrid composite structure with self-healing interphase is presented in Figure 15. When load is applied to the composite after damage, the outer ply SMA apply a compressive force which produces crack closure and clamping. Interfacial delamination is become hindered by the stitching and bridging effect of fibrous reinforcement that produce crack mitigation and shielding. The matrix alloy is now become partially molten at the healing temperature to reverse damage induced plasticity and provide crack welding via LMP.

4. Conclusions

Engineering of interface/interphase in copper zinc/aluminum lamellar composite structure is established through unidirectional pressing. Lamellar composite structure is obtained in 3, 5 plies symmetrical arrangements. The way these foils stacked as well as hipping parameters induce macroscopic aspects as well as microscopic aspects. Transient liquid phase bonding (TLP) implies that equilibrium state of the joint region is solid. The route, by which solidification can be achieved, can involve up to five related phenomena. These are the diffusion kinetics of alloying elements, materials flow, Transient liquid phase bonding (TLP), shape memory alloy and self-healing kinetic of micro fibrous interphase. The diffusion into the joint metals is mainly temperature, pressure and time dependent. Brass-Aluminum system can be TLP bonded at 540°C produces new emerging shape memory alloy and hybrid composite. The possibility of hybridization introduced with new emerging Cu-Zn-Al shape memory ally (outer ply), controlled micro fibrous interphase and delocalized central Cu-enriched whiskers and particulate in aluminum enriched matrix. Crack mitigation, and toughening mechanism established by stitching and bridging effect of micro Fe-Cr fibrous within aluminum matrix interphase. New nontraditional self-healing interphase with shape memory alloy hybrid composites have been designed to address the need for self-repairable high-strength structural materials.

References

1. Hurley S, MMCs Find Broad Range of Niche Markets, 1995, ***Met. Bull.*** Mon., pp. 54–56.

2. Otsuki M S, Kakehashi S, Kohno T, "Powder Forging of Rapidly Solidified Aluminum Alloy Powders and Mechanical Properties of Their Forged Parts", in: Advances in Powder Metallurgy, vol. 2, Metal Powder Industries Federation, Princeton, NJ, 1990, pp. 345–349.

3. Lu L, Lai M O, Ng C W, "Enhanced Mechanical Properties of An Al Based Metal Matrix Composite Prepared Using Mechanical Alloying", ***Mater. Sci. Eng.*** 1998, A 252 (2), pp. 203–211.

4. Sankar R, Singh P, "Synthesis of 7075 Al/SiC Particulate Composite Powders by Mechanical Alloying", ***Mater. Lett.*** 1998, 36, pp. 201–205.

5. Soni P R," Mechanical Alloying- Fundamentals & Applications", Press: Cambridge International Science Publishing, 2001, 6-23 p.

6. Benjamin J S , Volin T E," The Mechanism of Mechanical Alloying, ***Metall. Trans.*** 5, 1974, pp.1929–1934.

7. Gilman P S, Nix W D, "The Structure and Properties of Aluminum Alloys Produced by Mechanical Alloying: Powder Processing and Resultant Powder Structures", ***Metall. Trans.*** 1981, A 12, pp. 813–823.

8. Naiqin Zhao, Philip Nash, Xianjin Yang, "The Effect of Mechanical Alloying on SiC Distribution and The Properties of 6061 Aluminum Composite", ***Journal of Materials Processing Technology,*** 2005, 170, pp. 591.

9. Kirkby, Eva L.; Rule, Joseph D.; Michaud, Véronique J.; Sottos, Nancy R.; White, Scott R.; Månson, Jan-Anders E.," Active repair of self-healing polymers with shape memory alloy wires", Behavior and Mechanics of Multifunctional and Composite Materials 2007. Edited by Dapino, Marcelo J.. Proceedings of the SPIE, Volume 6526, article id. 65261E (2007).

10. Oleg Neikov, Stanislav Naboychenko, Irina Mourachova , Victor Gopienko ,Irina Frishberg, Dina Lotsko, Handbook of Non-Ferrous Metal Powders, Technologies and Applications; Press: Elsevier, 2009, 267-283 p.

11. Wang, C. C.; Ding, Z.; Purnawali, H.; Huang, W. M.; Fan, H.; Sun, L., "Repeated Instant Self-healing Shape Memory Composites", ***Journal of Materials Engineering and Performance***, Volume 21, Issue 12, pp.2663-2669, 12/2012.

CARBON FIBER REINFORCED COMPOSITES and MODELING & SIMULATIONS

Advanced Composites for Aerospace, Marine, and Land Applications II
Edited by: T. Sano and T.S. Srivatsan
TMS (The Minerals, Metals & Materials Society), 2015

SURFACE MODIFICATION OF CARBON FIBER POLYMER COMPOSITES AFTER LASER STRUCTURING

Adrian S. Sabau[1], Jian Chen[1], Jonaaron F. Jones[2], Alexandra Hackett[2], Gerald D. Jellison[1], Claus Daniel[1], David Warren[1], Jackie D. Rehkopf[3]

[1]Oak Ridge National Laboratory, Oak Ridge, Tennessee, 37831
[2]University of Tennessee, Knoxville, Tennessee, 37996
[3]Plasan Carbon Composites, Bennington, VT 05201

Keywords: laser, interference, carbon fiber, polymer matrix composites

Abstract

The increasing use of Carbon Fiber-reinforced Polymer matrix Composites (CFPC) as a lightweight material in automotive and aerospace industries requires the control of surface morphology. In this study, the composites surface was prepared by ablating the resin on the top fiber layer of the composite using an Nd:YAG laser. The CFPC specimens with T700S carbon fiber and Prepreg - T83 resin (epoxy) were supplied by Plasan Carbon Composites, Inc. as 4 ply thick, 0/90° plaques. The effect of laser fluence, scanning speed, and wavelength was investigated on the removal rate of the resin without an excessive damage of the fibers. In addition, resin ablation due to the power variation created by a laser interference technique is presented. Optical property measurements, optical micrographs, 3D imaging, and high-resolution optical profiler images were used to study the effect of the laser processing on surface morphology.

Introduction

The increasing use of Carbon Fiber Polymer Composite (CFPC) as a lightweight material in automotive and aerospace industries requires more effective and reliable joining techniques. The surface treatment is a critical step governing the quality of adhesive bonded joints. Composites contain mold releases and current state-of-the art surface preparation for composites is typically accomplished through multiple mechanical/chemical processes, which lack the speed and repeatability needed for high-throughput manufacturing. These traditional surface preparation methods of composite surfaces leave a resin-rich surface layer on the composite, which is susceptible to the nucleation of cracks that may limit its structural integrity. The surface also contains chemicals, such as mold releases and contaminants, which often require removal prior to adhesive bonding. Niino et al. [1] used a 308nm XeCl excimer laser to induce surface micromodifications onto polymers and carbon fiber composites. The use of an excimer laser treatment, selective removal of the organic matrix without any degradation of fiber reinforcements, was investigated by Galantucci et al. [2] and Benard et al. [3, 4]. Warren et al. [5] also used a KrF excimer laser to process glass fiber-reinforced polymer matrix composites. Significant glass fiber damage was found due to the ablation process, as the glass fibers acted as a lens focusing the laser energy to the backside of the fibers and thus produced microgrooving of the fiber along the side of the fiber farthest away from the incident direction. Lima et al. [6] used a 20 W of a Nd:YAG pulsed laser to producing fiber optical guidelines in carbon fiber reinforced polymer (CFRP) composites using laser texturing and machining.

In this study, the CFPC surfaces were modified by ablating the resin-rich surface layer and the resin surrounding the top fiber layers of the composite using a Q-switched Nd:YAG [7]. In the second section of this paper, the experimental laser procedures, surface morphology, and optical characterization of the as-received CFPC are presented. The CFPC specimens with T700S carbon fiber and Prepreg - T83 resin (epoxy) were supplied by Plasan Carbon Composites, Inc. Optical measurements were performed on several samples. In the third section of this paper, the effect of scanning speed, wavelength, and or number of laser pulses on the surface condition was investigated. Optical micrographs and 3D surface profiles, which were taken using a Keyence VHX-2000 digital microscope, were also used to characterize the effect of laser structuring.

Materials and Optical Properties

The CFPC specimens with T700S carbon fiber and Prepreg - T83 resin (epoxy) were supplied by Plasan Carbon Composites, Inc. A WYKO 9100 surface profilometer was used to obtain surface roughness data for the as–received CFPC. Wyko surface profilometer systems are non-contact optical profilers that use two methods to measure a wide range of surface topographical features. Phase-shifting interferometry mode allows the measurement of smooth surfaces and small steps, while vertical scanning interferometry mode allows the measurement of rough surfaces and steps up to several mm high. The surface profile shown in Figure 1a reveals that the as–received CFPC exhibits some grooves on its surface, approximately 9μm deep, which may be generated from a previous processing step such as rolling. The average roughness, or arithmetical mean deviation of the roughness profile, Ra, root-mean-square deviation of the roughness profile, Rq, and maximum height of the roughness profile, Rt, were estimated to be 0.349, 0.54, 11.59 μm, respectively.

In order to understand the laser ablation of the CFPC, optical measurements were conducted on several samples (Table 1). The goal was to characterize the absorption of laser energy at different wavelengths for carbon fiber and for carbon fiber plus epoxy resin and to then separate what part of the absorption was due to the fiber and what part was due to the epoxy resin. Four specimens were prepared by tightly wrapping T700S carbon fiber tows around stainless steel carriers. Two samples were then heated to 350°C in helium to volatize and remove the fiber sizing. One of those samples without sizing and one with sizing were then used for optical measurements without further preparation. The two remaining samples (one with sizing and one without) were mounted in epoxy specimen molds and polished to expose the fiber with care being taken not to damage the fibers. It was noted that there was minimal epoxy around the fibers due to the placement in the specimen containers, the high viscosity of the mounting epoxy and degree of the close packing of the fiber tows but that some epoxy was present. Additionally a sample of the 4-ply composite material was studied yielding fiber samples (Table 1).

Optical measurements were made of those samples. Specifically, the absorption properties of the resin and fiber were individually sought. Without considering the laser fluence variation with wavelength, the desired laser ablation wavelengths for the CFPC would be those at which the resin would exhibit high absorption. The diffuse reflectivity measurements were made on all samples using a Perkin-Elmer Lambda 900 spectrophotometer using an integrating sphere attachment. This was necessary because the samples were not specularly polished, requiring the integrating sphere attachment. Spectra were taken from 250 nm to 850 nm (4.96 to 1.46 eV).

Table 1. Sample descriptions used for the optical measurements.

Name	Physical Preparation	Sizing	Mounted in epoxy	Thermal Preparation
CF+sizing (No-resin)	T700S fiber ribbon tightly wrapped on a stainless steel carrier ~1.3x2.5 cm	Yes	No	None
CF-sizing removed (No-resin)	T700S fiber ribbon tightly wrapped on a stainless steel carrier (~1.3x2.5 cm)	No	No	Heated at 350°C in flowing He to volatilize sizing
CF+sizing mounted in epoxy	T700S fiber ribbon tightly wrapped on a stainless steel carrier (~1.3x2.5 cm) and then mounted in epoxy and polished to expose the fibers	Yes	Yes	None
CF-sizing mounted in epoxy	T700S fiber ribbon tightly wrapped on a stainless steel carrier ~1.3x2.5 cm and then mounted in epoxy and polished to expose the fibers	No	Yes	Heated at 350°C in flowing He to volatilize sizing prior to mounting in epoxy
CFPC	4-Ply Composite which includes epoxy resin	Yes	No	CFPC panel, cleaned on one side

Figure 1b shows the diffuse reflection from the 5 samples examined. The only sample that gave appreciable reflectance was the CFPC, which is the only one that contains resin, too. The other 4 samples consisted of carbon fiber ribbons, which are multi-layered bundles of fibers and can therefore trap a significant fraction of the light. That is, light that is incident upon the sample can easily pass past the surface layer and be reflected non-specularly off one of the underlying fibers, to be absorbed or scattered by another fiber. Therefore, the reflectance from these 4 samples cannot be taken as a fundamental property of the carbon fibers - as a bulk material, but rather as a property of the CF bundle. On the other hand, the reflectance from the CFPC does represent a fundamental property of the material, since the reflection occurs at the surface. The reflectance is high for light wavelengths greater than ~300 nm, and indicates that the material is optically metallic, i.e., would exhibit a high absorption coefficient. Below 300 nm, the reflectivity decreases significantly; this indicates a fundamental optical characteristic of the material. Additional, spectroscopic ellipsometry measurements, which are not discussed in detailed here, showed that the surface reflectivity is quite small (~1.6), which would result in a reflectivity ~5.3%, considerably lower than the reflectance measured using diffuse reflectance. Clearly, the carbon fibers of the composite material, which are below the surface, contribute significantly to the total reflectance, resulting in the higher value of reflectance for CFPC sample. In conclusion, diffuse reflectivity data indicate that the optical properties of the CF itself do not vary strongly with the wavelength while the resin affects the optical properties of the CFPC at wavelengths higher than 280 nm. Based on this reflectivity data, it would be desired that for an effective ablation of the resin, a laser wavelength of 266 nm would be desired. However, as shown in the next section, the wavelength selection would affect the laser power and this factor needs to be considered when selecting the wavelength.

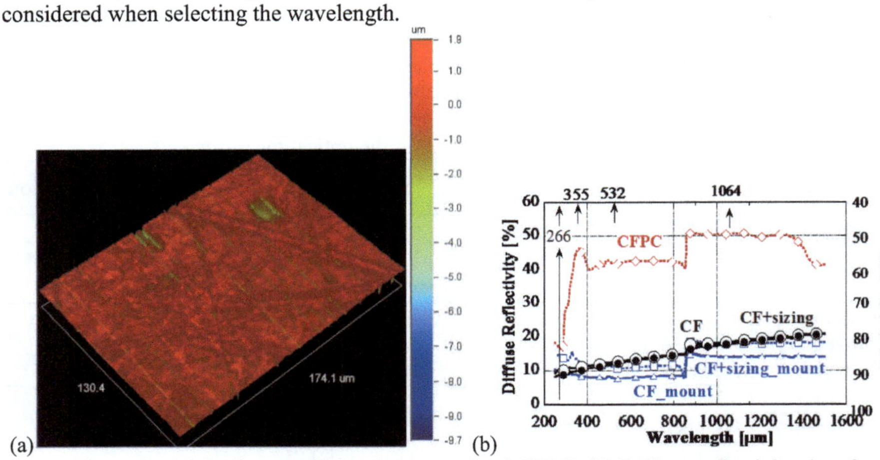

Figure 1. (a) Surface profile as the as-received CFPC. (b) Diffuse reflectivity data for the CFPC samples considered.

Experimental Procedures

A 10Hz Q-switched Nd:YAG laser (Quanta-Ray PRO 230, Spectra Physics) was used in this study. With only 10ns the laser pulse duration is very short, creating an extremely high power above 100MW, resulting in greater than 10^{12}K/s heating rates. The beam diameter was 8mm. The fundamental emission with the wavelength of 1064 nm was transformed to 355 nm and 266 nm using non-linear crystals. The maximum laser energy per pulse was 0.15 J at 355 nm and 0.038 J at 266 nm. This effect of the decrease in the laser power at the 266 nm with respect to that at the 355 nm wavelength had to be taken into account together with the optical properties of the CFPC materials, which were presented in the previous Section.

The laser interference power profile was created by splitting the beam and guiding those beams to the sample surface by overlapping each other with defined angles to each other (Figure 2). The coherent beams create an interference pattern instead of just adding their intensity. This allows a microscopic modulation and creates a light pattern without any loss of energy during the interference process. Patterns can be dot-, line-, and ring shaped [8-12]. The schematic setup of the laser setup is shown in Figure 2. The number of pulses was selected by a mechanical shutter. In some experiments, the laser pulse fluence was increased by using two identical focal lenses were used in each path of the splitted beam to focus laser beams from its original size of 8 mm to smaller sizes. The periodicity between power peaks, and laser-interference induced undulations, is defined by the wavelength, λ, and the angle, α, between the two beams, as : $d = \lambda / (2\sin(\alpha/2))$.

Other features of the structured surface morphology than those included in Table 1, include: (a) Undulation spacing of 0.5 - 50 mm, (b) Structured area of $0.27 cm^2$/shot, (c) scanning speed: 10,000 lines at a time, 79 million dots at a time, up to $162 cm^2$/min.

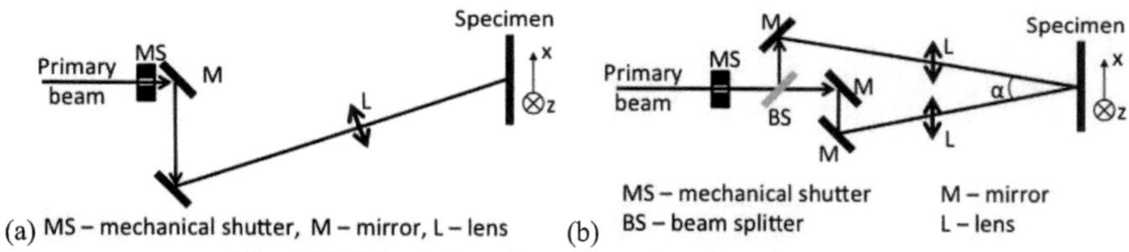

Figure 2. Schematic setup of laser interference system.

Table 2. Features and differences of proposed technique from conventional laser ablation techniques.

Feature	*One-beam laser*	*Advantages of laser-interference*
Patterns and shape	Shape of surface patterns identical to the laser spot shape	New patterns on CFPC can be dot- or line- shaped
Size of geometrical features	0.5 to 1 mm identical to the laser spot size	Feature size: 1 to 500 nm Feature density: 20 to 2,000 per μm

CFPC were laser-processed using both the one-beam (i.e., without the power interference) and two-beam laser interference setup. For the one-beam setup, two mirrors were also used in order to minimize the setup time to/from the two-beam laser interference setup, the incidence angle with respect to the normal to the sample surface was kept at 3° at close values to those for the laser interference setup. For the laser interference setup, the two beams were focused on a spot and each spot on the specimen surface was processed independently, i.e., firing a certain number of pulses after translating the specimen such that the next spot will be laser processed. For the traditional one-beam laser setup, the specimen was continuously translated at constant speeds while the laser operated at 10Hz.

Microstructure Analysis of the Laser Ablation and Structuring

In order to illustrate the laser structuring capability with the current optics setup, optical micrographs for a 316 stainless steel specimen are shown in Figure 3, in the conditions as-received and after laser structuring. The surface profile, which was obtained using a Keyence VHX-2000 digital microscope, Figure 3c, indicates a periodic spacing at the micron length scales and a submicron height of the laser-induced structures.

Due to the power decrease when selecting the wavelength, the first series of experiments were conducted at 355 nm wavelength. First, preliminary experiments were run to assess the effect of the laser operating parameters on the ablation of the resin, structuring of the resin, and structuring of the carbon fiber. These preliminary experiments indicated that operating the laser at maximum average power of 1.25-1.5W would be required in order to observe a change in the surface condition of the CFPC. Several cases were then run to assess the effect of number of laser pulses on the surface condition of the CFPC (Table 3). The laser fluence for the cases shown in Table 3 was approximately 0.25 J/cm^2. For some of the cases indicated in Table 3, optical micrographs are shown in Figure 4. At number of pulses per spot of 20 to 60, isolated spots of ablated resin were observed. The onset of the carbon fiber structuring was observed at 80 pulses per spot; indicated by striations on the carbon fiber. It has to be noted that no structuring in the resin itself was observed, indicating its complete ablation in those regions. As the number of pulses was increased, the ablated area was also increased (Figure 4 a and b).

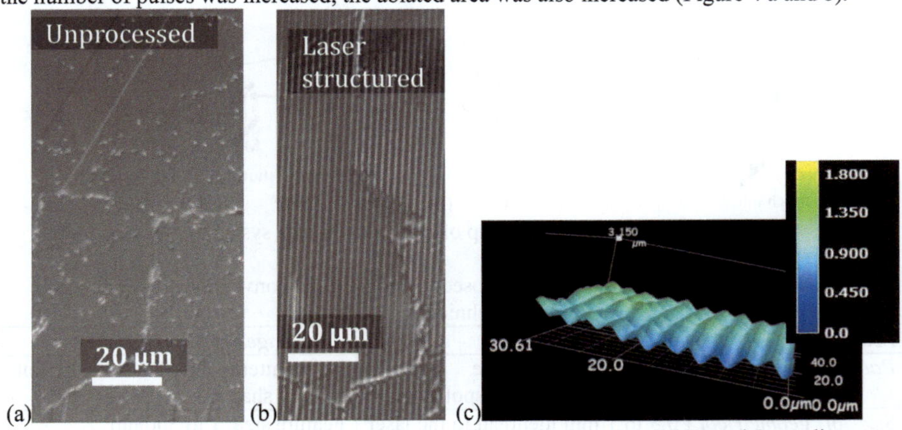

Figure 3. Laser structuring of 316 stainless steel at average power 1.5 W and 8mm diameter spot: (a) before laser structuring, (b) laser structuring, and (c) 3D surface profile after laser structuring.

The quality of the structuring, as evidenced by the sharpness of the striations shown in Figures 4c and 4d on the CF strands, was also found to increase with the number of laser pulses per spot. It has to be noted that based on these results at laser fluences of 0.25 J/cm^2, the number of pulses per spot may be considered quite excessive for some applications. Following this sensitivity study, a larger sample was prepared at 350 s per spot (laser fluence - 0.3 J/cm^2) and its surface was analyzed with a Keyence VHX-2000 digital microscope. The optical micrograph and corresponding 3D surface profiles are shown in Figure 5. The 3D surface profile shows the unevenness of the specimen surface, indicating macroscale features that cannot be observed in the optical micrograph.

Table 3. Experiments conducted to attain both resin ablation and structuring of carbon fiber at an average power of 1.25 W over an 8mm spot size at 355 nm wavelength.

Run No.	*No. pulses per spot*	*Resin ablation*	*CF structuring*

1	20	spots	-
2	40	spots	-
3	60	spots	-
4	80	spots	Onset
5	100	Increased area	Slightly enhanced
6	150	Increased area	Slightly enhanced
7	200	Increased area	enhanced
8	250	Large area	enhanced
9	300	Large area	Very good
10	350	Large area	Very good

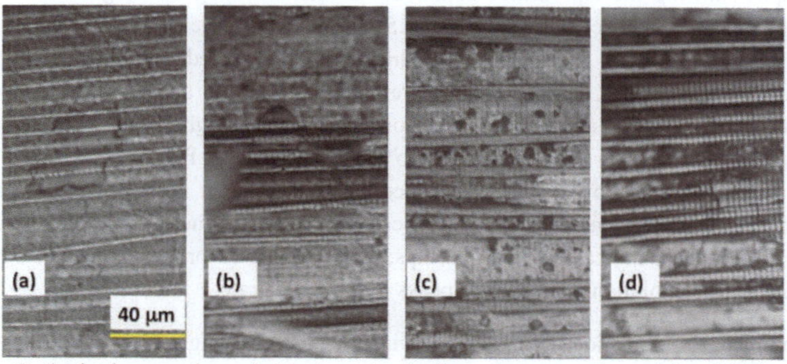

Figure 4. Optical micrographs and ablation of resin from the top surface of the CFPC using laser interference technique at 355 nm, average power 1.25 W, over an 8mm diameter spot after number of pulses of: (a) 60, (b) 200, (c) 250, and (d) 300 (frequency of pulses 10Hz).

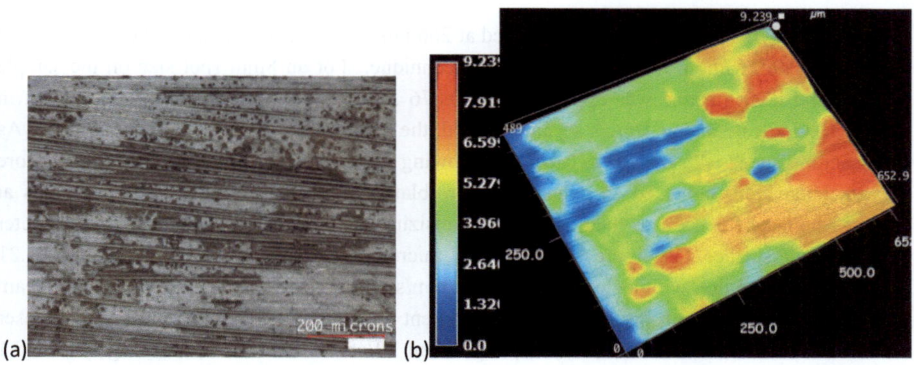

Figure 5. Ablation of resin from the top surface of the CFPC using laser interference technique at 355 nm, average power 1.5 W, over an 8mm spot diameter after 350 pulses per spot (35 s of exposure at 10Hz): (a) optical micrograph and (b) 3D surface profile [µm]. (500X magnification).

The next series of experiments was conducted with at one-beam traditional laser setup, i.e., without the laser interference technique. The one-beam setup was sought as a means to increase the productivity of the surface processing, i.e., as the idle times in the stop-and-go schedule would be avoided when the beam is simply translated across the sample surface. This second series of experiments aimed at identifying the scanning speed threshold for an effective resin ablation. Since the laser interference sensitivity conducted at 8mm spot diameter (laser fluences of 0.25-0.3 J/cm^2), indicated that an excessive number of pulses per spot, i.e., larger than 250-300, have to be used to attain an effective laser ablation and CF structuring, the laser fluence was increased by resizing the beam from 8 mm to 4 mm diameter at the sample surface. Optical micrographs showing the top surface of the CFPC after rastering the single laser beam at an average power 1.43 W at scanning speeds of 0.5, 1, and 3 mm/s are shown in Figure 6. At slow speeds (0.5 and 1 mm/s), most of the CF are clearly delineated. As the rastering speed is increased to 1 and 3mm/s, partially ablated regions between CF appear, as evidenced by the region between CF #5 and #6 (as numbered from the left to the right in Figure 6b and indicated by the ellipse drawn) and the region close to the right edge of Figure 6c. It was found by examination of the 3D profiles using a surface profilometer that most of the unclear regions along the CF direction are not simply out of focus regions but artifacts due to broken CF (Figure 7). The highest peaks indicate broken fiber tips; the profile of the broken CF can also be seen by the narrow profile evolution from highest peaks into the substrate.

The incomplete resin ablation became more evident as the rastering speed was further increased (Figure 8 for scanning speeds of 6.5, 7.5, and 8.5 mm/s). In Figure 9, the optical micrograph and corresponding 3D surface profile are shown at a laser fluence of 1.14 J/cm^2 (rastering speed of 0.5mm/s). The CF are clearly distinguished in the optical micrograph indicating a complete ablation of the resin on the top surface of CF. The 3D surface profile indicates the complete resin ablation between successive CF.

The last series of experiments was conducted at 266 nm wavelength with at one-beam traditional laser setup, i.e., without the laser interference technique. For an 8mm spot size on the sample surface, the laser fluence was approximately 0.076 J/cm^2. The optical micrograph shown in Figure 10a indicate that at 1mm/s scanning speed, the resin on top of CF is partially ablated. As the scanning speed is decreased to 0.2mm/s, allowing for an increase in the exposure time, more CF are uncovered, indicated an enhanced laser ablation. As for the scanning experiments at 355nm, the laser fluence was then increased by resizing the beam from 8, 4, and 2 mm diameter at the sample surface. In Figure 11, the optical micrograph is shown at a laser fluence of 1.21 J/cm^2 for the scanning with a slow speed of 0.1 mm/s of a surface specimen with a single beam setup at 266nm and 2mm laser spot. This excellent ablation of the resin was attained at laser fluence similar to that for the 355nm, i.e., larger than 1 J/cm^2 but at smaller scanning speeds (0.1 vs. 0.5mm/s).

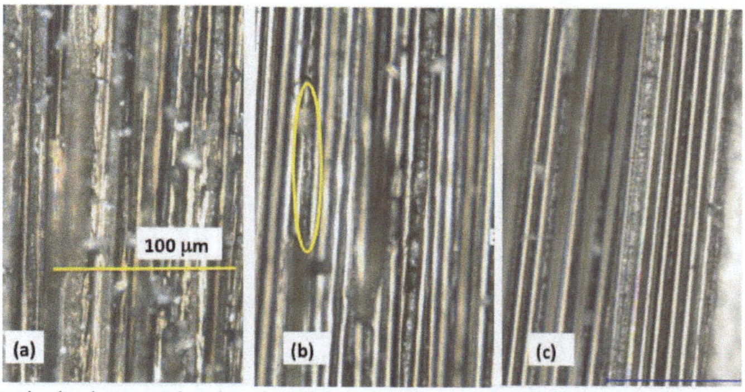

Figure 6. Optical micrographs showing the top surface of the CFPC as processed with a single laser beam at 355 nm, average power 1.36 W, over an 4mm diameter spot at scanning speeds of (a) 0.5, (b) 1, and (c) 3 mm/s.

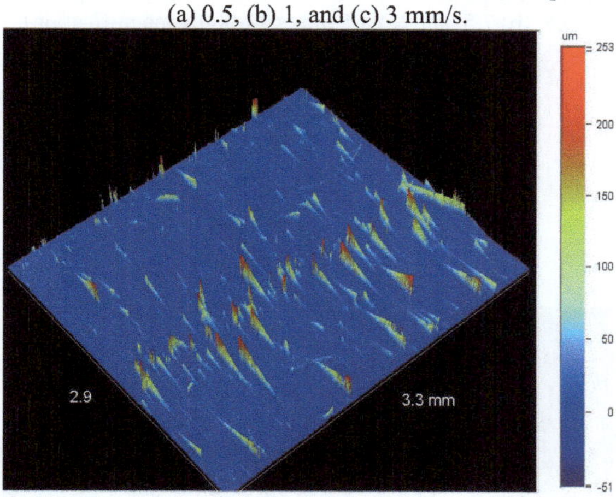

Figure 7. Surface profile obtained with WYKO 9100 surface profilometer. Highest peaks indicate the tips of broken fibers. (Laser operating parameters as those for specimen shown in Figure 6a but with an average power profile of 1.43W).

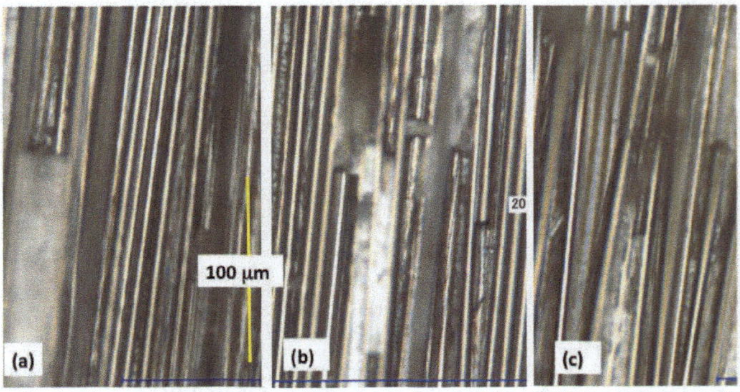

Figure 8. Optical micrographs showing the top surface of the CFPC as processed with a single laser beam at 355 nm, average power 1.36 W, over an 4mm diameter spot at scanning speeds of: (a) 6.5, (b) 7.5, and (c) 8.5 mm/s. ~~(1,000X magnification).~~

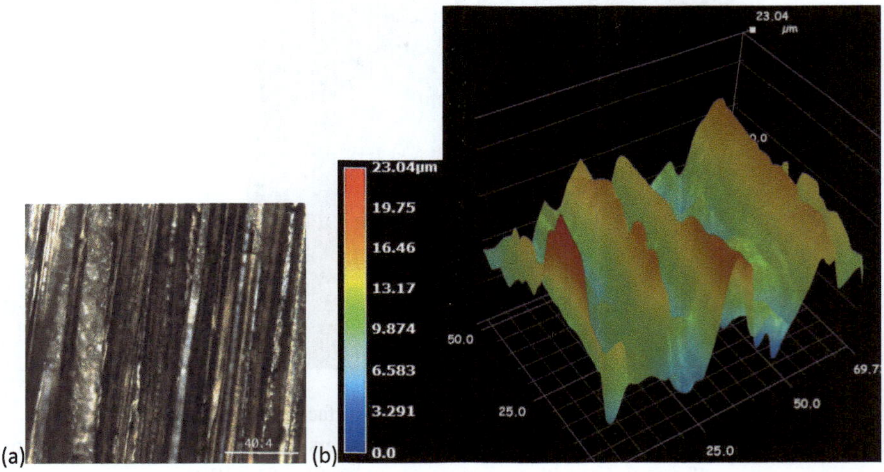

Figure 9. Ablation of resin from the top surface of the CFPC with a single laser beam at 355 nm, 1.43 W average power, over an 4mm diameter spot scanning speed of 0.5mm/s: (a) optical micrograph at 1,000X and (b) 3D surface profile [μm] at 5,000 X obtained with Keyence digital microscope.

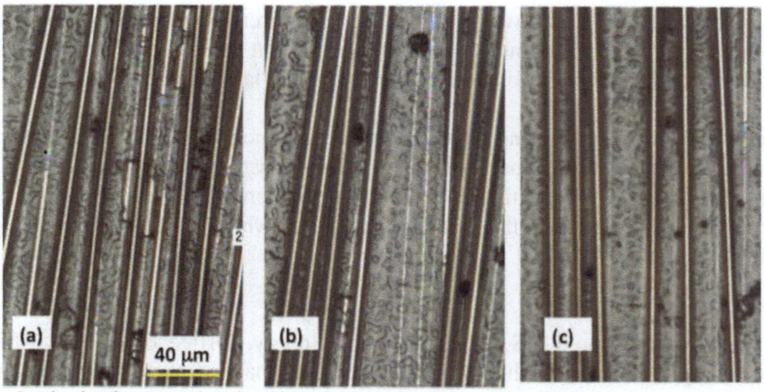

Figure 10. Optical micrographs showing the top surface of the CFPC as processed with a single laser beam at 266 nm wavelength, average power 0.382 W, over an 8mm diameter spot at scanning speeds of: (a) 1, (b) 0.4, and (c) 0.2 mm/s.

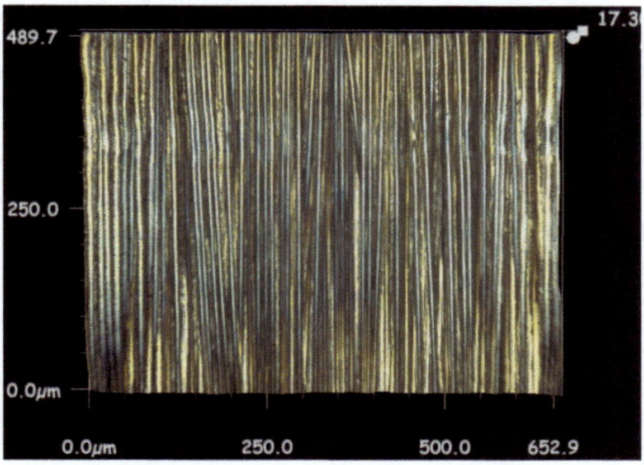

Figure 11. Optical micrograph showing an excellent ablation of resin with a single laser beam at 266 nm, 0.382 W average power, over an 2mm diameter spot at scanning speed of 0.1mm/s.

Conclusions

The present work introduced a novel surface pretreatment technique using laser interferometry produced by two beams of Nd:YAG laser operating at 355 and 266 nm of wavelength. Thus, the untreated smooth surface of the CFPC is replaced by a rough fiber surface. It was found that the resin affects the optical properties of the CFPC, optical properties of fiber themselves do not vary strongly with the wavelength, and sizing has a negligible effect on reflectivity. Although from optical considerations alone, the laser wavelength of 266 nm would be desired as the resin exhibits a larger absorptivity than those at other operating wavelengths, the laser power decrease had to be compensated by decreasing the laser spot in order to increase the fluence. At 266nm, laser ablation becomes difficult at larger spot sizes than 2mm due the inherent power loss 0.382 W. An effective resin ablation and structuring of the carbon fibers was attained at 355nm using the laser interference technique. An effective resin ablation was obtained by using the laser in a traditional fashion, i.e., without involving the laser interference technique. A certain fraction of carbon fiber was found to break during laser processing but the CF itself was not damaged. It was found that an effective resin ablation can be attained at laser fluences larger than 1 J/cm^2.

Future efforts will aim at optimizing the laser surface treatment of CFPC surface to minimize the surface preparation cleaning processes, increasing industrial acceptance for high volume applications, particularly for joining in automotive structures. This surface preparation is expected reduce the variability in the surface preparation, which is inherently in the grit blasting or abrasion techniques.

Acknowledgments

Research sponsored by the U. S. Department of Energy, Office of Energy Efficiency and Renewable Energy, Vehicle Technologies Program, as part of the Transportation Materials Program under contract DE-AC05-00OR22725 with UT-Battelle, LLC. The authors would like to thank John J. Henry Jr. of ORNL for obtaining the characterization of the surface profile using the Wyko surface profilometer.

References

1. H. Niino, M. Nakano, S. Nagano, H. Nitta, K. Yano, A. Yabe, Excimer Laser Ablation of Polymers and Carbon Fiber Composites, Journal of Photopolymer Science and Technology, Vol. 3 (1990), pp. 53-56.
2. Galantucci, LM; Gravina, A; Chita, G; Cinquepalmi, M, Surface treatment for adhesive-bonded joints by excimer laser, Composites Part A-Appl. Sci. and Manuf., Vol. 27, pp. 1041-1049, 1996.
3. Benard, Q., Foisa, M., Grisel, M., Laurens, P., Surface treatment of carbon/epoxy and glass/epoxy composites with an excimer laser beam, Int. J. of Adhesion and Adhesives, Vol. 26, pp. 543-549, 2006.
4. Benard, Q.; Fois, M.; Grisel, M.; Laurens, P.; Joubert, F., Influence of the Polymer Surface Layer on the Adhesion of Polymer Matrix Composites, J. of Thermoplastic Comp. Mat., Vol. 22, Pages: 51-61, 2009.
5. Warren, C.D., Paulauskas, F.L., and Bowman, R.G., "Laser Ablation Assisted Adhesive Bonding of Automotive Structural Composites," ORNL/CP-102637, ICCM-12, Paris France, July 3-9 (1999).
6. Lima, M. S. F.; Sakamoto, J. M. S.; Simoes, J. G. A.; Riva, R., Laser processing of carbon fiber reinforced polymer composite for optical fiber guidelines, Lasers In Manufacturing (LIM 2013), Physics Procedia, Ed. by Emmelmann C; Zaeh MF; Graf T; Schmidt M, Vol. 41, pp. 565-573, 2013.
7. C. Daniel, F. Mücklich, Z. Liu, Periodical micro-nano-structuring of metallic surfaces by interfering laser beams, Appl. Surf. Sci. 208-209 (2003) 317-321
8. C. Daniel, F. Mücklich, Quantification of periodical surface structures by white light interferometry, Prakt. Metallogr. 41 (2004) 6, 277-285.
9. F. Mücklich, C. Daniel, A. Lasagni, F. Yu, Direct Surface Patterning Induced by Interfering Laser Beams, Mat. Res. Soc. Symp. Proc. Vol. EXS-2 (2004) M11.1

10. F. Mücklich, A. Lasagni, C. Daniel, Laser Interference Metallurgy – Periodic Surface Patterning and Formation of Intermetallics, Intermetallics 13 (2005) 3-4, 437-442
11. F. Mücklich, A. Lasagni, C. Daniel, Laser Interference Metallurgy – using interference as a tool for micro/nano structuring, Int. J. Mater. Res. 97 (2006) 10, 1337-1344.
12. C. Daniel, Biomimetic structures for mechanical applications by interfering laser beams – more than only holographic gratings, J. Mater. Res. 21 (2006) 2098-2105

Advanced Composites for Aerospace, Marine, and Land Applications II
Edited by: T. Sano and T.S. Srivatsan
TMS (The Minerals, Metals & Materials Society), 2015

Comparing Strengthening Mechanisms of Vapor Grown Carbon Fiber vs. Titanium Carbide Reinforced Powder Metallurgy Titanium Metal Matrix Composites

Staub Franco [1], Katsuyoshi Kondoh [2], Junko Umeda [2], Hisashi Imai [2 1]

FrontierLab Program
University of California Irvine
Irvine, California, USA

[2] *Joining and Welding Research Institute*
Osaka University
Osaka, Japan

Keywords: titanium, carbon nanotubes, powder metallurgy, composite material

Abstract

In this experiment, TILOP-45 commercially pure titanium powder was mixed with vapor grown carbon fibers (VGCF) to form a 200 g 0.5 wt. % VGCF solution. After adding 0.15 grams of cle-safe oil, a rocking mill shook the sample at 60.0 Hz for 2 hours, resulting in satisfactory dispersion of VGCF on the titanium powder surface. The powder solution was compacted by spark plasma sintering (SPS) and hot extruded. The SPS temperature was set to either 800 °C or 1,000 °C and the pressure to 35 kN. Using an extrusion ratio of 13:1 and ram speed of 3 mm/s, the titanium billet, preheated to either 800 °C or 1,000 °C, was deformed to a 10 mm diameter rod. All four permutations of SPS and extrusion temperatures were tested. Microstructure, grain size, hardness, and oxygen/nitrogen/carbon content were observed. Also, a UTS experiment was done followed by SEM observations of the fractured surfaces.

1. Introduction

Each year energy demands, as well as emission standards that limit them, rise. One method by which we improve emission standards is by developing high specific strength materials. Titanium has long been recognized for his high specific strength, high corrosion resistance, and high-temperature resilience, making it a qualified candidate for a green material **[1-3]**. The majority of current titanium parts used in industry are made of the alloy Ti-6Al-4V. Though proven to be economical, the inclusion of rare earth metals in the alloy, such as vanadium (V) and niobium (Nb), inhibit many industrial applications. Resultantly, this research aims to form an economical, widely applicable titanium material with improved mechanical properties.

By dispersing vapor grown carbon fibers (VGCF), a commercial form of carbon nanotubes (CNT), with TILOP-45, a high-purity titanium powder, a titanium metal matrix composite (TMC) was produced. Powder metallurgical TMCs, as shown in previous studies by Kondoh et al. **[4,5]**, have demonstrated improvements in yield strength (YS) and Young's modulus (E) from pure titanium counterparts. This improvement in YS and E was mainly attributed to titanium carbide (TiC) particle reinforcement, C/N/O solid solution strengthening, and grain refinement. It has been established that carbon reacts readily with titanium above 882 °C up until concentrations of about 15 wt. % carbon **[6]**. For this study, a 0.5 wt. % carbon solution was prepared. To manipulate the TiC content, sintering and extrusions temperatures were set to 800 °C or 1,000 °C and samples of all temperature permutations were fabricated and investigated in detail. Also, low oxygen and nitrogen content titanium powder was chosen to lessen the strengthening effect of nitrogen and oxygen solid solute strengthening, further isolating the strengthening effects of TiC vs. that of intact VGCF.

2. Experimental Procedure

2.1 Powder mixture preparation

Commercially pure titanium powder (TILOP-45, Sumitomo Corp.) and vapor grown carbon fibers (VGCFs, Showa Denko) were used as starting materials. Characteristics of the as-received TILOP-45 and VGCFs are shown in Tables 1 and 2, respectively. The titanium powder was manufactured by the gas atomization method and had a maximum particle size of 45 μm. VGCFs, a type of multi-walled CNT, have a high aspect ratio with a diameter of approximately 150 nm and a length of 8 μm. Due to the morphology and the Van der Waal's forces between VGCFs, the nanotubes tend to tangle into aggregations. An effective mixing method was used to break up these VGCF aggregations and satisfactorily disperse the VGCF on the titanium powder surface. First, 1 g of VGCF and 199 g of TILOP-45 titanium powder were mixed in a container. Secondly, 0.15 g of cle-safe oil was added to the powder solution so the VGCF would adhere to the titanium. Next, 16 10-mm-diameter ZrO2 balls and a metal mesh were added to the container, which was subsequently shaken for 3 hours at 60.0 Hz on a rocking mill. As reference materials, samples of pure TILOP-45 powder were prepared under matching conditions. Micrographs of the mixed materials were taken by a field emission scanning electron microscope (FESEM, JSM- 6500F, JEOL).

Table 1 – Characteristics of the as-received pure titanium powder

Ti*	Fe*	N*	C*	O*	Particle Size (μm)
99.7	0.08	0.03	0.03	0.17	<45

*measured in wt. %

Table 2 – Specimen preparation conditions

	SPS			Pre-heating			Extrusion	
Specimen	Temp. (°C)	Time (min.)	Pressure (MPa)	Temp. (°C)	Time (min.)	Atmosphere	Die Temp. (°C)	Extrusion Ratio
800-800	800	30	35	800	3	Ar gas	400	13
800-1000	800	30	35	1000	3	Ar gas	400	13
1000-800	1000	30	35	800	3	Ar gas	400	13
1000-1000	1000	30	35	1000	3	Ar gas	400	13

2.2 Spark plasma sintering and hot extrusion

The powder mixtures were loaded into a cylindrical graphite die and consolidated with a spark plasma sintering system (SPS-1030S, SPS Syntex). The sintering temperature was set either to 800 °C or 1,000°C depending on the sample (see Table 2), with a heating rate of 20 °C/min. It was kept at the destination temperature for 30 minutes while being compressed at 35 MPa in a 5 Pa vacuum. The sintered billets had a diameter of 36 mm and a height of 43 mm. Before the hot extrusion, the billets were pre-heated to and kept either 800 °C or 1,000°C (see Table 2) for 3 minutes under an argon atmosphere. The billet was then immediately extruded using a hydraulic press machine (SHP-200-450, Shibayama) with a load capacity of 2,000 kN. The extrusion ratio and ram speed were set at 13:1 and 3.0 mm/s, respectively. The final diameter of the extruded rod was 10.0 mm. The pure titanium samples were also prepared under the same conditions as a reference material.

2.3 Materials characterization

The oxygen and nitrogen concentrations of the extruded materials were measured with an inert gas fusion nitrogen-oxygen determination instrument (TC-300, LECO). The samples were loaded into a graphite crucible and heated to 3,000 °C. The gases extracted during the fusion were directly analyzed after passing through a dust filter. The sample oxygen content was measured by two non-dispersive infrared analyzers (NDIR). Likewise, nitrogen content was determined by a thermal conductivity detector (TCD). The carbon concentrations were measured by an infrared absorption carbon analyzer (EMIA-221V20K, Horiba). The sample phase compositions were identified using x-ray diffraction (XRD-6100, Shimadzu). As for the microstructure and grain size, an optical microscope (BX-51O, Olympus) was used for lower magnifications, while a field-emission scanning electron microscope (JEM-6500F, JEOL) was used for high magnifications. Hardness data was collected using an MHV1000 Digital Vickers's Hardness Tester at the HV 0.2 setting,

3. Results and Discussions

3.1 Powder dispersion

The tubular structure of VGCFs have proven to be troublesome when dispersing them into other materials in past experiments. An extremely high tube length-to-diameter ratio coupled with attractive Van der Waals forces between carbon atoms leads to CNT tangles referred to as aggregations. To monitor how effectively the VGCF was being dispersed, SEM observations were conducted at 1-hour intervals during the mixing process. After the first hour on the rocking mill, many aggregations up to 20 μm in diameter remained. By the end of the third hour, every particle became evenly coated in VGCFs; and all that remained were about 10 μm-diameter VGCF aggregations on the surface of deformed powder particles (Figure 1). Time had a significant impact on the level of dispersion achieved. It also appears as if the spherical powder breaks apart aggregations more effectively than the rough morphology. This seems to be because dimples on the Ti powder surface house and shield an aggregation from the impact of another particle during the rocking process, preserving the aggregation. This would explain why aggregations were mostly found on concave surfaces. Overall, this dispersion method proved to be useful in breaking apart VGCF aggregations and dispersing them on spherical morphology powder particles.

Figure 1

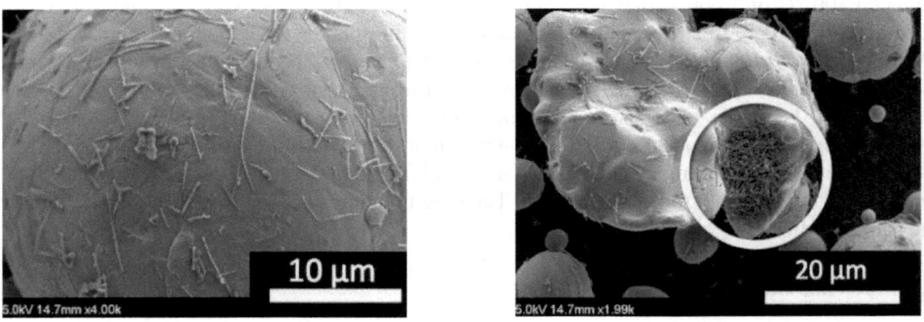

The left photo shows an even coating of VGCF on Ti powder surface. Every particle had this coating. The image on the right shows a VGCF aggregation on the dimple of a powder with an irregular morphology.

3.2 Material Properties

Variations in sintering and extrusion temperatures proved to vary sample properties. To observe the samples' mechanical properties, ultimate tensile stress (UTS) and Vickers's hardness tests were run. All 4 samples with VGCF experienced an increase in hardness, Young's Modulus, as well as yield strength (YS). Table 3 compares the mechanical properties of pure titanium samples with their respective VGCF filled samples.

Table 3 – Mechanical properties of extruded pure Ti materials and their composites with VGCF.

Sample name	0.2% YS (MPa)	UTS (MPa)	Elongation	E (GPa)	Hardness (HV)
800-800	364	505	36.3	88.4	204
800-800 w/ VGCF	593	669	34.6	113.1	287
800-1000	350	502	34.9	106.7	198
800-1000 w/ VGCF	466	585	26.3	112.6	242
1000-800	348	484	40.0	91.7	211
1000-800 w/VGCF	537	6359	30.8	111.8	266
1000-1000	355	496	30.6	112.8	202
1000-1000 w/ VGCF	418	550	28.7	124.4	229

Note: The first number in the sample name represents the SPS temperature, and the second number represents extrusion temperature

It is a rule of thumb in metallurgy that when the strength of a material is improved, ductility is sacrificed. The current material is no exception, though in some samples the ductility was much better preserved than in other samples. Most notably, the sample prepared at the lowest temperatures managed to increase in strength by 62.7% while only sacrificing 4.7 % of its elongation. There were several events responsible for this improvement in mechanical properties; O, N and C solid solute strengthening, grain refinement, TiC particle reinforcement, and VGCF fiber reinforcement. Next, we will discuss how, and to what extent, each of these played a role in improving mechanical properties.

3.3 Solid solute strengthening

As Table 4 shows, the O and N content did not change much among the samples. This was expected due to the high purity Ti powder used. Also, when the samples were sintered in vacuum (<5 Pa) and pre-heated in Ar gas atmosphere, enough care was taken to reduce atmospheric exposure, resulting in a higher purity material.

N and O atoms strengthen metals by integrating themselves in the crystal lattice of the Ti matrix. Up until a solubility limit, the solutes strengthen Ti proportionally to the wt. % O or N. From this, it can be observed that O and N solid solutes strengthening did not change the material properties upon adding VGCF. C levels, however, varied a little bit more. Of course VGCF is made up entirely of C, so it makes sense that the samples with 0.5 wt. % VGCF contained at least that amount of C. The reason C levels were read higher than 0.5 wt. % could be the result of contamination or improper calibration of the infrared absorption carbon analyzer. Regardless, the C content surpasses the previously measured solubility limit of 0.05 wt. % **[4, 5]**, which has been correlated to a strength increase of 20 MPa. Taking this into account, we can move on to other strengthening mechanisms.

Table 4 – O, N and C content of pure Ti materials and composites with VGCF.

Sample name	O (wt. %)	N (wt. %)	C (wt. %)
800-800	0.1293	0.0148	0.0061
800-800 w/ VGCF	0.1532	0.0156	0.6271
800-1000	0.1503	0.0184	0.0062
800-1000 w/ VGCF	0.1498	0.0136	0.5754
1000-800	0.1458	0.0141	0.0180
1000-800 w/ VGCF	0.1365	0.0147	0.6083
1000-1000	0.1406	0.0159	0.0000
1000-1000 w/ VGCF	0.1523	0.0147	0.5740

3.4 Grain refinement

Another method by which metals can be strengthened is by reduction in their grain size. This change in strength can be measured by the widely accepted Hall-Petch equation **[7, 8]**;

$$\sigma y = \sigma o + kyD^{-0.5} \quad (1)$$

where σy is the yield strength, σo is a constant stress, ky is a material constant, and D is the mean grain diameter. The constant stress σo and material constant ky for pure titanium are reported as 172.5 MPa and 18 MPa/mm$^{-0.5}$ **[9]**, respectively. By measuring the grain size of the pure titanium sample and comparing it to its respective sample with VGCF, we can calculate the change in yield strength of the material using the equation:

$$\Delta\sigma y = ky(D2^{-0.5} - D1^{-0.5}) \quad (2)$$

Table 5 shows the increase in yield strength in each of the materials using the above calculation method.

Table 5 – Changes in yield strength upon adding VGCF

Sample name	Strength improvement from pure Ti sample (MPa)
800-800 w/ VGCF	-9.1
800-1000 w/ VGCF	75.1
1000-800 w/ VGCF	-18.2
1000-1000 w/ VGCF	50.1

Figure 2

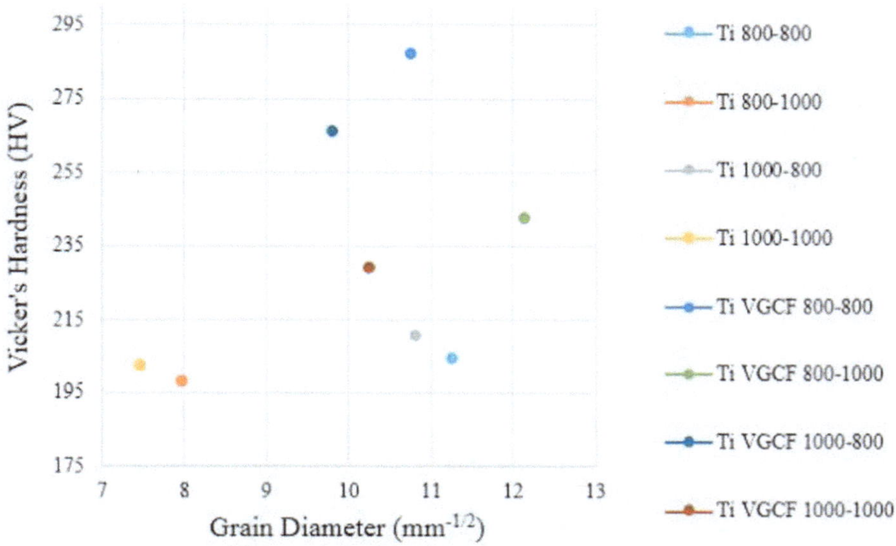

The Hall-Petch equation can also be used to compare the hardness to grain size. In the Hall-Petch chart (See Figure 2), all materials have been plotted in terms of hardness and the inverse square root of the average grain diameter. As can be witnessed, 800 °C extrusions did not affect grain size as much as the 1,000 °C extrusion did. This could be due to the fact that titanium undergoes a phase transition at 882 °C, which lies between the two aforementioned temperatures **[10]**. It has also been reported that β-titanium, which occurs above the transition temperature, is a more brittle metal than α-titanium **[11]**. It is possible that the overwhelming pressure exerted during the extrusion process shattered the β-titanium into many smaller grains, whereas the more ductile α-titanium was able to change shape without breaking into smaller grains. Be that as it may, it was also the low extrusion temperature materials that experienced a greater improvement in mechanical properties.

3.5 TiC particle reinforcement

There are several ways that C can reinforce a titanium metal matrix. Solid solutes were already discussed, however only 0.05 wt. % of the C from the VGCF was measured to go into solid solute formation. As for the rest of the C, we will assume it was divided between TiC particle and VGCF fiber reinforcements. TiC is an extremely hard ceramic that is formed as the result of Ti reacting with C above 920 °C **[12]**. Although the fabrication temperature was kept under 800°C for some of the samples, XRD results indicated all four VGCF samples showed traces of TiC (See Figure 3). The question is, how much of the VGCF reacted with the matrix to form TiC, and how much of it remained intact as CNT? To answer this question, one assumption was made: the wt. % of TiC is linearly proportional to the strength improvement it brings on. The same assumption was made for VGCF fiber reinforcement. In order to calculate the wt. % TiC, four micrographs (see Figure 4) were taken, one of each VGCF sample. From these images, the area percent TiC, appearing as a gray color, was calculated. Assuming uniformity in the 3rd dimension of the image, the volume percent was the same as the area percent. Finally, knowing the mass and volume of the entire sample, as well as the density of TiC, it was possible to calculate the wt.%. From the micrographs, it can be seen that the more time the sample spent at 1,000 °C, the more of the VGCF had reacted to form TiC. Despite room for human error, this calculation method produced remarkably accurate results.

Figure 3

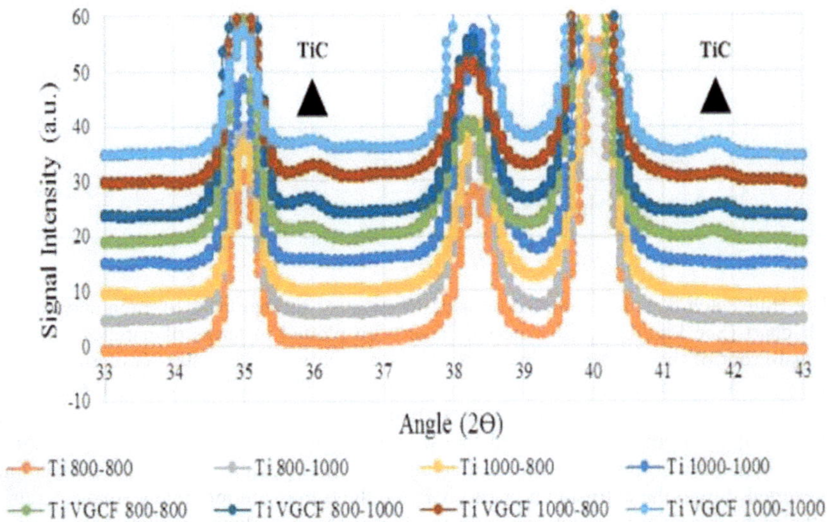

Figure 4 – Micrographs of VGCF samples

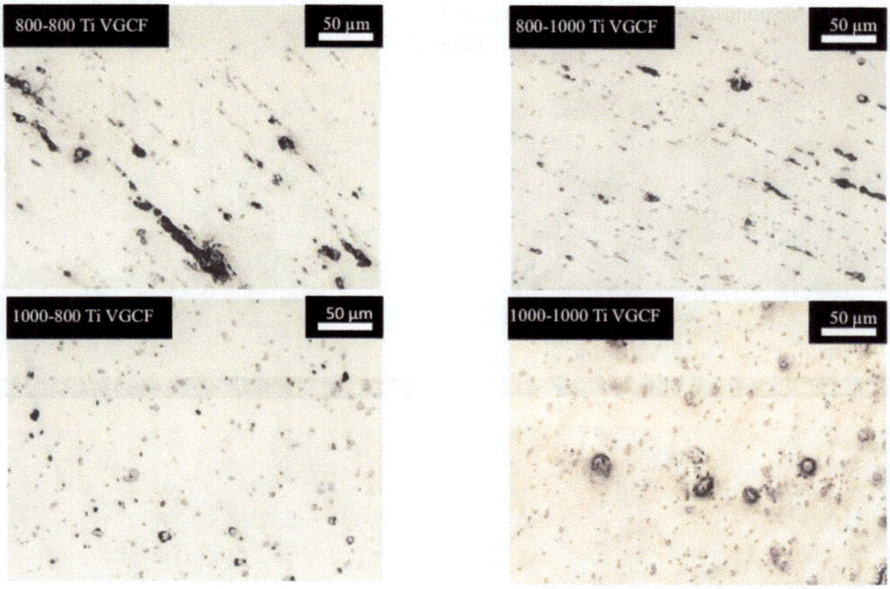

3.6 VGCF fiber reinforcement

In order to calculate the wt. % VGCF, the wt. % TiC and wt. C solid solute were subtracted from the total 0.5 wt. % C. Table 6 charts out the various wt. percentages of different forms of carbon in the TMC. This should further clarify that the longer the sample was heated to 1,000 °C, the more TiC formed.

Table 6 – wt.% of each material calculated in VGCF samples

	800-800	800-1000	1000-800	1000-1000
VGCF wt. %	0.393	0.309	0.157	0.01
TiC wt. %	0.057	0.061	0.293	0.44
C solid solute wt. %	0.05	0.05	0.05	0.05

While both TiC and VGCF result in an improvement in mechanical properties, the method of reinforcement varies. TiC particles work to stop dislocations by a pinning effect, whereas VGCF fiber reinforcements seem to rely on a load transfer effect. When a stress is applied to the matrix, the load jumps from the matrix to the individual fibers, which have a much higher tensile strength. SEM observations were performed on the fractured surface of the UTS test specimens. Bundles of nanotubes on opposing fracture surfaces, nanotubes in the dimples of the surface, as well as individual nanotubes with a clean break where they tore were captured on the SEM images (see Figure 5). From these pictures, it appears likely that some of the load was transferred to the VGCF. It is also important to note that the amounts of unreacted VGCF seen on each sample corresponded roughly to wt. % calculated.

Figure 5

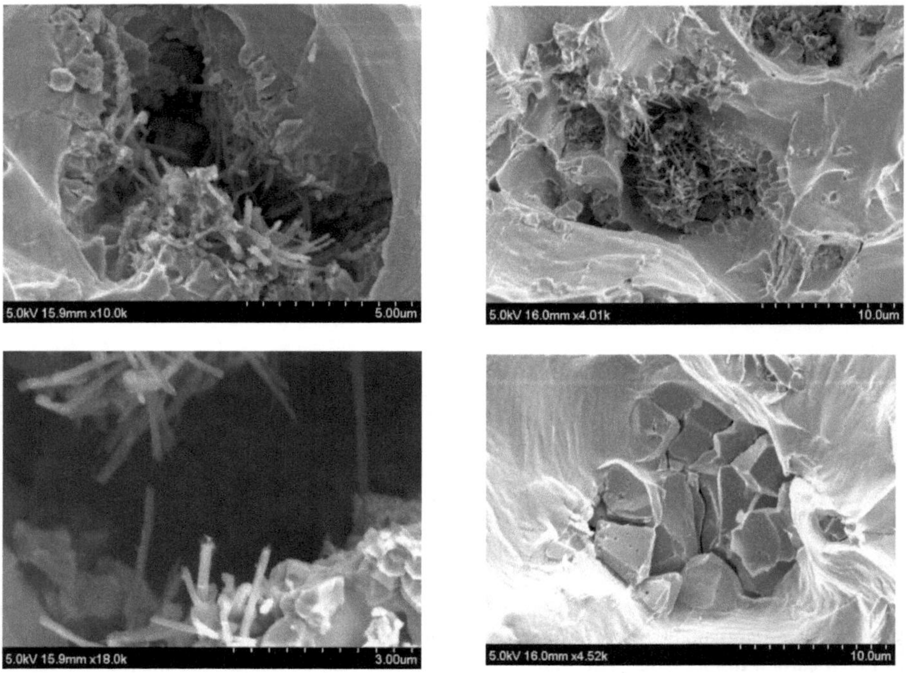

The top two images show large amounts of VGCF in dimples of the 800-800 sample. The bottom left image of the 800-800 sample shows a highly magnified image of a carbon nanotube that has fractured in the middle. Several tubes like this could be found in the matrix, suggesting at least some level of load transfer was achieved. The bottom right image is of a small TiC colony found in the 1000-800 sample.

3.7 Overview of strengthening mechanisms

From all the calculations above, the strengthening mechanisms were organized into the figure below (Figure 6). The light blue area represents the intrinsic strength of the Ti matrix by itself. Hence, the pure samples consist of just a light blue bar, with the corresponding VGCF filled sample to their right.

The samples extruded at 800 °C improved the most in YS. Oppositely, the samples prepared at 1000 °C did not seem to be affected much by VGCF or TiC reinforcements.

Rather, their primary strengthening mechanism was through grain refinement. Because this paper focuses on the strengthening mechanisms through TiC and VGCF reinforcements, the 800 °C extruded materials will be further examined.

Figure 6

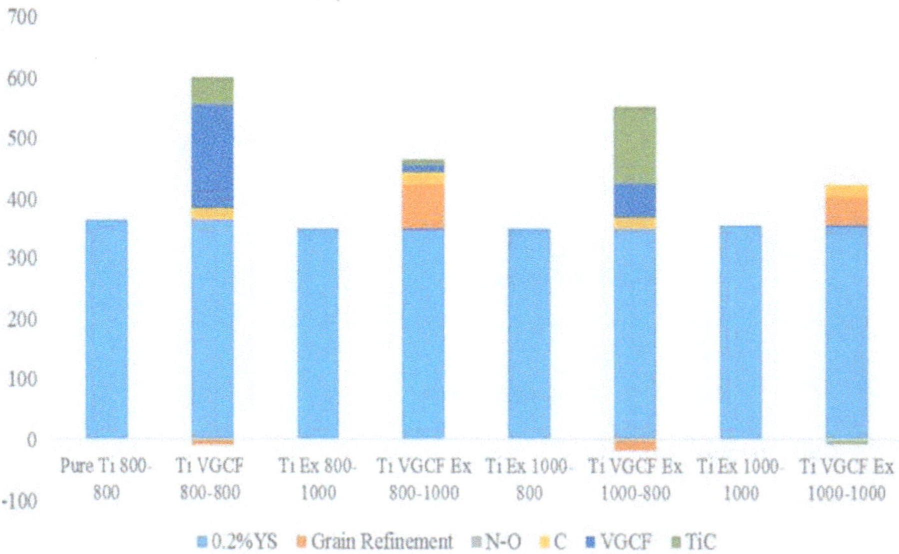

There were two primary strengthening methods established for the two samples extruded at 800°C. The sample sintered at 800 °C contained more unreacted VGCF, which boasted a slightly superior YS improvement. The main difference between these two materials, however, lies in their elongation properties. As demonstrated in the UTS chart (see Figure 7), the sample sintered at 800 °C preserved its ductility more successfully than the sample sintered at 1,000 °C. The 800°C sintered sample only dropped 4.7% in elongation, while the 1,000 °C sintered sample dropped a whopping 23.1%. Also, of the three UTS tests run with the 1,000 °C sample, one of the bars broke prematurely, only reaching a 21.9% elongation before failing. Contrarily, the 800 °C sample demonstrated consistency in ductility with all three test samples, varying only by 4.1% elongation. As long as manufacturing process temperatures are kept under 800 °C, VGCF can be preserved within a Ti matrix. It appears as if VGCF fiber reinforcement is a promising way to reinforce Ti matrices without compromising ductility.

Figure 7

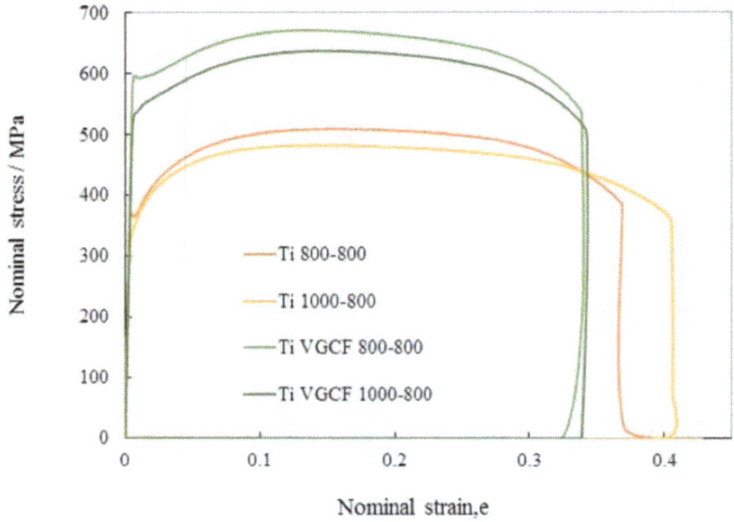

4. Conclusions

There are several conclusions to be made about the various materials fabricated during this experiment. First off, the spherical morphology of TILOP-45 coupled with the rocking mill seemed to effectively break down VGCF aggregations. Also, the cle-safe oil worked well as a coating to which the nanotubes were able to adhere. Overall, the VGCF dispersion method was successful.

As for the material properties of the four TMCs tested, only two samples proved to be promising. The samples extruded at 800 °C increased significantly in strength. One of these samples contained more VGCF and the other contained more TiC. The sample containing higher TiC levels also experienced a heavier drop in ductility, whereas the sample with unreacted VGCF experienced an almost negligible drop in elongation. Assuming that the strengthening effect is proportional to the wt. % of VGCF and TiC, it can be concluded that VGCF fiber reinforcement is more effective in strengthening titanium than TiC is. From SEM observations, there were a significant amount of VGCFs preserved as long as the preparation temperatures were kept at 800°C or below. Perhaps there is still room to increase the VGCF content. If the assumption that more VGCF results in a higher YS improvement, then it would certainly be worth looking into higher wt. % VGCF solutions.

Acknowledgements

I would like to thank the members of Kondoh laboratory for their infinite guidance and support. Also, I would like to thank everyone with the FrontierLab program and the Joining and Welding Research Institute (JWRI) at Osaka University.

References

[1] Leyens C, Peters M. Titanium and titanium alloys: fundamentals and applications. Weinheim: WILEY-VCH Verlag GmbH & Co. KGaA; 2003

[2] Luütjering G, Williams JC, Gysler A. Titanium engineering materials and progresses. 2nd ed. Berlin: Springer; 2007. 379.

[3] Boyer RR, Welsch G Collings EW. Materials properties handbook: titanium alloys. Metals Park, OH: ASM International; 1994.

[4] Li S, Sun B, Imai H, Kondoh K. Powder metallurgy Ti-TiC metal matrix composites prepared by in situ reactive processing of Ti-VGCFs system. ***Science*** Direct Vol. 61. 2013. 216- 228.

[5] Li S, Sun B, Imai H, Mimoto T, Kondoh K. Powder metallurgy titanium metal matrix composites reinforced with carbon nanotubes and graphite. ***Science*** Direct Vol 48. 2013. 57-66.

[6] Emamiam A, Corbin SF, Khajepour A. In-Situ Deposition of Metal Matrix Composite in Fe- Ti-C System Using Laser Cladding Process. Metal, Ceramic and Polymeric Composites for Various Uses: Chapter 2. 2011.

[7] Hall EO. Deformation and ageing of mild steel. ***Proc Phys Soc*** 1951;64B:747-53

[8] Petch NJ. The cleavage strength of polycrystals. ***J Iron Steel Inst*** 1953;173:25

[9] Kobayashi Y, Tanaka Y, Matsuoka K. Effect of forging ratio and grain size on tensile and fatigue strength of pure titanium forgings. ***J Soc Mater Sci Jpan*** 2005; 54:66-72

[10] Korbmacher D, Glensk A, Grabowski B, Neugebauer J. High accuracy ab initio prediction of phase transitions in Ti and TiNb alloys. Max-Planck-Institut für Eisenforschung GmbH 2009

[11] Donachie M Jr. Titanium: A Technical Guide. ASM International December 2009. Ch 12:102

[12] Cadoff I, Nielsen, J.P. Institute of Metals Division – Titanium-Carbon Phase Diagram. AIME Transactions. 1/1/54. 1564

Advanced Composites for Aerospace, Marine, and Land Applications II
Edited by: T. Sano and T.S. Srivatsan
TMS (The Minerals, Metals & Materials Society), 2015

MICROSTRUCTURAL CHARACTERIZATION OF FATIGUE DAMAGE OF CFRP IN THE VERY HIGH CYCLE FATIGUE REGIME

Daniel Backe[1], Frank Balle[1], Dietmar Eifler[1]

[1]Institute of Materials Science and Engineering, University of Kaiserslautern
P.O. Box 3049, 67653 Kaiserslautern, Germany

Keywords: CFRP, VHCF, ultrasonic fatigue, cyclic 3-point bending, failure mechanisms of CF-PPS

Abstract

Structural components in aircrafts made of carbon fiber reinforced polymers (CFRP) are often subjected to 10^8 and more loading cycles during their time in service. To obtain a comprehensive knowledge about the fatigue behavior of CFRP in the very high cycle fatigue (VHCF) regime and the corresponding failure mechanisms at loading cycles higher than 10^8, a new Ultrasonic Testing Facility for cyclic 3-point bending at 20 kHz has been developed. The system enables VHCF experiments up to 10^9 loading cycles in twelve days without unacceptable heating of the specimens. The investigated material is an aircraft qualified carbon fiber fabric reinforced polyphenylensulfide (CF-PPS). Constant load amplitude tests have been realized to characterize the fatigue behavior. Detailed light optical and SEM investigations have been performed in defined fatigue states and after failure or finally after reaching 10^9 cycles. Based on these results the failure mechanisms of CF-PPS in the VHCF regime were documented and analyzed.

Introduction and motivation

Carbon fiber reinforced polymers (CFRP) are getting more and more used for highly loaded lightweight structures and become increasingly important for the aircraft or automotive industry. To ensure an efficient and economic use of CFRP a detailed knowledge of their monotonic and especially fatigue properties in the case of very long lifetime or high frequently loaded components is required. Even current CFRP structures are subjected to cyclic loading with more than 10^8 cycles [1] corresponding to 20 years and more in service. This range of more than 10^7 cycles is called Very High Cycle Fatigue (VHCF) regime [2] corresponding to low cyclic load amplitudes. At present, only rare fatigue data of the VHCF behavior of CFRP are available due to the long test times to carry out VHCF experiments. The few data available about very high cycle fatigue of CFRP were obtained by long-term experiments with testing frequencies of 3 to 100 Hz [3-6]. For example, a VHCF experiment up to 10^9 cycles, with a conventional servo hydraulic testing system at a frequency of 5 Hz, would take at least 6.3 years. To realize VHCF experiments up to 10^9 cycles in only 12 days, a new Ultrasonic Testing Facility (UTF) for cyclic three-point bending has been developed at the Institute of Materials Science and Engineering (WKK), University of Kaiserslautern, Germany. This UTF, called "UltraFAST-WKK-Kaiserslautern", works with a frequency of 20 kHz and a permanent contactless online monitoring via infrared thermography, laser Doppler vibrometry and confocal chromatic distance control. To verify the oscillation behavior of the CFRP specimens at 20 kHz as well as for calibrating the load amplitudes, measurements with a 3D-Scanning Laser Vibrometer (3D-SLV) were performed. In this investigation the fatigue behavior and the failure mechanisms of a commercially available carbon fiber fabric reinforced polyphenylensulfide (CF-PPS) was analyzed in the VHCF regime.

Experimental setup for cyclic 3-point-bending at ultrasonic frequencies

All the experiments for investigating the fatigue behavior of CF-PPS in the VHCF regime have been realized on a new in-house developed Ultrasonic Testing Facility (UTF) for cyclic three point bending, as shown in figure 1.

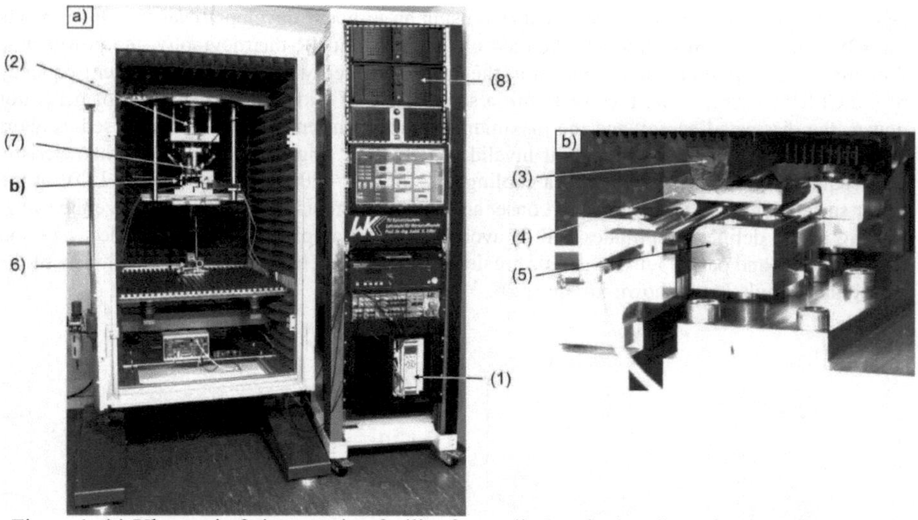

Figure 1. (a) Ultrasonic fatigue testing facility for cyclic 3-point bending of CFRP; (b) Detail of the variable shoulder unit

To avoid damage on the polymer by thermal effects every fatigue experiment is divided into ultrasonic pulse and pause sequences. Accordingly, the fatigue tests are carried out between ambient temperature and well below the glass transition temperature of polyphenylensulfide with $T_g \approx 90°C$ [7]. In addition, the specimens are permanently cooled with dry compressed air.
The developed UTF works as follows:
The digital high power ultrasonic generator (1) transforms the system voltage of 50 Hz into a high frequency alternating voltage of 20 kHz. This alternating voltage is relayed to the ultrasonic resonance system (2) which consists of a converter, a booster and a loading device (3). By using the inverse piezoelectric effect, the converter transforms the high frequency voltage into a mechanical oscillation of the same frequency [8]. This oscillation is amplified by the booster and mechanically transmitted to the loading device. The loading device finally converts and transmits this oscillation up to 60 μm displacement amplitude to the CFRP specimen (4) which is supported by a variable shoulder unit (5). This concept of high-frequency mechanical oscillation generation is already well known from fatigue testing machines for metals [9-12]. However as distinct from these metal fatigue systems, the CFRP specimen is not a positive part of the ultrasonic resonance system. Therefore a specific specimen design is required to achieve the first bending eigenmode of the CFRP specimen precisely at the resonance frequency of the UTF. Furthermore the specimen has to be deflected with more than the occurring oscillation amplitude (60 μm in maximum) to guarantee a permanent contact to the loading device during ultrasonic loading. The deflection is measured precisely in the middle of the specimen's lower surface via a confocal distance sensor. The superimposed cyclic displacement amplitude of the specimen during the fatigue loading with 20 kHz is permanently measured by a 1D-Laser Doppler Vibrometer (LDV) with a frequency of 500 kHz (6). The data is analyzed online and used for process control. The surface temperature as well as the planar temperature distribution of the CFRP specimens is monitored by infrared thermography (7) with a measuring rate of 120 Hz. Three computer systems (8) are integrated for controlling the UTF and for data acquisition. Two real-time computer systems control the exact timing of the pulse pause sequences. All measured

signals are observed, calculated and analyzed using an in-house developed National Instruments LabVIEW software. In addition to the data of the LDV and IR-thermography, the power-time function of the digital ultrasonic generator, the dissipated energy [13, 14] and the eigenfrequency of the CFRP specimens are recorded with a sample rate of 1 kHz. The maximum of generator power, the dissipated energy and the maximum of specimen temperature are also used as abort criterion for the VHCF experiments if invalid values occur. Figure 2a shows two characteristic load sequences of $t_{US} = 200$ ms with a cooling time of $t_{cool} = 2000$ ms measured by LDV at the lower surface of the specimen. A short onset and offset regime is realized to maximize the cyclic loading at the defined amplitudes and to avoid damaging amplitudes during the decay period between pulse and pause. Further details are described in Heinz et al. [14]. A detailed view of the sinusoidal stimulation is shown in Figure 2b.

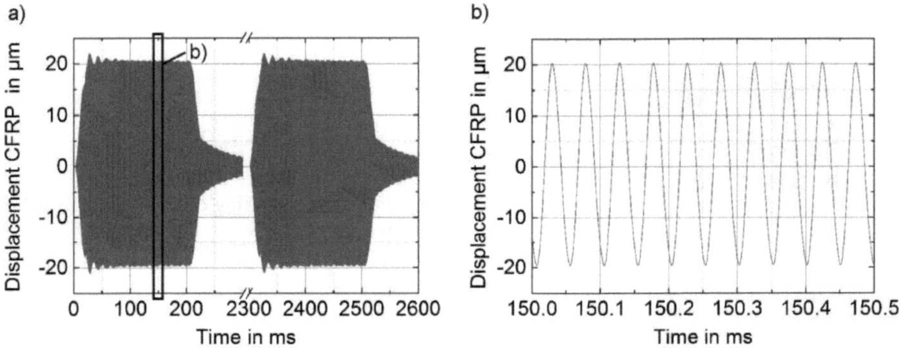

Figure 2. (a) Ultrasonic impulses measured at the lower surface of the CFRP specimen by LDV; (b) Detailed section of one ultrasonic pulse

Base material and specimen design for VHCF-experiments

A commercially available carbon fiber fabric reinforced polyphenylensulfide (CF-PPS) manufactured by Bond-Laminates GmbH (Brilon, Germany) has been characterized for its fatigue behavior in the VHCF regime. PPS is a semi-crystalline thermoplastic material with a glass transition temperature (T_g) of about 90°C and a melting temperature (T_{pm}) of around 290°C [7, 15]. In addition it offers a high stiffness, high chemical resistance and service temperatures of up to 200°C [7, 15]. Therefore CF-PPS is increasingly used especially in the aircraft industry. The laminate is 4 mm thick and is built up of 19 layers of twill 2/2 carbon fiber fabric with an area weight of 200 g/m^2. The fiber volume fraction of the laminate has been determined to 54.8% and the density to 1.54 g/m^3. An optical micrograph of the surface edge of one CF-PPS specimen with the alternating stacking sequence of 0°- and 90°-fiber layers can be seen in figure 3a. No manufacturing related defects like voids or even small delaminations have been detected in the initial state by computer tomography and ultrasonic inspection [16]. Nine elastic constants have been determined to build up an orthotropic simulation model. The simulations, implemented in Abaqus CAE, were necessary to adjust the frequency of the first bending eigenmode of the CFRP specimen to the resonance frequency of the ultrasonic resonance system at 20 kHz. The resulting specimen geometry for carrying out VHCF experiments is illustrated in figure 3b.

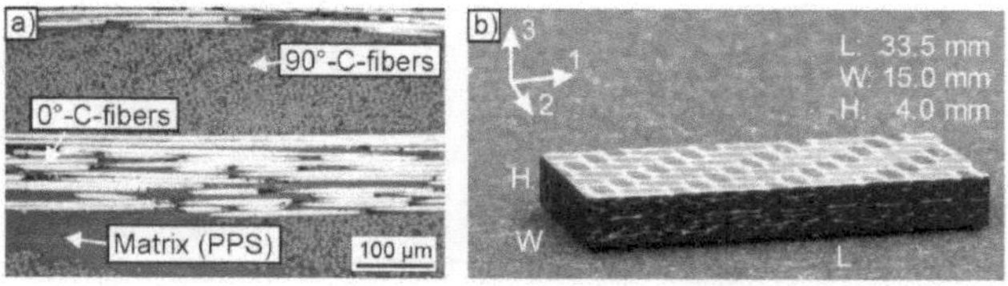

Figure 3. (a) Light optical micrograph of CF-PPS; (b) Geometry for CF-PPS specimen for VHCF experiments

The monotonic mechanical properties were determined in tensile and bending tests according to DIN EN ISO 527, DIN EN ISO 14125 and DIN 65148. Selected values are summarized in table 1 for the 11- and 22-direction (according to figure 3b) as well as for 13- and 23-direction.

Table I. Selected monotonic properties of CF-PPS

	Young's Modulus in GPa	Ultimate Tensile Strength in MPa	Flexural Strength in MPa		Shear Strength in MPa
11-dir.	57.6 ± 2	659 ± 36	590 ± 10	13-dir	37.7 ± 0.7
22-dir.	57.6 ± 1.6	585 ± 36	605 ± 17	23-dir.	35.4 ± 0.5

Selected results

3D-Scanning Laser Vibrometry

To determine the strain distribution and the absolute strain values high resolution non-contact measurements using a 3D-Scanning-Laser-Vibrometer (3D-SLV) were performed. Furthermore the oscillation behavior of the CF-PPS specimen has been analyzed in the initial state and after cyclic loading up to 10^9 cycles. The measurements were carried out during the ultrasonic pulses at 20 kHz in cooperation with Polytec GmbH (Waldbronn, Germany). In Figure 4a the sinusoidal oscillation of the CF-PPS specimen in the first transversal bending eigenmode at different instants of time is showed. The experimental results clearly confirm the computed results carried out with the orthotropic material model and also visualize the real sinusoidal oscillation measured by single spot laser vibrometry (see figure 2a). To avoid fatigue damage, this experiment has been carried out at nanometer amplitudes scale. Figure 4b visualizes the shear strain distribution at oscillation amplitude of 20 µm. The maximum of shear strain is clearly recognizable in the area between the shoulders and the loading device. In the upper and in the lower outer fiber, respectively, the highest tension and compression strain was measured, see figure 4c. Based on the detected strain values measured with this method, the corresponding shear-, tension- and compression stresses could be calculated using Hooke's law for orthotropic materials. Consequently the cyclic load amplitudes during the VHCF-experiments could be determined for all relevant displacement amplitudes.

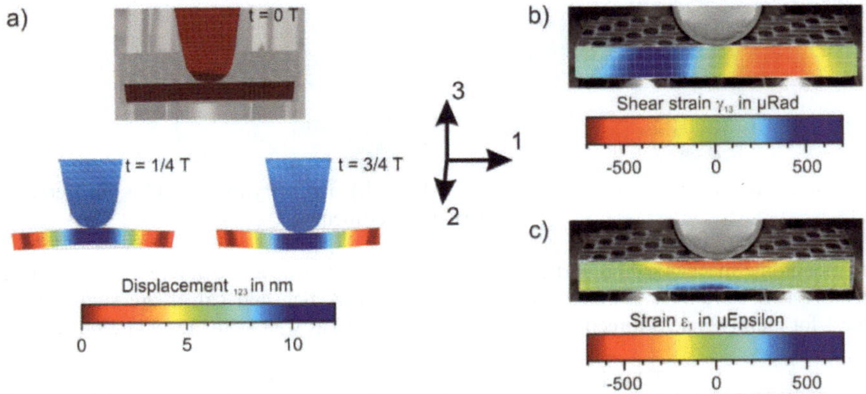

Figure 4. (a) Oscillation of the CF-PPS-specimen in the first transversal bending eigenmode at the working frequency of the UTF at about 20270 Hz, (b) Shear strain distribution of a CF-PPS specimen at 20 µm oscillation amplitude, (c) Tensile strain distribution of a CF-PPS specimen at 20 µm oscillation amplitude

Constant amplitude tests

The VHCF behavior of carbon fiber fabric reinforced polyphenylensulfide up to an ultimate number of cycles of 10^9 has been investigated in cyclic 3-point bending tests at constant load amplitudes and a frequency of 20.27 kHz. The S-N curve, shown in figure 5, summarizes the fatigue results of CF-PPS between 10^8 and 10^9 cycles. The cyclic shear stress amplitude $\tau_{a,13}$ is plotted versus the number of cycles to delamination N_{del}. The ratio of the maximum cyclic shear stress ($\tau_{a,13} + \tau_m$) during the fatigue experiment to the maximum monotonic interlaminar shear stress (τ_{ILSS}) is plotted on the right ordinate.

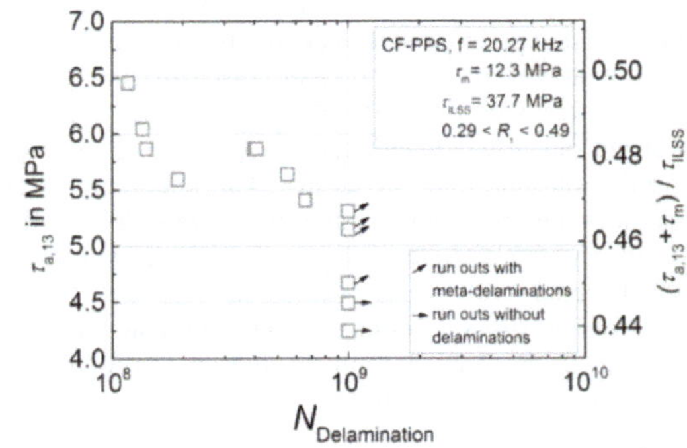

Figure 5. S-$N_{delamination}$-diagram in the VHCF range ($N > 10^8$) for CF-PPS

All VHCF-experiments have been realized with a monotonic mean stress of $\tau_m = 12.3$ MPa corresponding to a static deflection of 100 µm and a load ratio between $R_\tau = 0.29$ and $R_\tau = 0.49$. The cyclic shear stress amplitudes $\tau_{a,13}$ varied between 4.25 and 6.45 MPa. Specimen failure is defined by an overcritical heating due to delaminations or a significant decrease of the eigenfrequency which is caused by stiffness reduction. The $S\text{-}N_{delamination}$-curve clearly shows a decrease of the bearable shear stress amplitude between 10^8 and 2×10^8 as well as after 4×10^8 cycles. Run outs have been achieved at cyclic shear stress amplitudes between 4.25 and 5.3 MPa. Nevertheless even the specimens who reached the ultimate number of cycles showed a shear stress induced fatigue damage in the area between the shoulders and the loading device. The run outs under a cyclic shear stress level of 4.5 MPa indicated just fiber-matrix debonding and transversal cracks but no delaminations at all. Between a cyclic shear stress amplitude of 4.5 and 5.3 MPa micro- and meta-delaminations were detected in the specimens which reached the ultimate number of cycles. However, these delaminations did not cause an overcritical heating or a significant change in oscillation behavior. No fatigue failure up to 10^9 cycles occurred at specimens claimed with a maximum shear stress of up to 45% of the maximum interlaminar shear strength. Specimens claimed with up to 50% of the maximum interlaminar shear strength showed no failure up to 10^8 cycles.

<u>Online thermography</u>

The temperature distribution as well as the maximum temperatures on the surface edges of the specimen are detected online by IR-thermography during every VHCF experiment. All the measured temperatures of the specimen's surface are visualized and stored with a sample rate of 120 Hz. Based on the IR-thermography data the LabVIEW software is able to interrupt the fatigue experiment immediately if invalid temperatures ($T_{max} = 45°C \approx 50\%\ T_g$) occur due to fatigue damage. Figure 6 shows the temperature distribution on the length side of a fatigue damaged specimen after 4×10^8 cycles with a cyclic shear stress amplitude of $\tau_{a,13} = 5.8$ MPa.

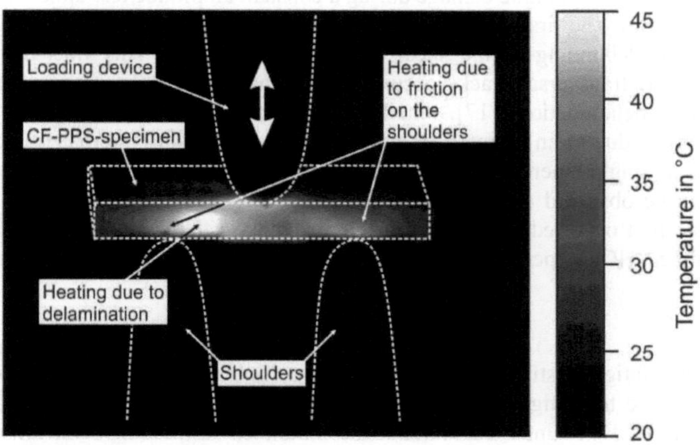

Figure 6. Characteristic surface temperature distribution of a fatigued CF-PPS specimen ($N = 4 \times 10^8$, $\tau_{a,13} = 5.8$ MPa)

The slight increase in temperature directly over the shoulders is caused by friction and characteristic for every specimen. For the specimen shown in Figure 6 the temperature increased by 20°C during the ultrasonic pulses the left shoulder and the loading device and caused an interruption of the experiment. The delamination increased the internal friction of the CF-PPS specimen which consequently caused a significant heating to a temperature of up to 45°C. Accordingly the experiment was interrupted and the specimen was investigated microscopically.

Microscopic investigations and failure mechanisms

To analyze the failure mechanisms of CF-PPS in the VHCF regime light optical and scanning electron microscopy (SEM) have been carried out. The constant amplitude tests have been interrupted after a defined number of cycles or in case of obvious fatigue damage for detailed microscopic inspection. Up to ten interruptions have been realized for one constant amplitude test. Figure 7 shows an overview of merged light optical micrographs of the lateral edge of a CF-PPS specimen after $N = N_{del} = 1.9 \times 10^8$ cycles and a shear stress amplitude of 5.6 MPa. The positions of the shoulder unit and the loading device are marked schematically. A detailed optical micrograph of the area of the maximum shear stress is presented. A macro-delamination over two carbon fiber fabric layers in the middle of the micrograph is clearly visible. Furthermore lots of transversal cracks in the 90°-fiber layers and meta-delaminations are noticeable.

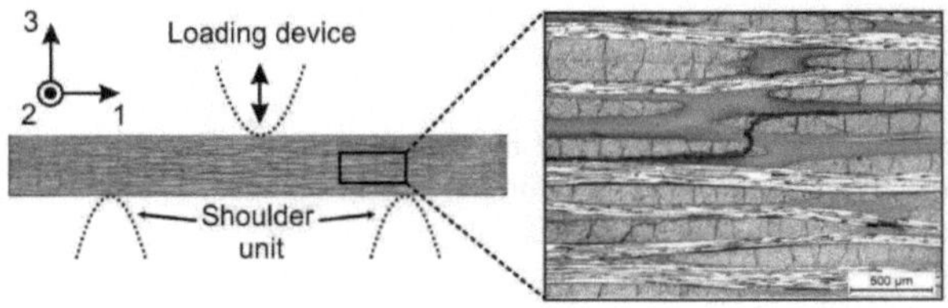

Figure 7. Light optical micrograph of the length side of a fatigued specimen and detail of the maximum shear stress area ($\tau_{a,13} = 5.6$ MPa, $N_{del} = 1.9 \times 10^8$ cycles)

The observed proceeding fatigue damage during a constant amplitude test up to 10^9 cycles can be described as follows: The first documented fatigue damage starts after 10% of the entire lifetime with fiber-matrix-debonding in the shear stress affected zones. This fiber-matrix-debonding increases and first transversal cracks occur. These transversal cracks propagated and grow to micro- and meta-delaminations [17]. Finally a complete delamination of two layers stops the VHCF experiment due to an increase of temperature by increasing internal friction. Beside the progress of one single fiber-matrix-debonding to a complete delamination, new fiber-matrix debonding can be observed on other positions in the area of maximum shear stress. Also not necessarily the first occurred meta-delamination growths to the delamination which causes the final abort of the VHCF-experiment.

Conclusions

A new ultrasonic fatigue testing facility for cyclic 3-point bending at 20 kHz has been developed at WKK to analyze the fatigue behavior of carbon fiber reinforced polymers in the very high cycle fatigue (VHCF) regime. To validate the simulated oscillation behavior of the fatigue specimens as well as to calibrate the fatigue loads precisely, 3D Scanning Laser Vibrometry measurements have been performed. Using the new developed testing facility a carbon fiber fabric reinforced polyphenylensulfide has been characterized for its fatigue behavior in the VHCF regime. For this reason constant amplitude tests up to an ultimate number of cycles of 1×10^9 have been carried out. The experiments were monitored and controlled by 1D laser Doppler vibrometry and IR-thermography. Using the IR-thermography the VHCF experiments were interrupted if invalid temperatures were measured in the area of maximum shear stress. The results show a significant decrease of the bearable shear stress amplitude after 10^8 cycles and a second time after 4×10^8 cycles. By interrupting the constant amplitude tests at defined fatigue states or in case of critical heating, the failure mechanism of CF-PPS in the VHCF regime from first fiber matrix-debonding over transversal cracks up to micro-, meta- and finally complete delaminations could be documented and analyzed by detailed microscopic analysis.

Acknowledgment

The authors would like to thank the German Research Foundation (DFG) for the financial support in framework of the priority program 1466 - Infinite Life.

References

1 C. Bathias, P. C. Paris, *Gigacycle Fatigue in Mechanical Practice* (CRC Decker, New York, 2005)

2 H. Mughrabi, "On "multi-stage" fatigue life diagrams and the life-controlling mechanisms in ultrahigh-cycle fatigue," *Fatigue and Fracture of Engineering Materials and Structures*, 25 (2001), 755 - 764

3 R. A. Abeles Couillard, P. Schwartz, "Bending fatigue of carbon-fiber-reinforced epoxy composites strands," *Composites Science and Technology*, 57 (1997), 229 - 235

4 A. Hosoi et al., "Interaction between transverse cracks and edge delamination considering free-edge effects in composites laminates," *Proc. of 16th International conference on composite materials*, Kyoto, 2007

5 A. Hosoi et al., "High-cycle fatigue characteristics of quasi-isotropic CFRP laminates over 10^8 cycles (Initiation and propagation of delamination considering interaction with transverse cracks)," *International Journal of Fatigue*, 32 (2010), 29 - 36

6 A. Hosoi et al., "Quantitative evaluation of fatigue damage growth in CFRP laminates that changes due to applied stress level," *International Journal of Fatigue*, 33 (2011), 781 - 787

7 G. W. Ehrenstein, *Polymeric Materials* (Carl Hanser, Munich, 2001)

8 J. Rotheiser, *Joining of Plastics* (Carl Hanser, Munich, 2004)

9 S. Stanzl, "A new experimental method for measuring life time and crack growth of materials under multi-stage and random loadings," *Ultrasonics*, 19 (1981), 269 - 272

10 C. Bathias, "Piezoelectric fatigue testing machines and devices," *International Journal of Fatigue*, 28 (2006), 1438 - 1445

11 S. E. Stanzl-Tschegg, "Ultrasonic Fatigue. Encyclopedia of Materials," *Science and Technology*, 2001, 9444 - 9449

12 M. Koster, G. Wagner, D. Eifler, "New measuring methods for the fatigue assessment of metals in the VHCF regime," *Proc. of Fourth International Conference on Very High Cycle Fatigue (VHCF4)*, 2007, 137 - 142

13 S. Heinz, F. Balle, G. Wagner, D. Eifler, "Innovative Ultrasonic Testing Facility for Fatigue

Experiments in the VHCF Regime," *Materials Testing*, 54 (2012), 750 - 755

14 S. Heinz, G. Wagner, D. Eifler, "Innovative piezoelectric testing facility for fatigue experiments in the VHCF regime," *Proc. of Fifth International Conference on Very High Cycle Fatigue (VHCF5)*, 2011, 479 - 484

15 G. W. Ehrenstein, G. Riedel and P. Trawiel, *Thermal Analysis of Plastics* (Munich, 2004)

16 U. Rabe et al., "Nonlinear Ultrasonic Testing of Carbon Fibre Reinforced Plastics in the Very High Cycle Fatigue Regime," *Proc. of Meetings on Acoustics (POMA), 16 (2012)*

17 S. Daggumati, et al., Fatigue and post-fatigue stress-strain analysis of a 5-harness satin weave carbon fibre reinforced composite. *Composites Science and Technology*, 74 (2013), 20 - 27

Advanced Composites for Aerospace, Marine, and Land Applications II
Edited by: T. Sano and T.S. Srivatsan
TMS (The Minerals, Metals & Materials Society), 2015

EFFECTIVE CREEP RESPONSE and UNIAXIAL TENSION BEHAVIOR OF LINEAR VISCO-ELASTIC POLYMER COMPOSITES

Tian Tang and **Sergio D. Felicelli**

Department of Mechanical Engineering
The University of Akron
Akron, OH 44325, **USA**

Keywords: effective creep response; macroscopic stress-strain behavior; linear viscoelastic polymer composite; Micromechanics; VAMUCH

Abstract

A micromechanics model was developed to characterize the effective creep response and macroscopic stress-strain behavior of linear viscoelastic polymer composites. The linear viscoelastic behavior of polymer was modeled by hereditary integral. The rate-dependent uniaxial tension behavior of the polymer composites can be calculated using the effective creep compliance coefficients predicted by the proposed model through the hereditary integral for the composites. Numerical examples were used to demonstrate the capability of the proposed model. All calculations were accomplished in the time domain, hence the Laplace transform and inversion commonly used for linearly viscoelastic composites are not needed in this theory.

1. Introduction

Composites containing linear viscoelastic constituents, such as polymer matrix composites, have been increasingly utilized in many engineering applications. As a consequence of the viscoelasticity of the polymer, such composites exhibit evidently viscoelastic behavior, which means that the magnitude of stress and strain are dependent on time and temperature. The creep of linear viscoelastic composites is one of main factors that affects their dimensional stability and their long-term strength, which in turn seriously restricts their structural applications [1, 2].

Numerous micromechanics models have been established to describe the linear viscoelastic behavior of polymer composites [3, 4]. Most of existing approaches adopted the Laplace transform and Laplace inversion, where the correspondence principle was applied [5-12]. Due to the difficulties in accurately and efficiently evaluating the Laplace inversion integral, some studies such as Brinson and Lin [13] and Fisher and Brinson [14] employed the finite element method to analyze the two-phase and three-phase viscoelastic composites in the Laplace transformed domain, respectively, and verified the results with the Mori–Tanaka model [15, 16]. In order to avoid the difficulties associated with the evaluation of Laplace inversion integral, Lahellec and Suquet [17] developed scheme operates directly in the time-domain by introducing an effective viscous strain (or effective internal variable) in each phase of the composite. Furthermore, commercial finite element software ABAQUS was employed to predict the creep response [18] of fiber reinforced polymer matrix composites considering a unit cell subjected to periodic boundary conditions.

The focus of this paper is to develop a micromechanics model to predict the effective creep compliance and macroscopic stress-strain behavior of polymer composites. This model is established on the basis of the framework of VAMUCH (variational asymptotic method for unit cell homogenization) [19].

2. Constitutive Equations of Effective Behavior of Linear Viscoelastic Materials

Effective Creep Compliance

According to the Boltzmann superposition principle and small deformation assumption, the constitutive equations for the linear viscoelastic material can be described in the time domain in the following way,

$$\varepsilon_{ij}(t) = \int_{-\infty}^{t} S_{ijkl}(t-\tau)\dot{\sigma}_{kl}(\tau)\mathrm{d}\tau \tag{1}$$

where $S_{ijkl}(t)$ is the creep compliance; $\dot{\sigma}_{kl}(\tau)$ is the stress rate; $\varepsilon_{ij}(t)$ is the strain tensor.
The creep is the tendency of a material to move slowly or deform permanently under constant mechanical stress, which means

$$\sigma_{kl}(t) = \begin{cases} 0 & t < 0 \\ \sigma_{kl}^{\mathrm{cst}} & t \geq 0 \end{cases} \tag{2}$$

Where "cst" stands for constant value. Eq. (2) implies: $\lim_{t\to-\infty}\sigma_{kl}(t)=0$.
Then applying the integral by parts to Eq. (1), we can obtain

$$\varepsilon_{ij}(t)=\left(S_{ijkl}(0)+\int_0^t\frac{\partial S_{ijkl}(t-\tau)}{\partial(t-\tau)}\mathrm{d}\tau\right)\sigma_{kl}^{\mathrm{cst}}=S_{ijkl}(t)\sigma_{kl}^{\mathrm{cst}} \quad (3)$$

From Eq. (3) one can see that the instantaneous strain values are dependent on the instantaneous values of creep compliance coefficients rather than history effects when the linear viscoelastic materials are subjected to constant loading.
The Eq. (3) can be rewritten as

$$\sigma_{ij}^{\mathrm{cst}}=C_{ijkl}(t)\varepsilon_{kl}(t) \quad (4)$$

where $C_{ijrs}(t)S_{rskl}(t)=I_{ijkl}$ with I_{ijkl} being fourth-order identity tensor. Note that $C_{ijkl}(t)$ is not the same as the stress relaxation stiffness C_{ijkl}^{sr} defined in $\sigma_{ij}(t)=\int_{-\infty}^{t}C_{ijkl}^{sr}(t-\tau)\dot{\varepsilon}_{kl}(\tau)\mathrm{d}\tau$.

The Eq. (4) may be derived from the following transient potential energy density functional,

$$U(t)=\frac{1}{2}C_{ijkl}(t)\varepsilon_{ij}(t)\varepsilon_{kl}(t) \quad (5)$$

such that

$$\sigma_{ij}^{\mathrm{cst}}=\frac{\partial U(t)}{\partial\varepsilon_{ij}(t)} \quad (6)$$

The effective properties of linear viscoelastic composites can be defined as the follows,

$$\bar{\varepsilon}_{ij}(t)=\int_0^t S_{ijkl}^{*}(t-\tau)\dot{\bar{\sigma}}_{kl}(\tau)\mathrm{d}\tau \quad (7)$$

$$\bar{\sigma}_{ij}^{\mathrm{cst}}=C_{ijkl}^{*}(t)\bar{\varepsilon}_{kl}(t)\quad\text{or}\quad\bar{\varepsilon}_{ij}(t)=S_{ijkl}^{*}(t)\bar{\sigma}_{kl}^{\mathrm{cst}} \quad (8)$$

$$\frac{1}{\Omega}\int_\Omega\frac{1}{2}C_{ijkl}(t)\varepsilon_{ij}(t)\varepsilon_{kl}(t)\mathrm{d}\,\Omega=\frac{1}{2}C_{ijkl}^{*}(t)\bar{\varepsilon}_{ij}(t)\bar{\varepsilon}_{kl}(t) \quad (9)$$

where "over-bar" indicates variables which are used in the macroscopic analysis of homogenized materials and Ω is the volume of unit cell, such that

$$\bar{\sigma}_{ij}^{\mathrm{cst}}=\frac{1}{\Omega}\int_\Omega\sigma_{ij}(t)\,\mathrm{d}\Omega \quad (10)$$

$$\bar{\varepsilon}_{ij}(t)=\frac{1}{\Omega}\int_\Omega\varepsilon_{ij}(t)\,\mathrm{d}\Omega \quad (11)$$

Although the externally applied mechanical stresses are kept constant as $\bar{\sigma}_{ij}^{\mathrm{cst}}$, the local stresses $\sigma_{ij}(t)$ and local strain $\varepsilon_{ij}(t)$ within unit cell vary with time due to the creep or stress relaxation of linear viscoelastic polymer.

Superscripts "*" in Eqs. (7-9) denote the effective properties whose calculations are determined by the micromechanics model one employs, Note that $C^*_{ijrs}(t)S^*_{ijrs}(t) = I_{ijkl}$ with I_{ijkl} being the fourth-order identity tensor.

3. Macroscopic Stress Strain Behavior

After obtaining the effective creep compliance coefficients $S^*_{ijkl}(t)$, the macroscopic stress-strain behavior of linear viscoelastic polymer composites subjected to constant loading rate $\dot{\bar{\sigma}}_{kl}(\tau)$ loading can be computed using Eq. (7) in the following way

$$\bar{\varepsilon}_{ij}(t) = \left(\int_0^t S^*_{ijkl}(t-\tau)\mathrm{d}\tau\right)\dot{\bar{\sigma}}_{kl} \tag{12}$$

With

$$\dot{\bar{\sigma}}_{kl} = \frac{\bar{\sigma}_{kl}(t)}{t} \tag{13}$$

Although the overall stress rates of the composite can be kept constant, the constituents respond differently with time, and the stress and strain fields within constituents vary with different rates at different times due to the different rates of creep in the materials at different times. The effects of loading rate on the macroscopic stress-strain behavior of the linear viscoelastic composites can be investigated using Eqs. (12) and (13).

4. Micromechanics Formulations for Effective Coefficients

In order to calculate the effective creep compliance coefficients $S^*_{ijkl}(t)$ of linear viscoelastic composites, we need to firstly develop a micromechanics model to compute the material matrix of effective coefficients $C^*_{ijkl}(t)$, which is the inversion matrix of $S^*_{ijkl}(t)$.
The micromechanics model for the calculation of effective $C^*_{ijkl}(t)$ of linear viscoelastic composites was constructed on the basis of the variational asymptotic method for unit cell homogenization (VAMUCH) [17]. Following a similar derivation procedure as in Yu and Tang [17], we can obtain a variational statement which will govern the micromechanics model of linear viscoelastic composites. The variational statement can be expressed as minimizing the following functional

$$\Pi_\Omega = \frac{1}{2\Omega}\int_\Omega C_{ijkl}(t)\left[\bar{\varepsilon}_{ij}(t) + \chi_{(i|j)}\right]\left[\bar{\varepsilon}_{kl}(t) + \chi_{(k|l)}\right]\mathrm{d}\Omega \tag{14}$$

subjected to periodic constraints. Here, χ_i is the commonly called fluctuating function, $\bar{\varepsilon}_{ij}(t)$ is the global transient strain tensor.
Let us introduce the following matrix notations:

$$\bar{\epsilon} = \lfloor \bar{\varepsilon}_{11}(t)\ \ 2\bar{\varepsilon}_{12}(t)\ \ \bar{\varepsilon}_{22}(t)\ \ 2\bar{\varepsilon}_{13}(t)\ \ 2\bar{\varepsilon}_{23}(t)\ \bar{\varepsilon}_{33}(t)\rfloor^T \tag{15}$$

$$\epsilon_1 = \begin{bmatrix} \frac{\partial}{\partial y_1} & 0 & 0 & 0 \\ \frac{\partial}{\partial y_2} & \frac{\partial}{\partial y_1} & 0 & 0 \\ 0 & \frac{\partial}{\partial y_2} & 0 & 0 \\ \frac{\partial}{\partial y_3} & 0 & \frac{\partial}{\partial y_1} & 0 \\ 0 & \frac{\partial}{\partial y_2} & \frac{\partial}{\partial y_3} & 0 \\ 0 & 0 & \frac{\partial}{\partial y_3} & 0 \end{bmatrix} \begin{Bmatrix} \chi_1 \\ \chi_2 \\ \chi_3 \end{Bmatrix} \equiv \Gamma_h \chi \tag{16}$$

where Γ_h is an operator matrix and χ is a column matrix containing the three components of the fluctuation functions. χ can be descretized using finite elements as

$$\chi(x_i; y_i) = S(y_i)\boldsymbol{\chi}(x_i) \tag{17}$$

Where S representing the shape functions (in assemble sense excluding the constrained node and slave nodes) and $\boldsymbol{\chi}$ column matrix of the nodal value of the fluctuation functions for all active nodes. Substituting Eqs. (15-17) into Eq. (14), we obtain a discretized version of the functional as

$$\Pi_\Omega = \frac{1}{2\Omega}(\boldsymbol{\chi}^T E \boldsymbol{\chi} + 2\boldsymbol{\chi}^T D_{h\varepsilon}\bar{\epsilon} + \bar{\epsilon}^T D_{\varepsilon\varepsilon}\bar{\epsilon}) \tag{18}$$

Where

$$E = \int_\Omega (\Gamma_h S)^T D (\Gamma_h S) \mathrm{d}\Omega \qquad D_{h\varepsilon} = \int_\Omega (\Gamma_h S)^T D \mathrm{d}\Omega \qquad D_{\varepsilon\varepsilon} = \int_\Omega D \, \mathrm{d}\Omega \tag{19}$$

with D as the 6×6 material coefficient matrix condensed from the fourth-order tensor $C_{ijkl}(t)$. Minimizing Π_Ω in Eq. (18), we obtain the following linear system:

$$E\boldsymbol{\chi} = -D_{h\varepsilon}\bar{\epsilon} \tag{20}$$

Since the fluctuation function $\boldsymbol{\chi}$ is linearly proportional to $\bar{\epsilon}$, such that

$$\boldsymbol{\chi} = \boldsymbol{\chi}_0 \bar{\epsilon} \tag{21}$$

Substituting Eq. (21) into Eq. (18), we can compute the transient potential energy density of the UC as

$$\Pi_\Omega = \frac{1}{2\Omega}\bar{\epsilon}^T(\boldsymbol{\chi}_0^T D_{h\varepsilon} + D_{\varepsilon\varepsilon})\bar{\epsilon} = \frac{1}{2}\bar{\epsilon}^T D^* \bar{\epsilon} \tag{22}$$

where D^* is a 6×6 material matrix containing effective coefficients of the fourth-order tensor $C^*_{ijkl}(t)$.

5. Numerical Results

In order to demonstrate the capability of the proposed model, it was employed to calculate the time dependent effective creep compliance coefficients $S^*_{ijkl}(t)$, which were obtained from the inversion matrix of $C^*_{ijkl}(t)$, and the macroscopic stress strain behavior of fiber reinforced linear viscoelastic matrix composites. The fibers are assumed to be linear elastic and isotropic while the matrix exhibits linear viscoelastic behavior. The fibers are of circular shape and in square array. The creep compliance of linear viscoelastic polymer matrix in the time domain is expressed by a power-law model [12]

$$M_0(t) = D_0 + D_1 t^n \tag{23}$$

where $D_0 = 0.065\ \text{GPa}^{-1}$ is the initial compliance, and $D_1 = 0.016286\ \text{GPa}^{-1}$ and $n = 0.2297$ are material parameters determined by experiments. In this study, the values of D_0 and D_1, and n are calculated from the Fig. 13 in Naik et al. [18], who computed the creep compliance coefficients of pure viscoelastic polymer based on the shear and bulk moduli defined by a Prony series expansion. The creep compliance matrix of isotropic and linear viscoelastic polymer is given by

$$[S_{ij}(t)] = M_0(t) \begin{bmatrix} 1 & -\nu & -\nu & 0 & 0 & 0 \\ -\nu & 1 & -\nu & 0 & 0 & 0 \\ -\nu & -\nu & 1 & 0 & 0 & 0 \\ 0 & 0 & 0 & 2(1+\nu) & 0 & 0 \\ 0 & 0 & 0 & 0 & 2(1+\nu) & 0 \\ 0 & 0 & 0 & 0 & 0 & 2(1+\nu) \end{bmatrix} \tag{24}$$

where ν is the Poisson's ratio of the polymer and kept as ν=0.4 in this study.
The fiber reinforced composites are orthotropic materials and its effective creep compliance matrix can be expressed as

$$\begin{Bmatrix} \bar{\varepsilon}_{11}(t) \\ \bar{\varepsilon}_{22}(t) \\ \bar{\varepsilon}_{33}(t) \\ \bar{\varepsilon}_{23}(t) \\ \bar{\varepsilon}_{12}(t) \\ \bar{\varepsilon}_{13}(t) \end{Bmatrix} = \int_0^t \begin{bmatrix} S^*_{11}(t) & S^*_{12}(t) & S^*_{13}(t) & 0 & 0 & 0 \\ S^*_{12}(t) & S^*_{22}(t) & S^*_{23}(t) & 0 & 0 & 0 \\ S^*_{13}(t) & S^*_{23}(t) & S^*_{33}(t) & 0 & 0 & 0 \\ 0 & 0 & 0 & S^*_{44}(t) & 0 & 0 \\ 0 & 0 & 0 & 0 & S^*_{55}(t) & 0 \\ 0 & 0 & 0 & 0 & 0 & S^*_{66}(t) \end{bmatrix} \begin{Bmatrix} \dot{\bar{\sigma}}_{11}(\tau) \\ \dot{\bar{\sigma}}_{22}(\tau) \\ \dot{\bar{\sigma}}_{33}(\tau) \\ \dot{\bar{\sigma}}_{23}(\tau) \\ \dot{\bar{\sigma}}_{12}(\tau) \\ \dot{\bar{\sigma}}_{13}(\tau) \end{Bmatrix} \mathrm{d}\tau \tag{25}$$

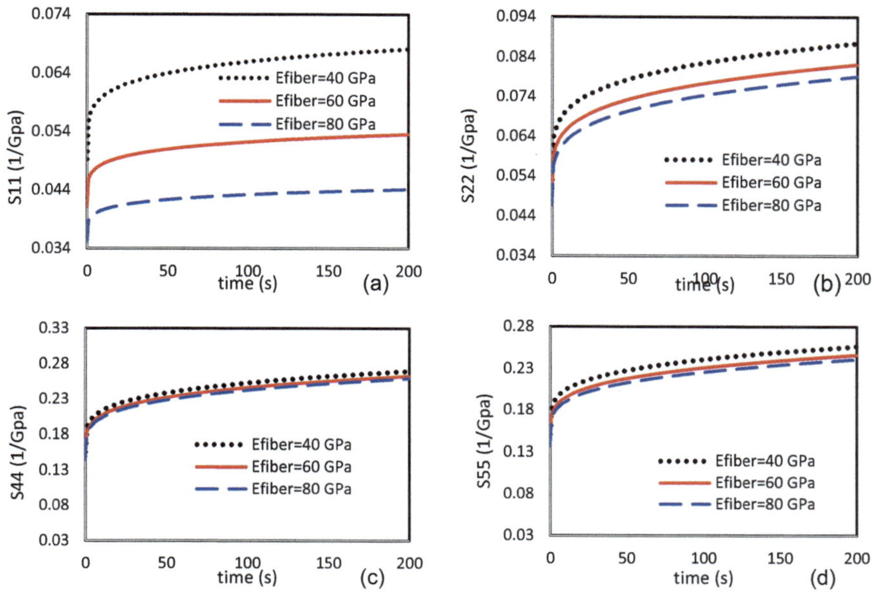

Figure 1. Variation of the effective creep compliance coefficients with respect to time for different Young's modulus of the fibers: (a) $S^*_{11}(t)$; (b) $S^*_{22}(t)$; (c) $S^*_{44}(t)$ (d) $S^*_{55}(t)$.

5.1 Effective creep compliance coefficients

In this section, the effective creep compliance coefficients were calculated by increasing the Young's modulus of the fibers from 40 to 80 GPa while the volume fraction of the fibers is kept constant as vof = 0.2. Fig. 1 shows that the variation of the effective creep compliance coefficients with respect to time for different Young's modulus of the fibers. It is as expected that the creep compliance coefficients increased with the decrease of the Young's modulus of the fibers.

5.2 Macroscopic stress-strain behavior

The macroscopic stress-strain behaviors of polymer composites can be predicted using Eq. (12) and (13). We calculated the macroscopic uniaxial tension behavior of fiber reinforced composites with the volume fraction of the fibers being vof=0.2 in longitudinal direction and transverse direction, respectively. Perfect bond is assumed to be between the fibers and the matrix. The predictions are presented in Fig. 2. The macroscopic uniaxial tension behaviors demonstrate that the required stress increases with the increase of the stress rate, namely, the faster the stretching, the larger the stress required.

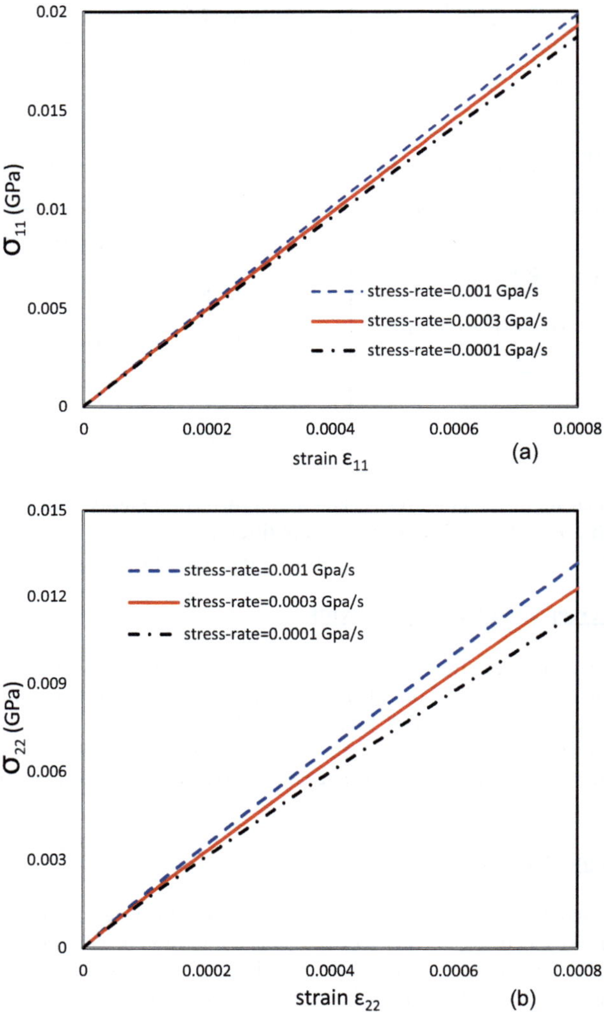

Figure 2. **Macroscopic stress-strain curves of fiber reinforced linear viscoelastic polymer matrix composites with the fiber volume fraction vof=0.2 for different stress rates: (a) longitudinal tension; (b) transverse tension.**

6. Conclusions

We have developed a micromechanics model being capable of efficiently predicting the effective creep compliance and macroscopic stress-strain behavior of linear viscoelastic polymer composites. Since all calculations are performed in the time domain the traditionally used Laplace transform and inversion are not needed. This model is applicable to polymer composites having any shapes of reinforcements although a fiber reinforced composites was used as a numerical example in this work. The present modeling technique can be readily extended to nonlinear viscoelastic behavior of polymer composites.

References

[1]. Raghavan, J. and Meshii, M. (1998). Creep of polymer composites. ***Composites Science and Technology***, **57** (12):1673-1688.

[2]. Barbero, E. J. (1994). *Construction: Applications and Design*, Lubin's Handbook of Composites, 2nd edn.

[3]. Nemat-Nasser, S. and Hori, M. (1993). *Micromechanics: overall properties of heterogeneous materials*. Amsterdam: Elsevier Science Publishers, Amsterdam.

[4]. Hashin, Z. (1983). Analysis of composite materials-a survey. ***Journal of Applied Mechanics***, **105**:481-504.

[5]. Christensen, R. M. (1979). *Mechanics of composite materials*. Dover Publications, Inc. Mineola, New York.

[6]. Barbero, E. J. and Luciano, R. (1995). Micromechanical formulas for the relaxation tensor of linear viscoelastic composites with transversely isotropic fibers, ***International Journal of Solids and Structures***, **32** (13): 1859-1872.

[7]. Yancey, R. N. and Pindera, M-J. (1990). Micromechanical analysis of the creep response of unidirectional composites, ***Journal of Engineering Materials and Technology***, **112**: 157-163.

[8]. Li, K, Gao, X. L., and Roy, A. K. (2006). Micromechanical modeling of viscoelastic properties of carbon nanotube-reinforced polymer composites. ***Mechanics of Advanced Materials and Structures***, **13**: 317-328.

[9]. Wang, Y. M. and Weng, G. J. (1992). The influence of inclusion shape on the overall viscoelastic behavior of composites, ***Journal of Applied Mechanics, Trans. ASME***, **112**: 157-163.

[10]. Haasemann, G. and Ulbricht, V. (2009). Numerical evaluation of the viscoelastic and viscoplastic behavior of composites, ***Tchnishe Mechanik***, **30** (1-3): 122-135.

[11]. Megnis, M., Varna, J., Allen, D. H., and Holmberg, A. (2001). Micromechanical modeling of viscoelastic response of GMT composites, ***Journal of Composites Materials***, **35**: 849-882.

[12]. Yancey, R. N. and Pindera, M.J. (1990). Micromechanical analysis of the creep response of unidirectional composites, ***Journal of Engineering Materials and Technology***, **112**:157-163.

[13]. Brinson, L. C. and Lin, W. S. (2003). Comparison of micromechanics methods for effective properties of multiphase viscoelastic composites, ***Composites Structures***, **41**:353-367.

[14]. Fisher, F. T. and Brinson, L. C. (2003). Viscoelastic interphase in polymer-matrix composites: theoretical models and finite element analysis, ***Composites Science and Technology***, **61**:731-748.

[15]. Mori, T. and Tanaka, K. (1973). Average stress in matrix and average elastic energy of materials with misfitting inclusions, ***Acta Metallurgica***, **21**:571-574.

[16]. Benveniste, Y. (1987). A new approach to the application of Mori-Tanka's theory in composite materials, ***Mechanics of Materials***, **6**:147-157.

[17]. Lahellec, N. and Suquet, P. (2007). Effective behavior of linear viscoelastic composites: A time-integration approach, ***International Journal of Solids and Structures***, **44**: 507-529.

[18]. Naik, A., Abolfathi, N., Karami, G., and Ziejewski, M. (2008). Micromechanical viscoelastic characterization of fibrous composites, ***Journal of Composites Materials***, **42**:1179-1204.

[19]. Yu, W. and Tang, T. (2007). Variational Asymptotic Method for Unit Cell Homogenization of Periodically Heterogeneous Materials, ***International Journal of Solids and Structures***, **44**: 3738-3755.

Advanced Composites for Aerospace, Marine, and Land Applications II
Edited by: T. Sano and T.S. Srivatsan
TMS (The Minerals, Metals & Materials Society), 2015

MOLECULAR DYNAMICS ASSESSMENT OF SMALL MOLECULE DIFFUSION IN MEDICAL PLASTICS

David M. Saylor, Sudabeh Jawahery, and Christopher Forrey

Center for Devices and Radiological Health
FDA
Silver Spring, MD 20993 USA

Keywords: diffusion, polymers, molecular dynamics, memory kernel

Abstract

The ability to accurately predict solute diffusivity in polymeric matrices has far–reaching implications, particularly in the capacity of predicting the behavior of polymer matrix composites used in medical devices. We have recently demonstrated that it is possible to use all-atom molecular dynamics simulations to acquire these values. However, as the size of the solute increases or polymer matrix becomes more glassy, the simulation time required becomes prohibitively long, and the observed dynamics are restricted to the sub–diffusive regime. Thus, we seek to explore potential relationships between the sub-diffusive and macroscale (Brownian) behavior. Here, we have characterized the dynamics of a model system within the context of generalized Langevin dynamics. Specifically, we compute the Laplace frequency dependence of the memory kernel $\Gamma(s)$ that quantifies non–trivial drag and random forces in the observed dynamics. Our observations span the inertial, sub-diffusive, and Brownian regimes over a wide range of temperatures. We find in all cases, that $\Gamma(s)$ is constant as $s \rightarrow 0$ or ∞, suggesting ideal dynamics, *i.e* "instantaneous" drag force and white noise, in these regimes. For intermediate values of s, $\Gamma(s)$ exhibit power–law relationships with exponents that increase monotonically with system temperature. These observations may serve as a basis to link the sub–diffusive and macroscale dynamics in these systems.

1. Introduction

Composite materials integrated into medical devices can significantly improve device performance and functionality. For example, drug-polymer composites coated on stents have demonstrated a marked improvement in therapeutic outcomes [1]. The successful implementation of these coatings depends intrinsically on small molecule diff sion, which dictates both the composite structure that forms during manufacturing and the rate of drug egress after implantation [2]. Further, biocompatibility of medical plastics, in general, is governed by the diff y of manufacturing by-products and other additives ubiquitous in these materials [3]. We have recently shown, for the fi time, that the diff y of relatively large solute molecules (> 50 Da) in polymer matrices can be accurately predicted using all-atom molecular dynamics (MD) simulations [4]. However, as the size of the solute increases or polymer matrix becomes more glassy, the simulation time required to reach the diffusive (i.e. Brownian) regime becomes prohibitively long.

While ever increasing computing resources will be helpful in overcoming this restriction, it will likely be decades before diff y can be calculated quickly and accurately for glassy systems and systems with larger diffusants. However, it may be possible to circumvent this limitation if a relationship between the "fast" sub-diffusive and "slow" diff e dynamics can be established. For example, work by Cicerone et al. suggests that the high frequency $O(ps^{-1})$ beta–relaxations in polymer matrices correlate with solute diffusion rate [5]. Further, quantitative relationships between the segmental relaxation time, which is intimately related to diffusivity, and fast dynamics have recently been re- ported [6, 7]. Here, we investigate the potential link between sub-diffusive dynamics and Brownian diff within the context of generalized Langevin dynamics. Specifically, we characterize the so-called memory kernel that captures temporally non–local drag forces and random forces with non-trivial correlations that are associated with the observed dynamics.

In this proceedings, we probe the character of the memory kernel that describes the dynamics of a model solute molecule in a polymer matrix determined by MD simulations over a wide range of times and temperatures. Specifically, we investigate systems com- prised of 0.10 w/w tetrahydrofuran (THF) in a poly(styrene-b-isobutylene-b-styrene) (SIBS) tri–block copolymer matrix. We start by describing the MD simulations and generalized Langevin equation (GLE) on which our analysis is based in the following Methods section. We then demonstrate that these data enable us to specify the evolution of the memory kernel over nearly fi e orders of magnitude in time in the subsequent Results and Discussion section. Finally, a brief overview of our finding is provided in the Summary section. We note that the timeframe of these simulation encompasses the inertial, sub-diffusive, and, for the highest temperatures, Brownian dynamical regimes, and that this characterization serves as an initial step towards developing a quantitative relationship between sub-diffusive and macroscopic character of small molecule diffusion in polymer matrices.

2. Methods

2.1 Molecular dynamics (MD) simulations

The details of the MD simulations have been previously reported elsewhere [4]. Briefly, a set of initial amorphous structures consisting of 0.10 w/w THF in a SIBS matrix (1565 total atoms per structure) were constructed using Materials Studio software from Accelrys. These structures were exported to LAMMPS [8], open source software made available by Sandia National Laboratory, for subsequent replication, density equilibration, and production runs. The generalized Amber force–field (GAFF) was employed for all molecular dynamics simulation stages within LAMMPS. For each individual structure imported to LAMMPS, a larger system was constructed by replicating each unit lattice 3 times in each dimension, generating a super-lattice with 27 times as many atoms as the original unit cell. The newly constructed super-lattice possessed a periodicity corresponding to the original unit cell dimensions. The structural memory of the replication process was erased by heating and annealing the super-lattice from 310K to 550K over one million molecular dynamics steps and then annealing at 550K for four million molecular dynamics steps. Elimination of super-lattice periodicity was monitored by following the decay of the peak in the Fourier Transform of the system density corresponding to the original unit cell. Following this initial annealing stage, the larger systems were each cooled to a unique temperature, namely 310, 370, 430, 490, and 550, K. Each system was then subjected to a six million step equilibration in the isobaric-isothermal (NPT) ensemble to eliminate remaining high energy structural features and to determine equilibrium density for subsequent production runs in the canonical (NVT) ensemble. Final structures generated for production runs were fully atomistic, containing 42255 atoms each. Production runs utilized a time-step of 4 fs, such that a total of 2.5×10^6 steps were required to achieve 10 ns of simulation time. Vibrational motion of hydrogen atoms was explicitly considered. Center–of–mass trajectories for the 297 individual THF molecules in each system were sampled every 40 fs. These trajectories were then used to determine the average mean-square displacement $\langle r^2 \rangle$ as a function of time, t. Based on these calculations, we can define a diffusivity, D, using the quantitative relationship originally derived by Einstein [9] that can be expressed as:

$$D = \frac{1}{6}\frac{\partial \langle r^2 \rangle}{\partial t}. \qquad (1)$$

We note that this quantity represents an effective diffusivity that corresponds to the macroscopic, Brownian value only in the limit $t \to \infty$.

2.2 Generalized Langevin equation (GLE)

We seek to describe the dynamics predicted by the MD simulations in terms of the generalized Langevin equation (GLE), which is given by:

$$M\frac{\partial V}{\partial t} = -\int_0^t \Gamma(t-\tau)V(\tau)d\tau + F(t), \qquad (2)$$

where M and V are the mass and velocity, respectively, of the diffusing molecule, $F(t)$ is the fluctuating force that is related to the memory kernel $\Gamma(t)$ by the fluctuation dissipation theorem $\langle F(t)F(\tau)\rangle = \Gamma(t-\tau)k_B T$, and k_B and T are the Boltzmann constant and system temperature, respectively. Following Ref. [10], we can multiply both sides of Eq. 2 by the initial velocity $V(0)$ and take the ensemble average, which yields:

$$M\frac{\partial C_V}{\partial t} = -\int_0^t \Gamma(t-\tau)C_V(\tau)d\tau, \qquad (3)$$

where $C_V = \langle V(t)V(0)\rangle$ is the velocity autocorrelation function. We can solve Eq. 3 by taking the Laplace transform and using the equipartition theorem $C_V(0) = 3k_BT/M$, which yields:

$$\widetilde{C_V} = \frac{3k_BT}{Ms + \widetilde{\Gamma}}, \tag{4}$$

where s is the Laplace frequency and the tilde indicates the Laplace transform of the given quantity. Finally, we recognize that the Laplace transform of mean square displacement and velocity autocorrelation function are related by $s^2\widetilde{\langle r^2\rangle} = 2\widetilde{C_V}$ which allows us to write:

$$\widetilde{\Gamma} = \frac{6k_BT}{s^2\widetilde{\langle r^2\rangle}} - Ms. \tag{5}$$

Based upon the above relation, we can directly specify the Laplace transformed memory kernel by numerically taking the Laplace transform of $\langle r^2\rangle$ computed from the MD simulations.

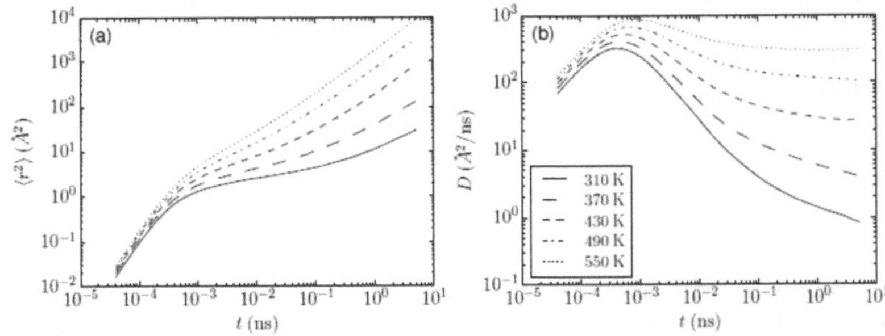

Figure 1. **Molecular dynamics simulation results for THF molecules in a SIBS matrix.**
a) **The time dependence of the mean square displacement of THF molecules at five different temperatures.**
b) **The corresponding *D* values as a function of time. Note that at 5 ns a converged value of *D* indicating diffusive (Brownian) behavior is only observed for $T \geq 430$ K.**

3. Results and discussion

Prior to exploring the character of the memory kernel, it is instructive to look at the temporal dependence of the mean squared displacement predicted by the MD simulations and corresponding impact on the convergence of the diffusivity of the solute molecules to the macroscopic value, *i.e.* as $t \to \infty$. Figure 1a depicts the relationship between $\langle r^2\rangle$ and time for the five simulation temperatures. The plot clearly illustrates the presence of the inertial regime at short times for all temperatures, T. This regime is characterized by the initial decay from a slope of two on the log–log plot (*i.e.* $\langle r^2\rangle \propto t^2$) associated with ballistic motion as the particles begin to interact with the surrounding matrix. This decay persists until around one picosecond, after which there is evidence of a sub–diffusive or localization regime for all T. This regime is characterized $\langle r^2\rangle \propto t^{\alpha}$, where a is an essentially constant value of less than one. We note that with increased T, the extent and duration of localization decreases dramatically, until it is barely perceptible at the highest $T = 550$ K. Finally, at longer times the solute molecules become delocalized, and the slopes of the log–log plot increase steadily until they reach the Brownian value of one. Over the timeframe covered by the simulations, Brownian behavior is only reached in systems with $T \geq 430$ K. In fact, previous results [4] suggest that at the lowest T(310 K), the time scales associated with the Brownian regime are $\geq O(1\,\mu s)$. Figure 1b depicts the manner in which the above observations manifest in the value of D computed using Eq.1. Inspection of the plot reveals that the D values increase substantively in the inertial regime. We note that this "super–diffusive" type behavior is expected because the slope of corresponding log–log plots in Figure 1a are in excess of one. The localization regime is then characterized by a marked decrease in D prior to asymptotically approaching the constant, macroscopic value. Using the data depicted in Figure 1, we next probe the character of the memory kernel, which characterizes the drag force imparted to the solute particles by the matrix, as a function of (Laplace) frequency and temperature. To accomplish this, we numerically Laplace transform the mean square displacement data, which can then be used along with Eq. 5 to compute the transformed memory kernel. The results of these calculations are shown in Figure 2. The figure reveals qualitatively similar behavior at all T, again with the three distinct regimes evident. For example, at high s, i.e. short t, the value of Γ becomes essentially constant, $\Gamma(s) = \Gamma_0$. This observation suggests that as $T \rightarrow 0$, the system dynamics are essentially ideal, i.e instantaneous drag force and delta- correlated (white) noise, with a magnitude given by Γ_0. This behavior is expected if one considers that for high frequencies, insufficient time has elapsed for correlations in the drag and random forces to manifest. In addition to being constant at high frequencies, $\Gamma(s)$ also approaches constant values at low frequencies, with $\Gamma(s) = \Gamma$ as $s \rightarrow 0$. This is characteristic of Brownian dynamics, where Γ / kBT is simply the inverse of the macroscopic D value. Finally, intermediate values of s exhibit a power{law relationships similar to those that have been observed in other constrained systems, such as protein conformational fluctuations [11].

Based on the observations illustrated in Figure 2, we can now quantify the memory kernel in all three distinct regimes. These data are summarized in Table 1 in terms of the normalized values, $\Gamma = \Gamma$ /kBT. The second column of the table shows the constant values of Γ_0 as $s \rightarrow \infty$. We note that these values are essentially linear with 1/T, implying a constant value of Γ_0. To characterize the power{law regime quantitatively, we have fit the linear portions of the curves to $\Gamma = \beta s^{\alpha}$. The resulting fit α and β values are presented in the third and fourth column of the table, respectively.

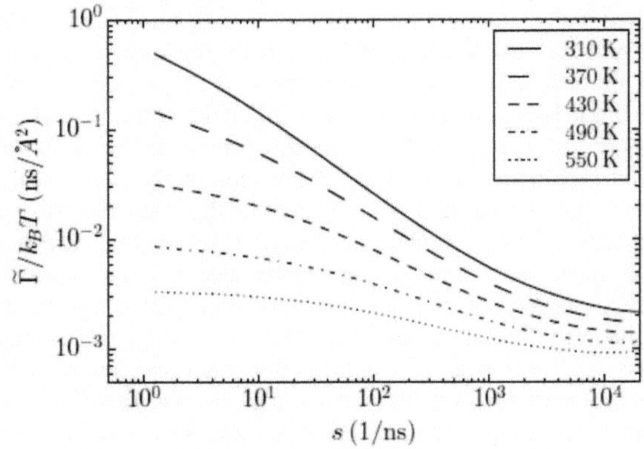

Figure 2. Laplace transformed memory kernels at different temperatures.

We find that α increases linearly with T over this range, while β decays exponentially. If Γ exhibits power law decay in the Laplace domain, it will also have power law behavior in the time domain, Γ / $t^{-\alpha-1}$. Thus, the exponents will become more negative with increasing T in the time domain. The fifth column highlights the values of Γ in the Brownian regime.

Table 1: Memory kernel characteristics as a function of temperature

T (K)	$\hat{\Gamma}_0$ (ns/Å^2)	α	β (ns$^{1-\alpha}$/Å^2)	$\hat{\Gamma}_\infty$ (ns/Å^2)	$\hat{\Gamma}_\infty/\hat{\Gamma}_0$
310	2.1×10^{-3}	-0.76	8.5×10^{-1}	10 (Ref. [4])	4.8×10^3
370	1.7×10^{-3}	-0.65	3.1×10^{-1}	–	–
430	1.4×10^{-3}	-0.50	7.9×10^{-2}	3.7×10^{-2}	2.6×10^1
490	1.1×10^{-3}	-0.35	2.0×10^{-2}	9.4×10^{-3}	8.5×10^0
550	9.2×10^{-4}	-0.24	6.6×10^{-3}	3.3×10^{-3}	3.7×10^0

We note that this regime is only reached for $T \geq 430$ K within the timeframe of the current simulations. However, previously [4] we carried out comparable simulations at 310 K to 1 μs that reached the constant, macroscopic value. Therefore, this value has also been included in the table for reference. As expected, we observe an exponential decrease in ΓΓ with increasing T covering nearly four orders of magnitude over the T range explored here. In the fifth column, we provide the ratio of the two regimes in which the dynamics are "ideal", as $s \to 0$ or ∞.

With the data in Figure 2 and Table 1, we have characterized quantitatively the memory kernel describing the dynamics of THF molecules in a SIBS matrix over a wide range of frequency an d temperature. In general, the dynamics exhibit ideal behavior at very high and very low frequencies, with an intermediate power–law regime. Although these observations are from a single system, albeit at multiple T, we suspect that this behavior is universal in locally constrained systems. Thus, specifying Γ_0, α, and β represents a reasonably robust metric for the sub-diffusive dynamics in a given constrained system. We are currently exploring links between this description of the sub–diffusive character of the dynamics, *i.e.* in the inertial and localization regimes, and the low frequency Brownian behavior on the basis of these data. If a quantitative link can be established, the fundamental constraint in using molecular dynamics to predict macroscopic diff coefficients of solutes in polymeric matrices can be circumvented. This would enable one to rapidly assess diffusion mitigated phenomena in polymer matrix composites.

4. Summary

We have characterized the memory kernel of the solute dynamics in systems comprised of 0.10 w/w THF in a SIBS matrix. The characterization is based on the predictions from 10 ns fully atomistic molecular dynamics simulations. We find that high frequency flow in the system do not appear to exhibit memory and are consistent with uncorrelated, white noise.

However, at lower frequencies, the solute molecules become localized and exhibit power–law fluctuation, consistent with observations in other con- strained systems. Given sufficient time, the molecules become delocalized and display ideal, Brownian behavior. Over the time frame of the simulations only systems with $T \geq 430$ K have had sufficient time to reach this regime. It is this intrinsic constraint that makes it challenging, if not impossible, to accurately predict macroscopic diffusion coefficients with molecular dynamics in systems with large solute molecules or glassy matrices. However, theories are emerging that link sub–diffusive and macroscopic dynamics in these systems.

We find that three parameters $\hat{\Gamma}_0$, α, and β represent a reasonably robust metric for the sub-diffusive dynamics in our systems, and perhaps all locally con- strained systems. We are currently exploring links between these quantities and the macroscopic, Brownian behavior that ensues after delocalization. If a quantitative link can be established, the fundamental constraint in using molecular dynamics to predict macroscopic diffusion coefficients of solutes in polymeric matrices can be circumvented. This would allow a wide range of potential issues associated with migration of solute molecules through medical device polymers to be rapidly assessed.

Disclaimer

The mention of commercial products, their sources, or their use in connection with material reported herein is not to be construed as either an actual or implied endorsement of such products by the Department of Health and Human Services.

References

[1] P.A Lemos, PW Serruys, and JE Sousa. Drug-eluting stents. ***Circulation,*** 107(24): 3003–3007, 2003.

[2] David M Saylor, Chang-Soo Kim, Dinesh V Patwardhan, and James A Warren. Diffuse-interface theory for structure formation and release behavior in controlled drug release systems. ***Acta Biomaterialia***, 3(6):851–864, 2007.

[3] R Kroes. The Threshold of Toxicological Concern Concept in Risk Assessment. ***Toxicological Sciences***, 86(2):226–230, May 2005.

[4] Christopher Forrey, David M Saylor, Joshua S Silverstein, Jack F Douglas, Eric M Davis, and Yossef A Elabd. Prediction and validation of diff coefficients in a model drug delivery system using microsecond atomistic molecular dynamics simulation and vapour sorption analysis. ***Soft Matter***, 2014, in press.

[5] Marcus T Cicerone and Jack F Douglas. β-Relaxation governs protein stability in sugar-glass matrices. ***Soft Matter***, 8(10):2983, 2012.

[6] L Larini, A Ottochian, C De Michele, and D Leporini. Universal scaling between structural relaxation and vibrational dynamics in glass-forming liquids and poly- mers. *Nature Physics*, 4(1):42–45, December 2007.

[7] David S Simmons, Marcus T Cicerone, Qin Zhong, Madhusudan Tyagi, and Jack F Douglas. Generalized localization model of relaxation in glass-forming liquids. ***Soft Matter***, 8(45):11455, 2012.

[8] S Plimpton. Motion of suspended particles on the kinetic theory. ***J Comp Phys***, 117:1–19, 1995.

[9] A Einstein. Motion of suspended particles on the kinetic theory. ***Ann. Phys***, 17(3): 549–560, 1905.

[10] T G Mason, Hu Gang, and D A Weitz. Diffusing-wave-spectroscopy measurements of viscoelasticity of complex flow ***JOSA A***, 14(1):139–149, January 1997.

[11] Wei Min, Guobin Luo, Binny J Cherayil, S C Kou, and X Sunney Xie. Observation of a Power-Law Memory Kernel for Fluctuations within a Single Protein Molecule. ***Physical review letters***, 94(19):198302, May 2005.

POSTER SESSION

Advanced Composites for Aerospace, Marine, and Land Applications II
Edited by: T. Sano and T.S. Srivatsan
TMS (The Minerals, Metals & Materials Society), 2015

EFFECT OF T6 TREATMENT ON THE HARDNESS OF CARBON NANOTUBE REINFORCED AA6061 ALUMINIUM ALLOY MATRIX COMPOSITES

Dilemando Nagle Travessa[1], Matheus Pianassola[1], Mirian Glícea S. Carneiro[1], Marcela Lieblich[2]

[1] *Instituto de Ciência e Tecnologia*
Universidade Federal de São Paulo,
ICT-UNIFESP. Rua Talim, 330, CEP 12231-280
São José dos Campos, SP, Brazil.

[2] *Centro Nacional de Investigaciones Metalurgicas*
Consejo Superior de Investigaciones Cientificas
CENIM-CSIC. Avda
Gregorio del Amo, 8. 28040, Madrid, Spain.

Keywords: Aluminium matrix composite, Carbon nanotubes, High energy ball milling, Hot extrusion, Heat treatment.

Abstract

The development aluminum matrix composites, using carbon nanotubes as strengthening dispersion phase, is of great interest for the transport sector, that continuously demands high strength-low density materials for efficient structures.

In the present work, multiwall carbon nanotubes (MWCNT) were used to reinforce the AA6061 aluminium alloy. 1 and 2% (in weight) of MWCNT were mixed to the alloy powder by high-energy ball-milling process. The blended powder was consolidated by hot extrusion. The obtained composite bars were solution heat treated and aged to T6 temper, and characterized by optical and scanning electron microscopy, and by hardening testing. A typical wrought microstructure free of defects was obtained. Hardness of the composites was substantially increased for 2% MWCNT addition, milled for 10 h. The better performance for the composites blended at a higher milling time was supposed to be due to a better MWCNT dispersion. The effect of aging on the composite hardness was evaluated, and it was found that no benefits for the composite hardness were obtained after solution heat treatment followed by T6 aging according to SAE AMS2772 specification, when comparing to the as-extruded hardness values. Possible reasons for such composites behavior are discussed.

Introduction

Over the past seventy years, aluminum alloys have dominated the market for the aerospace industry, due to a set of intrinsic properties and characteristics, as mechanical strength, corrosion resistance, low density, easy processing and moderate to low cost. However, in the latest decades, the strength level of the commercial Al alloys seems to be stalled at around 600 MPa of ultimate strength, and conventional thermo-mechanical processing seems to be unable to reach additional strengthening.

Carbon nanotubes (CNT) are an emerging and promising class of materials, exhibiting very interesting properties, among them, high strength and elastic modulus [1, 2]. Having nanometer dimensions, carbon nanotubes can be added to a set of ductile metallic materials in order to increase their strength by several mechanisms: stress fields development at the metallic reticulate, Orowan strengthening, and the classic load-transfer mechanism of composite materials [3]. However, as their dimensions are nanometric, van der Waals forces promote their agglomeration and make their homogeneous dispersion in composite materials very challenging.
High energy ball milling (HEBM) has been the most used route to produce CNT reinforced aluminum matrix composites (CNT-AMC), although the energy of the process could be deleterious to the CNT integrity. Fragmentation and amorphization are the main risks to which HEBMilled CNT are subjected. [4]. However, it is believed that the ductile aluminum particles can hold the CNT particles at initial milling stages, protecting them from direct impacts from the milling media. As a result, several authors have reported success on CNT dispersion in aluminum by this process, keeping their structural integrity [5-8].The final consolidation of the composites produced by HEBM can be obtained by several processes: hot extrusion, cold or hot pressing followed by conventional sintering, spark plasma sintering, cold or hot milling and plasma spraying are examples.

Most of the literature regarding the production of CNT-AMC is related to pure aluminum and a maximum of 5% by weight of the reinforcing phase [5-6, 8, 9-15]. Few works are related to precipitation hardenable aluminum alloys, like the AA2024 aklloy [7, 16-17]. However, the combined effect of precipitation hardening and CNT dispersion in strengthening such age hardenable aluminum alloys composite, is rarely evaluated. Kondoh and co-workers [18] investigated the effect of the heat treatment on the AA6063 alloy reinforced by multi wall carbon nanotubes (MWCNT) produced by spark plasma sintering and hot extrusion, and observed that after aging to T6 condition both yield and ultimate tensile strength decreased, when comparing to the as-extruded condition. The authors argue that Mg matrix depletion, as a consequence of Al_2MgC_2, MgO and/or $MgAl_2O_4$ phases formed around nanotubes, are responsible for the unsuccessful heat treatment. If such reaction occurs, less Mg atoms would be available for Mg_2Si precipitation.

In the present work, the effect of 1 and 2 % (in weight) of MWCNT reinforcing AA6061 AMC was evaluated by microhardness measurements, in the as extruded condition and after solution heat treatment followed by natural and artificial aging (T4 and T6 tempers, respectively). Optical and scanning electron microscopies were also employed to assess the microstructure development of the composites. AA6061 bars with no MWCNT reinforcement were also produced by the same processing route, for comparison purposes. It was found that the hardness was significantly increased after the addition of MWCNT to the alloy, in the as-extruded condition. However, subsequent heat treatment did not provide additional hardening for both unreinforced and reinforced material, when comparing to the as-extruded condition. Possible reasons for such behavior are discussed.

Experimental procedures

AA6061 alloy powder, produced by inert gas atomization by ALPOCO – United Kingdom – having particles lesser than 50 μm, was employed in the present work. The nominal chemical composition of the alloy, which complies with AMS-QQ-A-200/8 specification [19], is shown in Table 1.

Table 1. Nominal chemical composition of AA6061 alloy used in the present work.

Element	Weigth %
Si	0,45
Mg	0,96
Cu	0,27
Cr	0,16
Fe	0,15
Mn	0,0023
Al	Bal

Bundles of MWCNT (Baytubes® C150P) produced by chemical vapor deposition process by Bayer Materials Science – Germany – with a minimum purity level of 95%, were employed as reinforcing phase. Figure 1 shows a transmission electron micrography of a typical isolated nanotube, were an aspect ratio of around 100 can be observed.

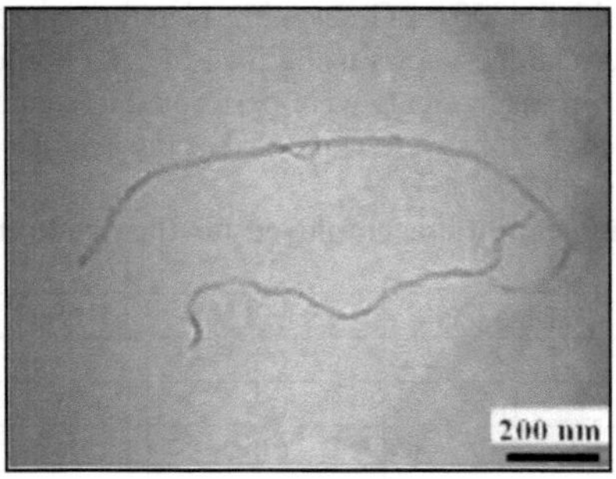

Figure 1: Aspect of isolated Baytubes® C150P, after ultrasonically assisted dissolution in acetone. For observation at the microscope, the MWCNT were captured from the solution by a carbon film covered Cu grid.

The AA6061 powder and the MWCNT were blended into a planetary ball mill Retsch PM4 in air atmosphere, using 250 cm^3 stainless steel vials and 20 mm diameter hard steel balls. A ball to powder ratio of 10:1 at 200 RPM were the main milling parameters employed, and 0,5% in weight of stearic acid was used as process control agent (PCA). Composites having 1 and 2% in weight of MWCNT were produced. For comparison purpose, bars without reinforcement were

also produced employing the same milling conditions. Milling times of 6 and 10 h were employed. For the composite milled for 10 h, the aluminum powder, bundles of MWCNT and PCA were simply loaded into the vials. For the composite milled for 6 h, a special mixing procedure was performed before loading into the milling vials. MWCNT was previously dispersed into acetone for 0,25 h under sonication. The AA6061 powder was added to this suspension with sonication for another 0,25 h. The mixture was subsequently heated under magnetic agitation until thick slurry was obtained. Final drying at 50 °C for 24 h was performed before adding the mixture to the milling vials, with the PCA and balls. The Table 2 resumes the main process parameters employed to obtain the composite powders.
Raman spectroscopy was performed on samples containing 2% of MWCNT in order to assess their structural quality after milling, when comparing to the pristine condition. A micro-Raman Renishaw microscope system 2000, employing an argon laser (514.5 nm wavelength) was employed on this characterization.

The powders obtained in the HEBM process were encapsulate into clean aluminum tubes and extruded at 450 °C, at 2 mm/s and with an extrusion rate of 28:1, resulting in 8 mm diameter bars.

Samples from the bars were cut, degreased and submitted to heat treatment according to SAE-AMS2772E specification [20]. Solution heat treatment was performed in air by heating at 530 + 5 °C for 1 h and quenched in water. For T4 temper, solubilized samples were allowed to age at room temperature for 96 h minimum. For T6 temper, T4 samples were aged at 175 + 5 °C for 8 h and cooled in air.

Metallographic examination was performed on samples in the as-extruded, T4 and T6 conditions. These samples were prepared by conventional grinding procedures and polishing with diamond past and colloidal silica suspension. Samples were observed in an Olympus BH-2 optical microscope and in a Hitashi S-4800 FEG scanning electron microscope, equipped with EDS microanalysis. Vickers microhardness tests using load of 1 kg applied for 10 s were performed on the samples. A total of 5 measurements were performed in each sample, and the highest and lowest values were discarded. Results were expressed as the average of the 3 remaining values.

Table 2. Milling conditions employed for the production of composites.

Sample	%MWCNT	MT	SM	RPM	B/P ratio
6061-NR-6h	-	6 h	yes	200	10:1
6061-1CNT-6h	1,0	6 h	yes	200	10:1
6061-2CNT-6h	2,0	6 h	yes	200	10:1
6061-NR-10h	-	10 h	no	200	10:1
6061-1CNT-10h	1,0	10 h	no	200	10:1
6061-2CNT-10h	2,0	10 h	no	200	10:1

NR= no reinforcement; %MWCNT in weight; MT= milling time; SM= Special mixing procedure of the powders before milling; B/P= balls to powder ratio.

Results and Discussion

The powders obtained after milling showed and homogeneous aspect, not sticking to the vials or the balls. The amount of PCA added was satisfactory. Raman spectroscopy presented in Figure 2 shows that typical D and G bands are present for the pristine nanotubes, as well as in the composite powders milled for 6 and 10 h. As the D band is frequently related to structural disorders of carbon based materials, the D to G band intensity ratio (ID/IG) is commonly used to monitor induced defects during nanotubes processing [21-22]. From Lorentzian fitting of the peaks, it seems that ID/IG ratio slight increase to the milled samples (around 1,10), when comparing to the pristine condition (1,05). It shows that small structural disorders are introduced to MWCNT during the milling process. It is also observed the presence of a shoulder at the right side of the peak of the pristine spectrum. This shoulder evolves to a more defined peak at around 1620 cm^{-1} as the milling time increases, and can be attributed to the D' band originating from a double resonance processes. The small increase of ID/IG observed suggests that MWCNT were minimally damaged during the HEBM process.

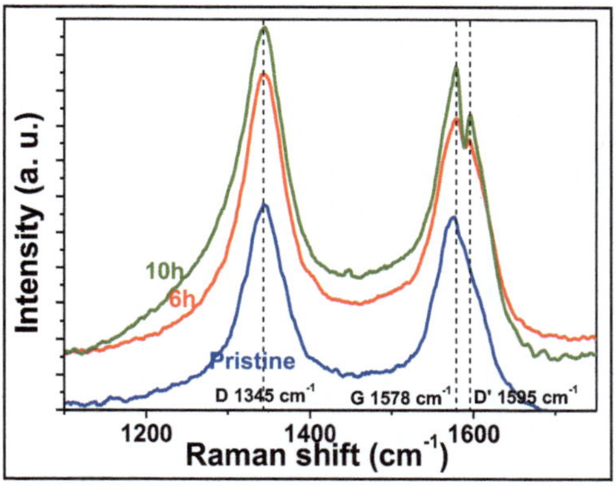

Figure 2. Raman spectroscopy of the samples containing 2% (weight) of MWCNT, after milling for 6 and 10 h. A spectrum from the isolated nanotubes in the as-received condition is also presented for comparison purpose.

Figure 3 shows the microstructure of the sample 6061-2CNT-6h in the as-extruded condition. This microstructure is typical for all as-extruded bars, and is similar to the microstructure of un-recrystallized wrought commercial products, except for the smaller size of insoluble phases. These insoluble precipitates consists mainly of Fe_3SiAl_{12} and Mg_2Si [23], and aligned with the extrusion direction. Their smaller size, when comparing to typical wrought commercial extruded products, is a consequence of the HEBM process. In 3A, few bigger particles are also observed. Higher magnification of such particles in 3B shows that the typical extrusion lines contour such particles, suggesting some sort of contamination probably from the milling media. Figure 4 shows the microstructure of the sample 6061-2CNT-10h in T4 condition, that is representative for all bars at both T4 and T6 heat treated conditions. This microstructure shows great similarity with the bars in the as-extruded condition, although visible Fe_3SiAl_{12} and Mg_2Si precipitates seem to be less abundant. This is consistent with the heat treated condition that dissolves some of these precipitates, re-precipitating into smaller particles during aging. From Figure 4, it is also possible to verify some voids arising after heat treatment. These voids, indicated by arrows in Figure 4, are aligned to the extrusion direction and can be a result of the expansion of entrapped gases, during solution heat treatment.

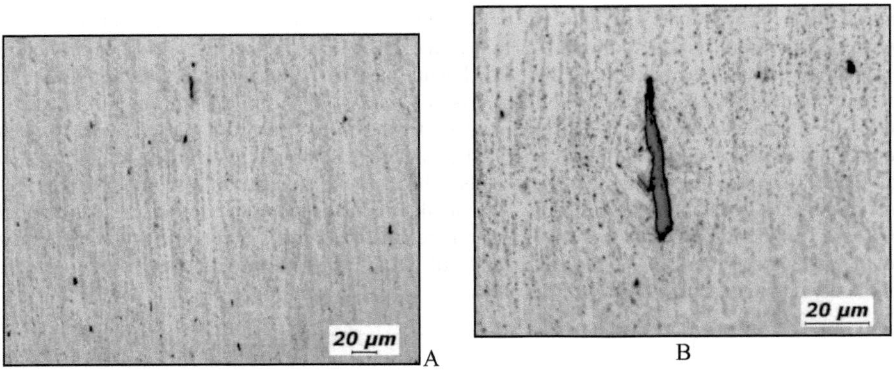

Figure 3. **Optical micrographs of samples containing 2% of MWCNT milled for 6 h (6061-2CNT-6h), in the longitudinal as-extruded condition: 3A- overall aspect showing fine precipitation; 3B- higher magnification showing extrusion lines around a bigger particle.**

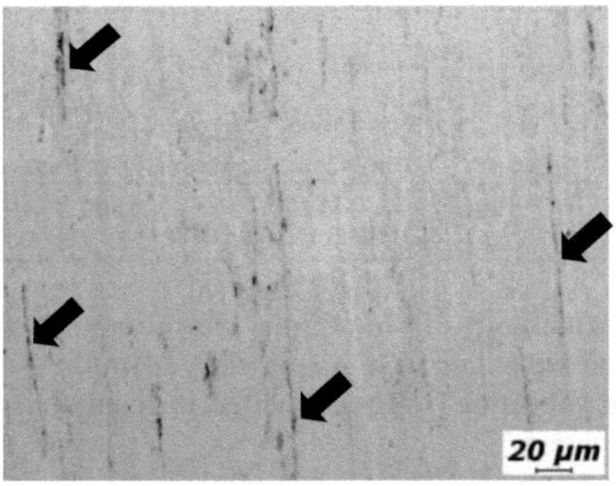

Figure 4. **Optical micrograph of sample containing 2% of MWCNT milled for 10 h (6061-2CNT-10h), in the longitudinal T4 condition. Some of the voids originating after solution heat treatment are indicated by arrows.**

Scanning electron microscopy analysis performed on the sample 6061-NR-6h shows in detail (Figure 5) one of the bigger particles mentioned in the Figure 3. Quantitative EDS microanalysis performed in such particle resulted in a composition similar to AISI304 series austenitic stainless steel, confirming that such particles are contaminants from the milling media (vials).

The results obtained from microhardness testing, preformed on samples in the as-extruded, T4 and T6 conditions, are presented in Table 3 and Figure 6. It can be easily observed the contribution of MWCNT additions to the hardness of the bars. This contribution is higher as the CNT content increases from 1 to 2% in weight. The increase of the hardness with the milling time is observed only in the composite bars. It seems that a better dispersion of MWCNT is obtained as the milling time increases from 6 to 10 h, resulting in better properties. When comparing to un-reinforced bars, the hardness of the samples 6061-1CNT-6h, 6061-2CNT-6h,

6061-1CNT-10h and 6061-2CNT-10h, increased by 20, 30, 57 and 68% respectively. This increase is remarkable, but surprisingly, after heat treatment, an additional contribution from precipitation hardening could not be observed at T4 and T6 conditions for any sample. In some cases, the hardness even decreased after heat treatment. Similar behavior has been observed in AA6063 alloy [18], in presence of CNT, although in this reference work the composites were not produced by HEBM. It was argued by the authors that alloying elements react with carbon, reducing the amount of solute in the matrix and thereby reducing the volume fraction of hardening precipitates that should be formed during the aging treatment. However, in the present work, even in the un-reinforced bars the hardness did not increase after heat treatment, when comparing to the as-extruded condition. It seems that microstructural benefits resulted from HEBM, as strain hardening, grain refinement and/or the development of a subgrain structure [24] could be present in the as-extruded condition and are lost during the solution heat treatment. The resulting decrease in hardness could compensate the precipitation hardening obtained by heat treatment. In fact, the un-reinforced bars showed an increase in hardness from T4 to T6 condition, what is expected to be a result from precipitation hardening.

The presence of hard reinforcing phases in ductile metals generates internal stresses [25] that can modify the precipitation kinetics. High defects concentration produced by HEMB process [24] can also account for similar behavior, due to the higher concentration of favorable nucleation sites for precipitation. It would be expected that, having higher vacancy concentration, HEBM processed alloys could accelerate the precipitation kinetics. However, it has been reported that a strong link can be established between vacancies and solute atoms, forming solute stable clusters that retard the precipitation [26-28]. In such conditions, the heat treatment procedures applicable to commercial alloys produced by conventional thermo mechanical processing, could lead to over or even under-aging of the composites produced by HEBM.

As far as precipitation hardening in HEBM processed materials is concerned, the literature also suggests that grain refinement reduces the potential for precipitation hardening in some alloys [3, 25, 29]. In this case, when the grain becomes too small, the grain volume to grain boundary ratio diminishes. As a consequence, precipitates could form at grain boundaries, were precipitation is more favorable, but less effective for hardening effects. This is another aspect to be investigated in order to assess the behavior of CNT reinforced aluminum alloys produced by HEBM during subsequent precipitation hardening heat treatments.

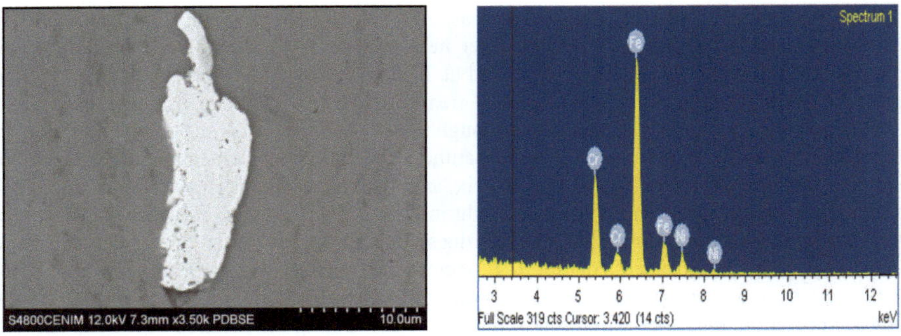

A B

Figure 5. **Scanning electron micrograph of the sample 6061-NR-6h, in the longitudinal as-extruded condition, in A. The particle in the center of the micrograph refers to stainless steel contamination from the milling vials, as confirmed by EDS microanalysis in B.**

Table 3: Vickers microhardness values obtained in the bars, in AE, T4 and T6 conditions

Sample	AE condition	T4 condition	T6 condition
6061-NR-6h	72,3 ± 3,1	54,7 ± 6,1	71,7 ± 2,9
6061-1CNT-6h	86,0 ± 3,5	68,7 ± 3,6	75,3 ± 4,9
6061-2CNT-6h	93,3 ± 0,6	79,0 ± 4,6	79,3 ± 3,8
6061-NR-10h	69,7 ± 1,9	53,5 ± 4,1	61,5 ± 4,3
6061-1CNT-10h	113,5 ± 4,4	100,3 ± 3,2	103,3 ± 6,2
6061-2CNT-10h	121,3 ± 1,2	119,7 ± 0,6	113,7 ± 3,1

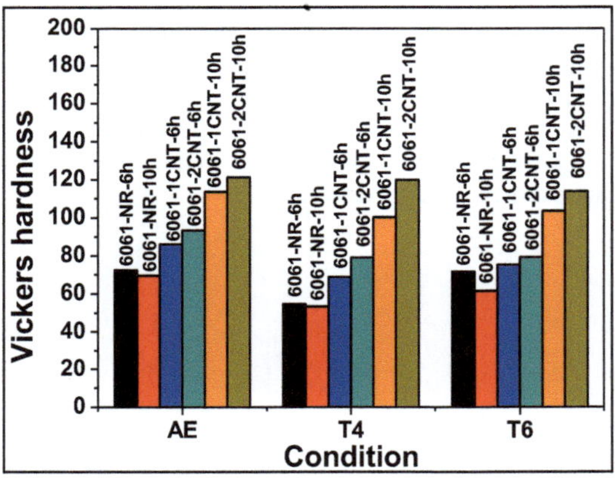

Figure 6: **Microhardness measurements of the composite and non-reinforced bars produced, in the as-extruded, T4 and T6 conditions.**

Conclusions

Composites of AA6061 alloy reinforced by MWCNT were successfully produced by the HEBM – Hot Extrusion processing route. The presence of MWCNT in the composite resulted in remarkable increase in hardness. Vickers hardness increased by 20 and 30% with additions of 1,0 and 2,0% of MWCNT (in weight) respectively, when comparing to the un-reinforced alloy produced by the same route. The benefits were even higher when the milling time was increased from 6 to 10 h, resulting in hardness increments by 57 and 68% for the same MWCNT additions. A better dispersion of nanotubes is supposed to occur with the higher milling time, resulting in higher strength of the composite.

The combination of MWCNT dispersion and alloy precipitation to the composite hardening was not observed in the present work, when heat treating the composites according to the procedures recommended for the commercial alloy [20]. It seems that benefits in microstructure obtained from HEBM process could be lost during solution heat treatment. Precipitation kinetics can be changed, and grain boundary precipitation could be favored in fine grain HEBM composites, making conventional heat treatment procedures of commercial alloys inapplicable in such materials. However, it has to be pointed out that the strength level reached in the as-extruded condition is already very promising for structural applications.

Acknowledgments

The authors are very thankful to researchers at Associated Laboratory for Sensors and Materials from the National Institute for Space Research (LAS-INPE – Brazil), for the Raman Spectroscopy analysis. Dilermando Nagle Travessa is very thankful to FAPESP for the founding related to an internship at CENIM-CSIC – Spain (process no. 2012-07831-2).

References

[1] Thostenson ET, Renb Z, Choua TW. Advances in the science and technology of carbon nanotubes and their composites: a review. Composite Science and Technology. 2001;61:1899-1912.

[2] Paradise M, Goswami T. Carbon nanotubes – Production and industrial applications. Materials. & Design. 2007;28:1477-89.

[3] Choi HJ, Shin JH, Bae DH. Grain size effect on the strengthening behavior of aluminum-based composites containing multi-walled carbon nanotubes. Composites Science and Technology. 2011;71:1699-1705.

[4] Poirier D, Gauvin R, Drew RAL. Structural characterization of a mechanically milled carbon nanotube/aluminum mixture. Composites: Part A. 2009;40:1482-89.

[5] Esawi AMK, Morsi K, Sayed A, Gawad AA, Borah P. Fabrication and properties of dispersed carbon nanotube–aluminum composites. Materials Science and Engineering A. 2009;508:167-173.

[6] Choi HJ, Kwon GB, Lee GY, Bae DH. Reinforcement with carbon nanotubes in aluminum matrix composites. Scripta Materialia. 2008;59:360-363.

[7] Deng C, Wang D, Zhang XX, Li AB. Processing and properties of carbon nanotubes reinforced aluminum composites. Materials Science and Engineering A. 2007;444:138-145.

[8] Bustamante RP, Guel IE, Madrid PA, Yoshida MM, Ramírez JMH, Sánchez RM. Microstructural characterization of Al-MWCNT composites produced by mechanical milling and hot extrusion. Journal of Alloys and Compounds. 2010;495:399-402.

[9] George R, Kashyap KT, Rahul R, Yamdagni S. Strengthening in carbon nanotube/aluminium (CNT/Al) composites. Scripta Materialia. 2005;53:1159-63.

[10] Morsi K, Esawi AMK, Lanka S, Sayed A, Taher M. Spark plasma extrusion (SPE) of ball-milled aluminum and carbon nanotube reinforced aluminum composite powders. Composites: Part A. 2010;41:322-326.

[11] Zhong R, Cong H, Hou P. F abrication of nano-Al based composites reinforced by single-walled carbon nanotubes. Carbon. 2003;41:848-851.

[12] Kwon H, Park DH, Silvain JF, Kawasaki A. Investigation of carbon nanotube reinforced aluminum matrix composite materials. Composites Science and Technology. 2010;70:546-550.

[13] Liao J, Tan M, Sridhar I. Spark plasma sintered multi-wall carbon nanotube reinforced aluminum matrix composites. Materials and Design. 2010;31:S96-S100.

[14] Laha T, Agarwal A. Effect of sintering on thermally sprayed carbon nanotube reinforced aluminum nanocomposite. Materials Science and Engineering A. 2008;480:323-332.

[15] Wang L, Choi H, Myoung J, Lee W. Mechanical alloying of multi-walled carbon nanotubes and aluminium powders for the preparation of carbon/metal composites. Carbon. 2009;47:3427-33.

[16] Deng CF, Zhang XX, Wang DZ, Lin Q, Li A. Preparation and characterization of carbon nanotubes/aluminum matrix composites. Materials Letters. 2007;61:1725-28.

[17] Deng CF. Zhang XX, Wang DZ, Ma XY. Calorimetric study of carbon nanotubes and aluminum. Materials Letters. 2007;61:3221-23.

[18] Kondoh K, Fukuda H, Umeda J, Imai H, Fugetsu B. Microstructural and mechanical behavior of MWCNTs reinforced Al-Mg-Si alloy composites in aging treatment. Carbon 2014; accepted manuscript – doi 10.1016/j.carbon.2014.01.013.

[19] Aerospace Material Specification. Society of Automobile Engineers – SAE. AMS-QQ-A-200/8: Aluminum Alloy 6061, Bar, Rod, Shapes,Tube, and Wire, Extruded. SAE International; 2007.

[20] Aerospace Material Specification. Society of Automobile Engineers – SAE AMS2772E: Heat Treatment of Aluminum Alloy Raw Materials. SAE International; 2008.

[21] Dresselhaus MS, Dresselhaus G, Saito R, Jorio A. Raman spectroscopy of carbon nanotubes. Physics Reports. 2005;409:47-99.

[22] Antunes EF, Lobo AO, Corat EJ, Trava-Airoldi VJ, Martin AA, Veríssimo C. Comparative study of first- and second order Raman spectra of MWCNT at visible and infrared laser excitation. Carbon. 2006;44:2202-11.

[23] MacKenzie DS, Totten GE. Analytical characterization of aluminium, steel, and superalloys. Boca Raton: CRC Press - Taylor & Francis Group; 2006.

[24] Suryanarayana C. Mechanical alloying and milling. Progress in Materials Science. 2001;46:1-184.

[25] Arakawa S, Hatayama T, Matsugi K, Yanagisawa O. Effect of heterogeneous precipitation on age-hardening of Al2O3 particle dispersion Al-4% mass Cu composite produced by mechanical alloying. Scripta Materialia. 2000;42:755-760.

[26] Edwards G A, Stiller K, Dunlop G L, Couper M J. The precipitation Sequence in Al-Mg-Si Alloys. Acta Materialia 1998; 46: 3893-3904.

[27] Niranjani V L, Hari Kumar K C, Subramanya Sarma V. Development of high strength Al–Mg–Si AA6061 alloy through cold rolling and ageing. Materials Science and Engineering A 2009; 515: 169–174.

[28] Pogatscher S, Antrekowitsch H, Uggowitzer P J. Interdependent effect of chemical composition and thermal history on artificial aging of AA6061. Acta Materialia 2012; 60: 5545–5554.

[29] Corrochano J, Lieblich M, Ibáñez J. On the role of matrix grain size and particulate reinforcement on the hardness of powder metallurgy Al-Mg-Si/MoSi2 composites. Composites Science and Technology. 2009;69:1818-24.

AUTHOR INDEX
Advanced Composites for Aerospace, Marine, and Land Applications II

SUBJECT INDEX
Advanced Composites for Aerospace, Marine, and Land Applications II

MIX
Papier aus verantwortungsvollen Quellen
Paper from responsible sources
FSC® C105338

If you have any concerns about our products,
you can contact us on
ProductSafety@springernature.com

In case Publisher is established outside the EU,
the EU authorized representative is:
Springer Nature Customer Service Center GmbH
Europaplatz 3, 69115 Heidelberg, Germany

Printed by Libri Plureos GmbH
in Hamburg, Germany